WITHDRAWN

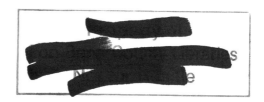

WITHDRAWN

Music and the Making of Modern Science

Music and the Making of Modern Science

Peter Pesic

The MIT Press
Cambridge, Massachusetts
London, England

© 2014 Massachusetts Institute of Technology

All rights reserved. No part of this book may be reproduced in any form by any electronic or mechanical means (including photocopying, recording, or information storage and retrieval) without permission in writing from the publisher.

MIT Press books may be purchased at special quantity discounts for business or sales promotional use. For information, please email special_sales@mitpress.mit.edu.

This book was set in Times by Toppan Best-set Premedia Limited, Hong Kong. Printed and bound in the United States of America.

Library of Congress Cataloging-in-Publication Data

Pesic, Peter.
Music and the making of modern science / Peter Pesic.
  pages   cm
Includes bibliographical references and index.
ISBN 978-0-262-02727-4 (hardcover : alk. paper)
1. Science—History. 2. Music and science—History. I. Title.
Q172.5.M87P47 2014
509—dc23
2013041746

10 9 8 7 6 5 4 3 2

For Alexei and Andrei

Contents

    Introduction   1

1   Music and the Origins of Ancient Science   9

2   The Dream of Oresme   21

3   Moving the Immovable   35

4   Hearing the Irrational   55

5   Kepler and the Song of the Earth   73

6   Descartes's Musical Apprenticeship   89

7   Mersenne's Universal Harmony   103

8   Newton and the Mystery of the Major Sixth   121

9   Euler: The Mathematics of Musical Sadness   133

10   Euler: From Sound to Light   151

11   Young's Musical Optics   161

12   Electric Sounds   181

13   Hearing the Field   195

14   Helmholtz and the Sirens   217

15   Riemann and the Sound of Space   231

16   Tuning the Atoms   245

17   Planck's Cosmic Harmonium   255

18   Unheard Harmonies   271

   Notes   285

   References   311

   Sources and Illustration Credits   335

   Acknowledgments   337

   Index   339

Introduction

Alfred North Whitehead once observed that omitting the role of mathematics in the story of modern science would be like performing *Hamlet* while "cutting out the part of Ophelia. This simile is singularly exact. For Ophelia is quite essential to the play, she is very charming—and a little mad."[1] If in the story of science mathematics takes the part of Ophelia, music might be compared with Horatio, Hamlet's friend and companion who helps investigate the ghost, discusses what may lie beyond their philosophies, sings the sweet prince to his rest, and tells his story.

This book will examine some significant moments in the relationship of music to science, especially those in which prior developments in music affected subsequent aspects of natural science. By investigating this direction of influence, we question the common presupposition that, however beguiling, music is conceptually derivative or secondary compared to other modes of thought or perception—an effect, rather than a cause. The examples considered in this book show the larger intellectual and cultural dimensions of music as a force in its own right.

By virtue of its special position in Greek natural philosophy, music occupied the perfect position to mediate between idealized mathematical objects and the world of experience.[2] Based on these ancient models, the continuing structures of learning mandated music's ensuing centrality as the "hinge" discipline connecting arithmetic, geometry, and the sensual world. This both reflected and moved the profound alterations that surrounded the birth of the "new philosophy" during the sixteenth and seventeenth centuries. My main contention is that, in whatever directions its interventions tended, from ancient until modern times music so deeply and persistently affected the making of science over so many historical vicissitudes that we should tell their stories jointly. Our awareness of how exactly music entered into the story can enlarge and deepen our understanding of the human and intellectual dimensions of science. In so doing, we may draw more closely together the study of "aural culture" and symbolic structures hitherto considered separate. This rapprochement calls for an enriched exploration of the felt dimensions of scientific experience, considered as a fully human activity vividly engaged with perception, feeling, and thought.[3]

As Horatio has ties to both Hamlet and Ophelia, music touches both natural philosophy and mathematics. Accordingly, we will examine several critical shifts of understanding in both these domains. Chapter 1 describes the most consequential move of all: ancient Greek natural philosophy connected music with mathematics and astronomy within a fourfold study, the *quadrivium*. This alliance involved both experiment and theory, and hence positioned music at the frontier between the worlds of physical sensation and ideal forms. During the subsequent fifteen centuries, music retained its central place among the mathematical sciences, an essential component of what became "liberal education," which explicitly continued a program the Pythagoreans began, Plato systematized, and Boethius transmitted to the West. These ancient developments are important not merely as historical background, buried as hidden sediment under the surface of modernity, but as continually active and reemergent forces that shaped and continue to shape "modern" science and mathematics.

Chapter 2 considers the status of these forces as they seemed to a preeminent fourteenth-century natural philosopher. Nicole Oresme used musical concepts as important elements in his reexamination of the cosmos, its possible cycles, and their relation to arithmetic and geometry. The case of Oresme shows how significant musical issues were well before the advent of the "new philosophy" of the sixteenth century. Oresme's contact with the "new music" (*ars nova*) led him to refute the simplest versions of cosmic harmony and to propose radical new alternatives that depend on the tension between arithmetic and geometry. The underlying musical and cosmological considerations allow us to deduce his own unstated conclusion in favor of geometry.

Disputes about the order of the planets had profound musical and cosmological implications, especially on the growing controversy over whether the seemingly immovable Earth could be understood to move, as the Pythagoreans held. Though he had presented strong arguments in favor of this view, Oresme himself finally accepted the geocentric account. Chapter 3 investigates the role of music in this cosmological controversy as it came to a head in the fifteenth century. The problem of a seemingly immovable yet moving center such as the Earth parallels the musical problem of changing the usually fixed modal center of a composition. Despite its proverbial impossibility, the theorist Heinrich Glarean drew attention to just such a shift of mode in Josquin des Prez's motet *De profundis*. If so, music might show how the immovable could move, after all. Indeed, *harmony* became the defining issue on which Copernicus and those following him phrased their arguments about the new cosmology. Musicians followed this controversy closely; in his 1588 polemic for a revival of ancient musical practice, Vincenzo Galilei was among the first Italians to defend the heliocentric view, many years before his celebrated son Galileo took up this cause, again under the banner of harmony.

Several decades before Vincenzo Galilei's surprising avowal, music influenced fundamental changes in the concept of number. Though irrational quantities had long been excluded from arithmetic and harmonics, sixteenth-century musical theory and practice called for irrational and rational quantities on an equal footing. Chapter 4 discusses three central figures in this story: the German mathematician Michael Stifel, who was the first person to use the phrase "irrational numbers" in the course of his exposition of music, but who then hesitated to grant those numbers full reality; Girolamo Cardano, celebrated physician-polymath, who gave such quantities even greater significance in his musical writings; and Nicola Vincentino, a composer obsessed with reviving ancient Greek quarter tones who found himself in need of what he called "irrational proportions" to define these unfamiliar intervals. Each of these three men was involved with practical music to a degree correlated with their respective reliance on irrational numbers.

Johannes Kepler, more than anyone, incorporated music into the foundations of his innovative astronomy. Chapter 5 relates his interest in musical practice to his novel approach to its theory, which moved him to reject algebraic results that contradicted musical experience. Kepler's search for cosmic polyphony points to Orlando di Lasso's *In me transierunt* as a moving expression of the "song of the Earth," down to the melodic spelling of the Earth's song. Kepler presents both cosmos and music as essentially alive and erotically active, based on his sexual understanding of numbers. The pervasive dissonance of the cosmic harmonies reflects the throes of war and eros. Like Oresme, Kepler realized the essential incompleteness of the cosmic music, which may never reach a final cadence, a universal concord on which the world-music could fittingly end. Kepler treats this as an indication of divine infinitude, inscribed in the finite cosmos.

René Descartes began his career writing about music, which affected his innovative natural philosophy throughout its development. Chapter 6 reads his correspondence with Marin Mersenne as tracing the interaction between musical, mathematical, and philosophical themes. Musical observations led to Descartes's initial observations of the overtones of vibrating strings, which in turn led to wider considerations of mechanics, motion in a vacuum, and eventually to his continuum theory of the universe. His theories emerged in constant dialogue with musical issues and problems.

Music was central to the natural philosophy of Mersenne. Chapter 7 begins with his musical arguments for heliocentrism, against the hermetist Robert Fludd. Mersenne used musical devices to make pioneering measurements of the frequency of vibrating strings and of the speed of sound. Mersenne's work on overtones both profited from and struggled with his musical preconceptions, as did his attempt to incorporate atomism into his account of vibrating bodies.

Though Isaac Newton walked out of the only opera he ever attended, music had a significant place in his work.[4] His early notebooks show his close study of music theory and his attention to matters of "elegancy" in practice. A decade later, he applied the musical scale to define the colors in the spectrum. Though he had initially assumed that color spanned a perfect octave from red to violet, chapter 8 discusses Newton's subsequent realization that it actually spanned a smaller interval. Yet he did not realize that the departure from an octave implied strong evidence for the wave nature of light.

Throughout his life, the great mathematician Leonhard Euler spent most of his free time on music, to which he devoted his first book. Chapter 9 shows how he reformulated the ordering of musical intervals, implying a mathematical basis for the greater "sadness" of minor chords, compared to major. For this purpose, Euler devised a "degree of agreeableness" that indexed musical intervals and chords. This work on numerical factorization and ratios immediately preceded his subsequent interest in number theory. Having devised a new kind of index, Euler was prepared to put forward indices that would address novel issues like the Königsberg bridge problem and the construction of polyhedra, basic concepts of what we now call topology. Euler also applied musical ideas and analogies with sound to the wave theory of light, as chapter 10 describes. He took the analogy with sound so far as to postulate light "overtones" and "undertones" based on musical theories, though undertones lacked any experimental justification. Euler's later musical writings include his reflections on "ancient" versus "modern" music through their use of different chords. He also used music as the centerpiece in his popular account of science.

Building on the work of Euler, Thomas Young advanced the wave theory of sound and light. Chapter 11 describes how Young found his way to music against the strictures of his Quaker milieu. His newfound passions for music and dance informed his studies of sound and languages. At many points, his understanding of sound influenced and shaped his approach to light, including the decisive experiments that established its wave nature. When Young turned to the decipherment of Egyptian hieroglyphics, he relied on sound and phonology. His final suggestions about the transverse nature of light waves again turned on the comparison with sound.

Those who followed Euler's wave theory of light often reengaged its relation to sound. The study of electricity and magnetism resonated with ongoing initiatives in light and sound, reflecting also wider philosophical ideas about the unity of nature. Chapter 12 examines the intertwined studies of electricity and acoustics by Georg Christoph Lichtenberg, Johann Ritter, and Ernst Chladni. The search to unify the forces of nature often relied on analogies with sound, which in turn looked to electricity for new tools. In the aftermath of Young's work, waves became a newly attractive explanatory approach to the problems of electricity. Building directly on Chladni's sound figures, Hans Christian Ørsted discovered the synthesis of what he called "electromagnetism." Ørsted brought a new unity to the two formerly separate forces of electricity and magnetism, advancing the unitive hopes of *Naturphilosophie,* the German Romantic tradition of natural philosophy. This dialogue between sound and electricity also affected Charles Wheatstone and Michael Faraday. Chapter 13 shows how their unusual collaboration led Wheatstone to discover telegraphy and Faraday to the intensive investigations of sound immediately preceding and preparing his discovery of electromagnetic induction.

Chapter 14 considers how Hermann von Helmholtz's studies of vision and hearing drew on his deep interest in music and art. The dialogue between these arts and their respective senses fed strongly into his investigations into the possible "spaces" of experience, which chapter 15 connects with the many-dimensional manifolds earlier considered by the mathematician Bernhard Riemann. Riemann's unfinished work on the mechanism of the ear affected Helmholtz's ensuing response, which used studies of sound and color to present an empirical basis for Riemann's hypotheses. Einstein drew on these results to shape his world-geometry.

At the same time, the study of spectra brought the music of the spheres to the atomic level. Chapter 16 explores the acoustical underpinnings of G. Johnstone Stoney and Johann Balmer's search for the order in elemental spectra. Balmer's basic formula for the spectral lines of hydrogen emerged from musical presuppositions and analogies. In the following years, Max Planck investigated a complex harmonium just before he began studying blackbody radiation in 1894. Though Helmholtz assumed that "natural" tuning would win out over the convention of equal temperament, Planck's performance experiments with choruses showed otherwise. Chapter 17 describes the relation between Planck's surprising musical findings and the "chorus" of resonators he subsequently introduced to determine the universal spectrum of black bodies.

Planck's colleagues, such as Werner Heisenberg, often considered their musical experiences to be formative of their relation to physics. As chapter 18 shows, even the unmusical Erwin Schrödinger found himself relying on musical analogies as he formulated his wave mechanics. The continuing development of string theory reengages the mathematics of vibration, though the reality of the strings rests on analogy built on analogy. The Pythagorean theme of harmony remains potent in contemporary physics, though its harmonies are more and more unhearable, ever more embedded in its mathematical formalism. Even so, the quest for these harmonies preceded and succeeded the profound changes in the "new philosophy" around the seventeenth century. Mathematics and physics, ancient as well as modern, have been and remain closely linked to essentially musical concepts, whose continuities may have been more significant than the changes generally ascribed to the "scientific revolution." To put it provocatively, that "revolution" may more nearly have been a phase in the restoration and augmentation of the ancient project of musicalizing the world than a change in the basic project of natural philosophy. I hope that bringing forward these overarching musical themes will allow us to see science in a new light, compared to standard accounts based on disruptive "revolutions" and "paradigm shifts."[5]

I have not attempted anything like a complete history of the connections between music and science but have chosen cases in which music led the way.[6] I have restricted myself to what we now call physics and mathematics, noting throughout the actual terms used by the actors themselves, terms generally quite different from ours.[7] About 1830, William Whewell advocated the use of the terms "scientist" and "physicist," but Michael Faraday did not care for them and called himself simply a philosopher, as did those who preceded him (sometimes qualifying their pursuit as "natural philosophy"). Using and assessing the terms and language of the historical actors is essential to approaching their meaning, which still requires a sustained, sensitive, and cautious comparison with our usage and its own presuppositions.

Restoring the name "philosophy" to physical science through the nineteenth century already establishes an important link to the ancient sources, whose fuller content we will explore. We will need to do much work to establish the full meaning of "music"; far beyond the current sense of music as particular specimens of fine arts in the sonic realm, the ancient concept of *mousikē* was far more inclusive of mathematical and philosophical studies. Though, in what follows, I perhaps should have used throughout this Greek term to emphasize its larger meaning, I decided to use our word "music," which was understood in that more inclusive way by many of the historical actors in this book. I beg the reader to bear in mind that older, larger meaning throughout. Then too, musical theory and practice (whether ancient or modern) lies within a broader realm that includes all kinds of sound; accordingly, we will often pass from music into that larger sonic world. The study of "aural culture" complements the "material culture" of science, its machines and devices, and "visual culture," its charts, diagrams, and illustrations.[8] Yet sound has generally been neglected, compared to sight and material objects.

In contrast, H. Floris Cohen's classic work showed the close connection between musical and scientific investigations during the first century of the new philosophy, as did pioneering work by Claude Palisca on musical humanism, by D. P. Walker on Kepler and Galileo, by Penelope Gouk on Bacon and Newton, and by Jamie Kassler on Hooke; Benjamin Wardhaugh extended these investigations into the succeeding century, including many neglected musical and scientific thinkers.[9] The whole field is enjoying a period of notable ferment, thanks to the exciting work of Alexandra Hui, Myles Jackson, Axel Volmar and others concerning the nineteenth and twentieth centuries. Emily Thompson and Jonathan Sterne examined the rich interactions between culture and technology in the twentieth century, in the interfaces between architecture, recording, and the sonic arts.[10] Hillel Schwartz's exuberant history of noise connected many facets of cultural and acoustic history.[11] Brigitte Van Wymeersch, Paolo Gozza, and Jairo Moreno reconsidered the wider implications of Descartes's writings on music; Veit Erlmann explored the deep connection between reason and resonance, in all its senses. Friedrich Kittler offered provocative links between ancient and modern, media and philosophy, music and mathematics.[12] I hope this volume will add some new avenues and approaches to this growing array of insights.

Attempts to make such broad-reaching connections should be circumspect. Consider, for instance, Erwin Panofsky's provocative argument that Galileo Galilei's artistic judgments, particularly his antipathy to mannerist art and its predilection for oval shapes, influenced his rejection of the elliptical planetary orbits Kepler proposed.[13] Yet surely Galileo's artistic views were one factor among many, not simple determinants of his scientific views.[14] Music was important to Galileo personally, not least through the influence of his musician-father Vincenzo, who may (as Stillman Drake suggested) have set him on his path to study nature by joining experiment with mathematics.[15] But lacking further hard evidence, Drake's speculation remains only that. Galileo's beautiful pages on music in his *Dialogues on Two New Sciences* show ways he applied his new science to music, rather than vice versa.[16]

Instead, I will present cases in which more substantive evidence shows the effect of music on science. Drake daringly claimed that music was "the father of modern physical science and mathematics was its mother"; I will make a more nuanced argument that music influenced the unfolding of science at many points and in different ways.[17] My studies point to diverse modes of interaction, both in kind and degree. Nor is musical influence merely "positive" in the sense of advancing what (from our perspective) became the received opinion, whose emergence I will view with some reservation. In particular, Newton's engagement with music arguably had a ambiguous effect on him, at least from the point of view of those who judge critically his role in the controversy between particle and wave theories of light.

Most of all, the case studies in this book bring forward the peculiar power of music, its autonomous force as a stream of experience and sensitivity, independent of language, capable of stimulating insights different than those mediated by visible representations and their attendant theoretical constructs. Compared to the realms of verbal and visible experience, we are only at the beginning of assessing and understanding the role in human experience played by music and sound. Examining the intimate relations between music and science may help us hear (not just see) them both more acutely.

Throughout the book, where I refer to various "♪ sound examples," please see http://mitpress.mit.edu/musicandmodernscience (please note that the sound examples should be viewed in Chrome or Safari Web browsers). See that link as well for further information on purchasing enhanced digital editions that will be available in a variety of formats. The text and examples are most easily and seamlessly available on the iBook version, available for iPad and Mac, in which you need merely touch a sound example to hear and see it.

1 Music and the Origins of Ancient Science

Music entered deeply into the making of modern science because it was already a central element of ancient philosophy. Greek concepts of number and cosmos were the foundations to which their successors looked, even when they turned toward new directions.[1] The ancient Greek word *mousikē* denoted all the activities of the Muses, vocal and instrumental art as well as the arts of poetry and dance, which the followers of Pythagoras then connected with their teaching that *all is number*, thereby also implying that *all is music*. This fundamental connection between music and mathematics had fateful consequences. Plato developed what Pythagoreans first named "philosophy" into a new kind of education that unified the study of arithmetic, geometry, music, and astronomy. Expressing the consonance of the primordial musical intervals, integers were separated from irrational magnitudes, setting arithmetic apart from geometry. Yet mathematical ratios shaped the physical world, as expressed in the mythical story of Pythagoras visiting a smithy: music was the meeting ground where the first experiments interrogated the mathematical underpinnings of experience. We retrace these deep connections by recapitulating their historical sequence.

Born on the island of Samos in the mid-sixth century B.C.E., Pythagoras himself remains so shadowy a figure that everything said about him is controversial. Even by the fourth century, the brotherhood who deified him had dispersed; modern historians no longer accept the traditional view that they founded Greek mathematics.[2] A century later, a few fragments remain from the writings of those who came to be called Pythagoreans: Philolaus, a contemporary of Socrates, and his student Archytas, whom Plato knew and admired.[3] Philolaus held that "all things, indeed, that are known have number: for it is not possible for anything to be thought of or known without this," underlining the primal status of number as the inescapable criterion of intelligibility. For him, music takes its place at the very center of the treatment of number and the cosmos. In Homer, *harmonia* has the literal sense of fastening together (the word is used to describe the fashioning of Odysseus's raft) or a covenant or agreement (such as the compromise Hector proposes to Achilles during their combat).[4] In Philolaus, *harmonia* has both the general sense of "locking together" like and unlike in the cosmos, "a unification of things multiply mixed," as well as specifically meaning an octave (♪ sound example 1.1). He lists more complex musical intervals that fit together to make the *harmonia* of an octave, which he also uses as a verb, to harmonize, literally meaning to "octavize": "Nature in the cosmos was harmonized from unlimiteds and limiters, both the whole cosmos and all things in it." *Cosmetein* means to put in order and thus beautify (hence "cosmetics"). The cosmos can be an ordered, intelligible whole only because it is *harmonized*, organized into octaves using alternate mixtures of what Philolaus called "limiters" and "unlimiteds," which came to be identified with odd ("limiter") and even ("unlimited") numbers.[5]

Several ancient accounts weave the moment of discovery into a famous, though problematic, myth. Dictating a simplified account for a noble lady in the second century C.E., Nicomachus gave the earliest extant version: Pythagoras was engrossed in trying to find "some instrumental aid for the hearing" that would be comparable to the measuring rod or compasses used by sight or the balances used for weight and touch. Then,

> happening by some heaven-sent chance to walk by a blacksmith's workshop, he heard the hammers beating iron on the anvil and giving out sounds fully concordant in combination with one another, with the exception of one pairing: and he recognized among them the consonance of the octave and those of the fifth and the fourth. ... Overjoyed at the way his project had come, with god's help [*kata theon*], to fulfillment, he ran into the smithy, and through a great variety of experiments [*peirai*] he discovered that what stood in direct relation to the difference in the sound was the weight of the hammers, not the force of the strikers or the shapes of the hammer-heads or the alteration of the iron which was being beaten.[6]

Writing as the minister of a barbarian king four centuries later, Boethius transmitted this story with some additions: "in order to test [*inquiriebat*] this theory more clearly, [Pythagoras] commanded the men to exchange hammers among themselves. But the property of the sounds was not contingent on the muscles of the men, but rather followed the exchanged hammers."[7] Thus, even though Pythagoras came to the smithy "by the favor of a god," what happened there was not a supernatural or miraculous event worked by divine power. Instead, we visit an everyday workplace and behold ordinary events revealed as wondrous. Pythagoras's epiphany led him not to worship or quiet contemplation but to human actions inquiring into that mysteriously unified sound. Had this been the scene of a divine revelation, what followed might be understood as inappropriate or even profane because Pythagoras actually tried to stop or alter the very thing he found most wonderful.

This test was the archetypal experiment, a trial (*peira*) in which Pythagoras *tests* the source of the wonderful sound through an *action* that attempts to alter it, rather than through purely verbal or rational means. In Homer, the verb *peiraō* can mean an assault that tests the enemy's strength, but also an attempt to gain information, to test someone's character or fidelity. In Boethius's version, Pythagoras tested the weight of the five hammers in the shop and found that four of them produced consonant intervals based on the ratios of their weights: the hammers whose weight was in ratio of 1:2 sounded an octave; those of 2:3, a perfect fifth; those of 3:4, a perfect fourth. Each consonance yoked an odd with an even number, corresponding to the "limiter" and "unlimited" Philolaus described. These ratios are successive pairs taken in the sequence 1:2:3:4, which Boethius also expresses in the sequence 6:8:9:12: within the octave (6:12 or 1:2), the fourth (6:8 or 3:4) and the fifth (6:9 or 2:3) find their place. This then implied, *without* adding any new information (or hammers), that between the interval of the fifth and the fourth emerges the ratio 8:9, later called a *tone* or *whole step* because it is the step between these two intervals (♪ sound example 1.2), which according to Nicomachus "was in itself discordant, but was essential to filling out the greater of these intervals."

While Nicomachus reconciles the discordant hammer as "essential" to the greater interval, Boethius tells that "the fifth hammer, which was dissonant with all, was rejected." This, too, should be taken as part of the foundational myth: *an experiment requires recognizing and dealing with dissonance*, the part of an experience that does not "sound together" with the rest. Clearly, there is peril here: how to decide what is "dissonance" that should be set aside, without throwing out some crucial piece of information? At this primal scene of Pythagorean science, confronting dissonance represents the price and also the potential danger of the knowledge achieved through the test and through *principled reconsideration or rejection of some experiences* in order that others may stand out more intelligibly. We shall return to the identity of this fifth hammer.[8]

Boethius notes that Pythagoras continued his examination after he left the smithy: he tested the pitches of strings of different lengths, some stretched by different weights; he tried pipes, "using some twice as long as others, as well as fitting in the other proportions" and glasses filled with different amounts of water by weight (figure 1.1). "Thus he made his belief complete by various experiments," for which Boethius now specifically uses the word *experientia*, whose literal meaning is "something lived through as a trial or even peril" (*ex-perire*). Yet a ten-pound hammer does not ring differently than a six-pounder (as you can hear for yourself in ♪ sound example 1.3); one wonders whether the smiths tried to set Pythagoras straight or whether he even talked to them. Perhaps the word *sphurōn* ("hammer") used in some texts was a misreading or corruption of *sphaira* ("sphere" or "disc").[9] If so, what Pythagoras may have heard were the pitches sounded by various-sized metal discs, which could conceivably have behaved in the numerical ratios recorded, whereas hammers could not. Still, the tale refers to the hammers as the smiths' tools, rather than objects forged in the shop.

Nevertheless, strings do behave as Boethius recounts in describing the "ruler" or monochord (*kanōn*) that Pythagoras developed (figure 1.2), which "is fixed and firm under the study of anyone."[10] A single stretched string mounted against a graduated ruler allows stopping the string at lengths that will realize various proportions; indeed, the proportions 1:2, 2:3, and 3:4 ring out their respective intervals as the story has them (♪ sound example 1.2) if the tension is held constant, but not otherwise. Simple pipes of the same diameter, material, and construction also will sound these intervals, as will water glasses, according to the Boethian story. The difficulties with some of these stories were probably known to

Figure 1.1
Images of the founders of music from Francinus Gaffurius, *Theorica musicae* (1492): Jubal, from the Bible (top left), along with images of Pythagoras (trying bells, glasses, strings, and pipes) and Philolaus.

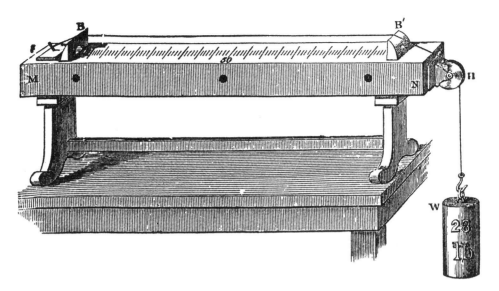

Figure 1.2
A monochord from John Tyndall, *Sound* (1871).

the great Alexandrian astronomer and music theorist Claudius Ptolemy (second century C.E.), who rejects the evidence given by reed pipes and flutes "or weights suspended from strings," as well as by "spheres or discs of unequal weight," and never mentions hammers, judging instead that the monochord (presumably at constant tension) "will show us the ratios of the concords more accurately and readily."[11] Following Ptolemy, we may speculate that the smithy story dramatized original findings with strings. No one may have thought to check whether hammers really behaved that way because it all seemed reasonable: how could hammers *not* obey the proportions already established for strings and pipes?

Here emerges another recurrent peril of experiment: taking a certain pattern, observed in one context, to dictate what "must" happen in another, seemingly analogous situation. None of these problems figured in Pythagorean lore, which presented the story as a miracle; number triumphs even in a smithy. The real wonder may be that numerical ratios can be clear for a simple string, however complex elsewhere. This underlying thread emerges more clearly if we return from Boethius's late Roman summary (written almost a thousand years after the earliest relevant texts) to consider what survives of the earliest Greek evidence.

The Pythagoreans called *pythmenes* ("base" or "foundation") two, three, and four the "first numbers" *because* they "produce the ratios of the concords," the primal consonances of octave, fifth, and fourth.[12] One, never considered a "number" in Greek mathematics, is both even and odd, the primal monad (*monos*, "solitary," "unique") out of which all the other numbers came. For Philolaus, the One was cosmologically as well as metaphysically central: "The first thing fitted together, the One, in the middle of the sphere, is called the hearth," the central fire around which all else revolves. Aristotle noted that, though "most people say that [the Earth] lies in the center," as he himself held, "the people in Italy who are called Pythagoreans speak in opposition to this. For they say that at the center is fire, while the Earth is one of the stars, and by traveling in a circle around the center makes night and day."[13] We will return to this Pythagorean notion in the following chapters.

Recounting Philolaus's views, Plato presents the same *harmonia* governing soul and body as regulates the cosmos. In his *Timaeus*, Plato asserts that the soul and the cosmos are both *made of music*, which explains why we are so moved by the primal cosmic harmonies: in them, we recognize the same numerical concords that ground our own being.[14] Archytas took Philolaus's ideas as the basis of what he called the four *mathēmata*, literally "learnable things": astronomy, geometry, "numbers" (arithmetic), and music. These four are united because "their concern is with the two primary forms of what is, which are sisters themselves," by which Archytas may have meant the realms of the visible and the audible.[15] In his *Republic*, Plato put these four at the center of the education that leads to what he called "philosophy," which he depicted as a journey from the shadowy illusions of perception to the full light of reality that shines beyond the dark cave we call "the world." This quest requires harmony in both body and soul, for which Plato prescribes the combined practice of *mousikē* and *gymnastikē*, which is not merely acrobatics or calisthenics but the living embodiment of *mousikē* in the moving body. He uses the expression "to rhythmize" as a synonym for education.[16]

Children learn these musical skills "by habit, not knowledge, imparting a kind of tunefulness by mode and gracefulness by rhythm," but those who aspire to guide and guard the human commonwealth must go further. To pass from habit to knowledge, they need "to learn to count," an ability (Socrates notes) even the famous warlord Agamemnon lacked, "that little matter of distinguishing one and two and three."[17] With his usual irony, Socrates may allude to the famous catalog of ships, which Homer includes as if the war chief were incapable of numbering his vast host, or may allude to Agamemnon's failure to understand the need to unite his army, rather than divide it by angering Achilles. In more senses than one, Agamemnon did not know what really counts.

Numbering here unites a basic sense of counting off with the larger judgment of what objects count as sufficiently separate to merit enumeration, which depends on context: Agamemnon and Achilles are two separate men, at odds whether or in what way they should constitute a single larger whole as allied Hellenes. Then too, in an era before widespread literacy, knowledge of calculation was far less widespread; the Greek notation for numbers lacked a zero and used the alphabet in ways that made numerical knowledge depend on basic literacy.[18] Plato takes these numerical foundations as important because they rely on fundamental distinctions between what is the same and what is other, what counts as one and what as two. Thus, though a child can parrot counting numbers, understanding them requires discernment.

This deeper knowledge leads to the four sister studies, each of which illuminates some essential aspect of *mathēmatikē*, the master art of grasping all learnable things. All four studies are necessary for the quest Socrates calls philosophy: each of these studies "appears to compel soul to use thought by itself for purposes of truth by itself."[19] Arithmetic relies on the most fundamental insight into being and otherness and how they can be combined into more or less inclusive wholes when considered as a one, or distinguished into the various numbers as many. The primal category of *multitude* is manifest in whole numbers, each one called in Greek an *arithmos*, meaning a countable multitude of countable things, an integer whose wholeness is essential to its integrity, its very essence. The word *arithmos* comes from the Indo-European root common to our word *rite*, as in the counting essential to the performance of sacred ritual, as well as the fundamental sense of rightness (in Sanskrit *ṛta*) or rhythm (in Greek *rhythmos*) underlying the cosmos as an ordered whole. A shape or pattern conveying motion, such as the fluid pose of a dancer or of a sculpted figure, was also called *rhythmos*, which the Romans commonly translated as *numerus*, number itself.[20]

Arithmetic also concerns how whole numbers can be connected by a ratio, a *logos*. In Homer, the verb *legein* means "to gather together," as when the grieving Achilles tells his comrades "let us gather up [*legōmen*] the bones of Patroklos" in preparation for his funeral rites.[21] By extension, this word for gathering or collecting also came to signify speaking, recounting, telling, and reasoning, implying that all these are deeply forms of *bringing together*, hence of *connected* expressions. In that sense, only by means of *logoi*—the primal relations between integers—do the counting numbers really become fully the object of accounts and reason, of *logos* manifest in what the Greeks therefore called *logic*. *Logos* also has the specific meaning of a musical interval, hence suggesting that, as *logoi*, musical intervals may be deeper even than the integers whose relation they express. One might daringly suggest that the intervals (such as 1:2, 2:3, 3:4) come before the integers themselves, which remain profoundly isolated until we express their relation. Can we understand the concepts of two or three only if we grasp each in relation to the unit of which they are implicitly composed (2:1, 3:1)? If so, arithmetic may implicitly rely on musical ratios to ground our awareness of number.[22]

In contrast, geometry deals with *magnitude* (*megathos*), which Plato (and Greek mathematicians in general) considered deeply different from *multitude* (*plēthos*). The Pythagoreans were credited with the crucial insight that showed the full extent of the distinction between arithmetic and geometry: in general, geometric lines cannot be expressed as any number or as any ratio of finite numbers. Most famously, the diagonal of a square is not commensurable with its side: if its side is a unit length, no ratio of whole numbers $m:n$ can express the length of the diagonal, called *irrational* (*alogon*, not having a *logos*) or *unspeakable* (*arrhēton*) because not expressible in terms of finite numbers. A beautifully terse contradiction emerges if one assumes hypothetically that such a ratio could be found, for if so, the "numbers" of that ratio would have to be simultaneously odd and even.[23] Our symbol $\sqrt{2}$ does not resolve this problem but merely gestures symbolically toward a "ratio" that is in fact no ratio of finite numbers but (as we would say) an infinite decimal, 1.41421356237 ... . It is, indeed, both even and odd, or neither.

Thus, Greek mathematics could never speak of "irrational numbers," as familiar as that terminology became after the sixteenth century. In Greek mathematical texts, numbers were by definition integers and did not include zero (a concept nowhere explicit in their mathematical texts) or one. Indeed, some Pythagoreans considered two not a number but a crucial intermediate, a mysterious dyad that bridges the utter solitude and uniqueness of the One and the multitude of the Many.[24] In his own way, Plato treats the dyad as "unlimited" by connecting pure Being—the One in its solitary splendor—and non-Being, leading to the variable, ever-changing multiplicity of the cosmos, in which Being and non-Being are not utterly separate but somehow interwoven into the structure of Becoming as we experience it.[25] Throughout his dialogues, Socrates and his friends examine the strange mixture of truth and story, *logos* and *mythos*, constituting the living stream of language and thought. They keep looking to the realm outside our dark cave where, if only in imaginative speech, we aspire to see the One and the other pure Forms or Ideas, each distilling the ultimate essence of a number or concept. Socrates suggests that "proceeding to the major and more advanced part of geometry tends to make it easier to behold the Idea of the Good," the highest Form, which many passages in Plato suggest may be identified with the One itself. As Socrates notes, everything in that realm of Forms tends to compel "soul to be turned round to that place in which the happiest of what is exists, which soul must in every way behold."[26]

Accordingly, Socrates pokes fun at geometers' use of phrases like "squaring," "applying," or "constructing," "as if all their words were for the sake of action" rather than "undertaken for the sake of knowledge," meaning the philosophic contemplation "of what always is, not of what sometimes comes to be and passes away." Though he touches on the practical uses of mathematics, which appeal to his companions, Socrates is much more interested in its "useless" aspects because it awakens and sharpens "an organ in the soul of every man which is purified and rekindled in these studies when it has been destroyed and blinded by other pursuits, an organ more worth saving than ten thousand eyes; for truth is seen by it alone."[27]

Given the exalted purity of what Socrates seeks to behold, he and the other Greek mathematical writers would have been amused, if not dismayed, by the breezy nonchalance with which later mathematicians speak of "irrational numbers," which would have seemed to them a sheer contradiction in terms, even nonsense: "nonnumerical numbers," in their terms, or "uncountable counting numbers." Socrates teases his young friends for being "as irrational as lines," an ironic judgment he likely would have passed on the modern mathematical proclivity to mix rational with irrational quantities. Socrates makes fun of fractions, which he does seem to have heard of: "For you surely know that if someone undertakes to divide One itself in speech, those who are skilled in these matters laugh and won't allow it. On the contrary, if you break [the One] up, they multiply it, taking care that one should never appear not one, but many parts."[28] He notes that we implicitly take any such supposed parts or fractions each to be one in itself, thus negating the premise that we had "broken" the One into many pieces.

For Plato, these counting numbers, with the One as their supreme source, stand as the touchstone of knowledge as such, the prime example showing the *possibility* of human reason, of knowing *anything* with certainty. We literally *count* on the difference between one and two and three as the most certain things we know, which indeed define knowledge itself. Thus, the Greek insistence on the utter distinction between number and magnitude is not merely terminological fussiness but the central point on which they grounded their search for truth. To admit, as moderns do, "irrational numbers," "imaginary numbers," or "surreal numbers" on equal terms with the integers implicitly rejects and ignores the fundamental insight of logic: *countable* entities cannot be confused with the endless divisibility possible (and necessary) for *uncountable* magnitudes, such as geometric lines.[29] By holding fast to this distinction, Euclid and Plato balanced the realms of the rational and the irrational, giving coordinate but separate domains to each, respecting both by never mixing them. This widely held set of assumptions will be of great importance at several points in this book, as they came to be challenged and replaced with the modern alternatives.

These fundamental mathematical premises are deeply grounded in musical findings. The primal Pythagorean ratios place music on the side of arithmetic, not geometry. For instance, if one tries to "hear" the irrational ratio formed by the relation of the diagonal of a square to its side by approximating that interval on two strings, the result is very close to the tritone, the interval later notorious as the *diabolus in musica* (♪ sound example 1.4). Ironically, a perfect *geometric* division corresponds to a highly *dissonant* musical interval, whereas the *arithmetic* division into simple ratios corresponds to the primal *consonances*. On this basis, it seems plausible that the fifth hammer that Pythagoras discarded as "dissonant with all" was irrational with respect to them; as Adrastus of Aphrodisias put it, the irrational is mere "noise [*psophos*]" that should not even "be called notes [*phthongoi*], but only sounds [*ēchoi*]." If so, the rejection of the dissonant hammer initiated the ancient separation between numbers and magnitudes, arithmetic and geometry.[30]

Indeed, the same fundamental problem plagues the division of any melodic interval. If, for instance, one tries to lay out equal tones (9:8) to fill out an octave, it turns out that five tones undershoot an octave but six tones overshoot it (♪ sound example 1.5); to create modes or scales, it is necessary to introduce some kind of "half tone" to fill out the missing space between five tones and an octave.[31] But, as the Greeks already realized, a perfectly equal division of a tone (as of an octave) would require the use of an irrational magnitude, as will become of crucial importance in chapter 4.

Box 1.1
The arithmetic, geometric, and harmonic means.

Arithmetic mean	*Geometric mean*	*Harmonic mean*
1:2:3	1:√2:2	1:2:4
2 is the arithmetic mean between 1 and 3, an equal difference from both: 2−1 = 3−2	√2 lies at an equal ratio (√2:1) between 1 and 2 because 1:√2 = √2:2	The *differences* 2−1= 1, 4−2 = 2 are in the same *ratio* as the terms, namely 1:2

Nevertheless, the Greek theorists found a way to include geometric proportion in music by introducing a *harmonic mean* that is a hybrid of the *arithmetic* and *geometric* means (box 1.1), positioning music *between* arithmetic and geometry. Like music, astronomy (the fourth of the sister sciences) is also poised between *arithmetic* proportions of the ratios between the movements of the heavenly bodies and their description using spherical *geometry*. Music and astronomy bridge the invisible realm of mathematical forms and the sensual realm of experience. Socrates argued that both music and astronomy use the experience of the senses to "summon or arouse thought," whereas many other modes of perception "do not summon thought to inquiry." Through beholding "intricate traceries in the heavens," astronomy "compels a soul to look upward," toward "what is and is invisible."[32] Music, in a different way, joins imperceptible numerical ratios with the perceptible intervals. Both music and astronomy connect the purely mathematical and the sensually perceptible. In what follows, both these sister sciences will play out their intermediary roles. But whereas the formative influence of astronomy in the development of science has long been acknowledged, the ways music has entered this story remain to be told.

These matters aroused deep and enduring controversy. The Pythagorean view of music as mathematical ratios was opposed by Aristoxenus, the great contrarian voice in Greek music theory, so famous that the Romans referred to him simply as "the musician." Where the Pythagoreans exalted reason over sensual judgment, Aristoxenus, like his teacher Aristotle, emphasized the fundamental role of the senses: "Through hearing we assess the magnitude of intervals, and through reason we apprehend their functions."[33] In reasserting the sensual, experiential character of music, Aristoxenus called into question the relation between music and mathematics. But Boethius, adhering to Pythagorean views, treated Aristoxenus briefly and dismissively. Because all musical study during the subsequent millennium relied on Boethius, Aristoxenus fell into obscurity until his texts were rediscovered in the sixteenth century with powerful effects on musical and mathematical thought, as we shall see in chapter 4.

Then too, Ptolemy took the side of the Pythagoreans against Aristoxenus. Ptolemy's *Mathematical Composition (Syntaxis)*, which Arabic scholars called *Almagest* ("The

Greatest"), synthesized the observational data of Babylonian, Egyptian, and Greek observers over centuries and presented a theoretical model that predicted the motion of the planets with an accuracy exceeding anything that had come before, which stood unchallenged for a millennium. At the same time, Ptolemy's *Harmonics* synthesized musical learning on a scale comparable to his astronomical work. Andrew Barker has emphasized the scientific accomplishment of this work, which bridges observational practice with theory in ways comparable with Ptolemy's *Almagest*.[34] Though the *Almagest* had been transmitted via the Arabs and received in the West by the twelfth century, the *Harmonics* only reentered the stream of Western music theory in the sixteenth century. Parts of this work were translated then, though its extant corpus was fully available only in the early seventeenth century. As we shall see, its reentry was consequential not just for the study of music.

Unlike his *Almagest*, Ptolemy's *Harmonics* was not transmitted in its entirety. Its final Book III is an intriguing torso that integrates certain astronomical topics, showing the explicit connection between these matters in his mind. In particular, Ptolemy connects the motions of the planets in his geocentric system with *changes* of musical modes, for which we use the modern term "modulation" to translate his term *metabolē*, whose primary connotations are both the transformations of metabolism (in which ingested foodstuffs are digested and changed into living flesh) and its political sense of *revolution*, change of regime. Ptolemy notes ways in which the "proper motions" of planets (moving closer or farther from their center) parallel musical change of mode.[35] This reflects ancient practices, especially the Athenian "New Music" of the late fifth century B.C.E.; Ptolemy's discussion shows the astronomical correlates of musical modulation.[36] Though the surviving text of his *Harmonics* breaks off at this point, enough survives to give a clear sense of the depth and range of the correlations between music and astronomy in his work. This connection will return in the controversies over heliocentric astronomy.

Plato's "fourfold way" of philosophical preparation set out arithmetic, geometry, music, and astronomy as higher studies to follow the initial "threefold way" of grammar, rhetoric, and logic, the linguistic basis on which all discourse rests. His *Republic* proposed the first utopia, the idealized "no-place" that, even if unrealizable in practice, could set a standard and pattern to which to aspire. His advocacy of the education of women and his critique of slavery remained controversial for millennia.

His radical educational proposals shaped the immediate future. Traditional Greek *paideia* had consisted of memorizing Homer and learning rudimentary arithmetic; Plato created "liberal education," worthy of the free born (*liberi*), as opposed to the rote training of slaves to perform their assigned tasks.[37] His vision everywhere haunts the modern university, even when it turns against the liberal arts to prefer utilitarian vocational training. Beginning with the Academy that Plato founded, his three- and fourfold ways became standard as education of the elite. This pattern was transmitted to the Romans and, via Boethius, to the West as the *trivium* and *quadrivium*; the church then used this plan of liberal education to form its clerics, gradually including rulers and nobility.

Thus, a continuous line of quadrivial studies goes all the way from Plato to about the eighteenth century, in the sense that educated persons were exposed to a unified curriculum of those four subjects, in which music (as mathematical harmonic science) was habitually studied in conjunction with the others. In the chapters that follow, we shall explore some of the consequences of this shared fourfold study, all the more powerful because it was the capstone of higher education, the central content of university learning. Even when the quadrivium is not specifically mentioned, we still need to remember that many educated persons up to about the time of Isaac Newton would have a shared experience of musical theory that was as much a part of their common fund of learning as was the basic study of arithmetic, geometry, and the basic linguistic arts of the trivium (considered so common that they underlie the connotations of "trivial"). Though we continue to share most of these studies as part of our elementary (and even higher) educations, music and perhaps also astronomy have fallen out. Not so for our predecessors, as we shall see.

2   The Dream of Oresme

For two centuries after he wrote it, Boethius's treatise on music was unavailable, seemingly lost in the "dark ages." Beginning in the ninth century, manuscript copies began appearing in ever-increasing numbers; Boethius became the principal source of music theory (and of arithmetic) long before Latin translations made Aristotle's writings directly accessible. In this intermediate period during which Aristotelian science remained relatively unknown, ancient musical theory continued to be taught.[1] In that sense, musical science maintained a continuity that other branches of natural philosophy generally had lost, in the absence of available ancient sources. After the renaissance of the twelfth century, during which Aristotle was translated by Richard of Moerbecke, the dialogue between astronomy, physics, and music could recommence more fully.

In that conversation, Nicole Oresme played an extraordinary role as the leading natural philosopher of the fourteenth century, remarkable both for the breadth and depth of his writings as well as for his penetrating questions and insights. Emerging from humble origins in Normandy, Oresme eventually became the Grand Master of the College of Navarre he attended, designed for students too poor to attend the University of Paris. His scholarly writings attracted the attention of the future King Charles V of France, for whom Oresme prepared translations and commentaries on several of Aristotle's major works, along with his own writings. The king eventually elevated Oresme to the bishopric of Lisieux, where he spent the final five years of his life. Study of his works gives us the opportunity to consider the transmission and reception of ancient natural philosophy as he reformulated its leading issues. Oresme was a probing and daring thinker who shows us the dimensions of the issues at stake before the advent of the "new philosophy" about two centuries later.

Oresme's writings and commentaries brought new life to the ancient texts. As would have been expected of an educated person of his time, he was versed in music as part of the quadrivium and wrote a book (now lost) on the division of the monochord.[2] Though he demonstrated fundamental contradictions in the received teachings about cosmic harmony, he gave that concept new dimensions. He investigated whether celestial motions were commensurable with each other or not, which bears on larger issues of celestial repeatability that he phrased in musical terms. In this, differences between arithmetic and geometry, music's sisters in the quadrivium, came forward musically, mathematically, and astronomically.

Oresme began by showing that the radii of the different celestial spheres cannot be expected to be commensurable with each other, based on Euclid's propositions about spheres circumscribing regular solids.[3] He drew on his older contemporary, Johannes de Muris, best known for theoretical writings on music, though Oresme went much further in examining the consequences. He concluded that the radii of different spheres are far more likely to be incommensurable than commensurable with each other, which still remains possible, though improbable.[4] But if two celestial motions are incommensurable, then any given initial position of the two bodies *will never recur*. His argument is simple; assume hypothetically the contrary, that their motions (say, their velocities) obey a certain ratio, say, 3:2. Then after six revolutions of the first body, both bodies will have regained their initial position, the first body having "lapped" the second body twice over. If the ratio were m:n, it would take $m \cdot n$ revolutions; if, on the contrary, their motions are *not* commensurable, m and n are not any finite numbers, and neither is $m \cdot n$. Hence, the initial positions of these two bodies would never recur even after an infinite time. If so, the cosmos will not return to any given initial configuration, continually assuming different states from any that came before.

On the other hand, Plato had described a cosmic cycle of 26,000 solar years, the "Great Year," after which the planets would return to their initial configuration.[5] Oresme concluded that no such recurrence was possible, hence no Great Year. He emphasized that his result disproves astrology: the impossibility of recurrence disallows the recurrent astral configurations on which astrological predictions depend. Learning this, he hoped that the ignorant would abandon astrological determinism and understand their free will.

At many points, Oresme's inclusion of musical matters plays a central role. His judgments reflected his interest in new music and in new ideas, including the speculation that the Earth might not be the center of the universe. More than century and a half before Copernicus, Oresme's writings give an invaluable view into the status of geocentric cosmology. Though he addressed this issue at several points in his earlier writings, his most extended treatment comes in his final work, *Le Livre du ciel et du monde* (1377), a translation and extensive commentary on Aristotle's *On the Heavens* written for Charles V (figure 2.1).

Oresme devotes considerable space to the suggestion that the Earth may not be at the center of the cosmos and might move, rather than remaining at rest, as Aristotle had argued. What Oresme calls the "Italian or Pythagorean" view places "the sphere of fire" at the middle, so that "the earth is a dark star moving in a circle around the center and … this is the cause of our nights and days." Explaining the heliocentric view, Oresme compares the planets moving clockwise around "the entire circumference of the wheel … just like people in a *carole*," a circular dance (*chorea* in Latin) that during the twelfth century was

Figure 2.1
An illuminated page from Oresme's *Le Livre du ciel et du monde* (1377, fol. 3r), showing God raising his right hand in benediction, holding in his left hand a circle of fire containing a symbolic representation of the globe, above which the inscription reads: "The senseless man shall not know; nor will the fool understand these things" (Psalm 91:7 [92:6]). In the margins, the heraldic swan, coat of arms, and motto *Le temps venra* ("the time will come") all indicate the patronage of the Duc de Berry.

still danced in church on special occasions.[6] Oresme presents a convincing account of this view, including the argument (often attributed to Copernicus and Galileo) that "we do not perceive motion unless we notice that one body is in the process of assuming a different position relative to another."[7] Given the prestige of Aristotle, the extent and sympathetic quality of Oresme's account of the heterodox "Pythagorean" alternative has been much remarked by scholars, some tempted to judge him a heliocentrist. He goes so far as to argue that it would be paradoxical, even preposterous, for the heavens to rotate diurnally, requiring them to travel at high speed to complete their daily revolution.

Oresme also confronted such scriptural passages as the famous miracle of the sun standing still for Joshua. Like many other scriptural interpreters, going back to Augustine, Oresme notes that the Bible "conforms to the customary usage of popular speech … where it is written that God repented, and He became angry and became pacified, and other such expression which are not to be taken literally. … Thus, we could say that the heavens, rather than the earth, appear to move with diurnal motion while the truth is the exact opposite."[8]

In the end, Oresme seems to draw back from advocating this extreme view, however strongly he had presented the arguments in its favor:

However, everyone maintains, and I think myself, that the heavens do move and not the earth: For God hath established the world which shall not be moved [Psalm 92:1], in spite of contrary reasons because they are clearly not conclusive persuasions. However, after considering all that has been said, one could then believe that the earth moves and not the heavens, for the opposite is not clearly evident. Nevertheless, at first sight, this seems as much against natural reason as, or more against natural reason than, all or many of the articles of our faith. What I have said by way of diversion or intellectual exercise can in this manner serve as a valuable means of refuting and checking those who would like to impugn our faith by argument.[9]

Oresme's statement might be read as a carefully balanced accommodation to common opinion, despite the powerful arguments against the geocentric view he had just presented. Perhaps he found those arguments privately persuasive but so disturbingly contrary to church teachings (and to common opinion) that he prudently overrode them. As an experienced ecclesiastic, Oresme may well have discerned the enormous doctrinal controversy that would ensue, were he to advocate heliocentric cosmology. When, a century and a half later, Copernicus espoused that view, the text of his *De revolutionibus* reveals his apprehension of ecclesiastical condemnation; the rhetoric of his dedicatory letter to Pope Paul III clearly aims to avert those dangers. One imagines that Copernicus was privately relieved that he could present his controversial theory from his deathbed, where he was supposed to have seen the first copy of his book, escaping any furor by disappearing into the hereafter.

If indeed Oresme privately rejected the geocentric view, he had no such escape route available; his rather tortuous formulation of his public position could be read as walking a tightrope between dishonesty to his intellect and imprudent disclosure. His quotation of the verse from the Psalms seems disingenuous, given his own dismissal of literal appeals to figurative language in the scriptures. Oresme presents his antigeocentric presentation as an "intellectual exercise" and mere "diversion," hence purely hypothetical and therefore not subject to the rigorous doctrinal scrutiny he may have anticipated, had he put forward those views as realistic representations of the cosmos. His allusion to "those who would like to impugn our faith by argument" may refer to long-standing controversies about the relation of reason to the mysteries of faith; as such, he tacitly seems to indicate the mathematically simpler motion of the Earth, compared to geocentric cosmology. His passing comment that the alternative cosmology may violate natural reason as much as "the articles of our faith" could be read as comparing the difficulties of the antigeocentric view with those already surmounted by Christian apologists. His implicit suggestion may be that heliocentrism strains human credulity no less than do the paradoxes of Christian doctrine.

But setting aside these more or less speculative possibilities, it is more likely that Oresme considered the matter finally undecidable, however probable the arguments he adduced for the Earth's motion and rotation. Likely affected by the wider skeptical and probabilistic currents in fourteenth-century natural philosophy, Oresme wrote that "I indeed know nothing except that I know that I know nothing" about natural knowledge, using the famous Socratic formulation to express the inadequacy of human opinion. As such, he distanced himself from the "real" explanations later claimed by Copernicus and Galileo because of his principled demurral from certainty, which also may have accorded conveniently with his desire to avoid doctrinal controversy.[10] Even so, we cannot consider his position merely timorous; he seems quite sincerely to have considered the foundations of natural philosophy ultimately to lie beyond human certainty—and it remains possible that he was right.

In trying to assess his real views, however, we should include his treatment of the musical context of astronomy that immediately precedes his discussion of geocentrism. His musical considerations provide additional and perhaps decisive evidence against the diurnal movement of the heavens and hence against geocentrism. Oresme follows Aristotle in considering that the celestial harmonies are not audible, but he takes the primal musical proportions quite seriously, arranging the harmonic ratios in a two-dimensional array, a figure he considers full of "very great mysteries" (figure 2.2, table 2.1). Note that the top row contains the successive powers of 2 (1, 2, 4, 8, 16, 32), while the leftmost column lists the powers of 3 (1, 3, 9, 27, 81, 243). Then the interior of the array lists the various products of these outer rows and columns precisely in accord with the modern rule for the terms of a matrix: the term in the nth row and mth column is given by nm. This may be the earliest "matrix" (though without using that modern term), in which Oresme lists all the various possible products that appear in musical theory, as he knows it. His diagram contains many blank cells, in addition to those in which he has noted numbers that appear explicitly in musical theory. By thus drawing attention to the other, heretofore unnoticed

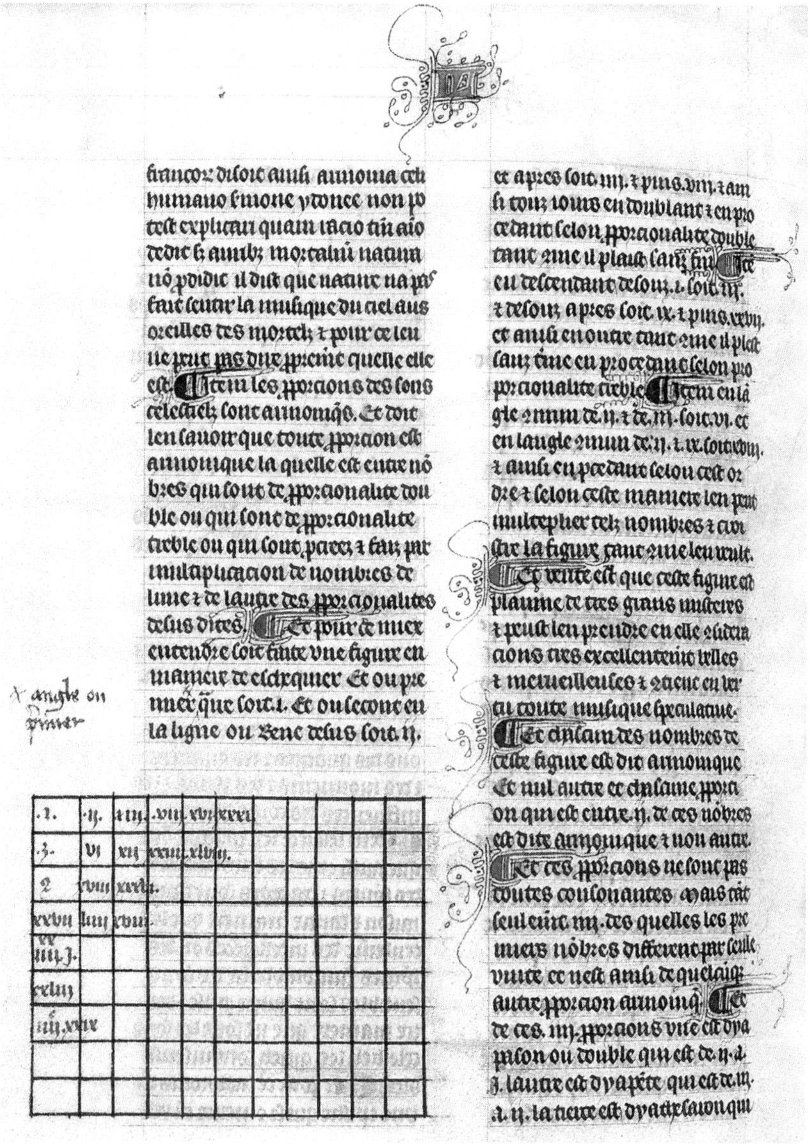

Figure 2.2
Oresme's diagram of the principal musical ratios, from *Le Livre du ciel et du monde* (1377, fol. 125v). (For his diagram in modern notation, see table 2.1 on the following page.)

Table 2.1
Oresme's diagram.

1	2	4	8	16	32				
3	6	12	24	48					
9	18	36							
27	54	108							
81									
243									
729									

possibilities, Oresme may be indicating implicit speculations about as yet unused "harmonic" possibilities. Less speculatively, his diagram indicates a nascent interest in pure combinatorics (the array of all possible products of a certain form, here powers of 2 and 3), as well as the possibility of visualizing them in such an array.

Oresme approvingly quotes Cassiodorus's sentiment (by then a commonplace) that human ears are too gross to perceive these celestial ratios, which also govern earthly music. Yet, along with Boethius and so many ancient authors, Oresme nonetheless goes on to consider even an inaudible "music" of the spheres as crucial to the cosmos. In so doing, he shows the continuing availability of music as the meeting ground between the suprasensual world of mathematics and the perceptible evidence provided by astronomy.

In fact, Oresme deploys music to solve a long-standing astronomical problem. In geocentric cosmology, there remained the question of which musical pitches should be assigned to the various heavenly spheres. In particular, does the "highest" sphere, that of the fixed stars, correspond to the lowest or the highest pitch of the celestial system, even assuming (as Oresme does, following Aristotle) that no audible sound results? Even though these spheres produce no sound grossly audible to our ears, Oresme still applies the musical language of relative pitch to describe the various possibilities. He uses the spheres' decoupling from ordinary processes of sound production to consider new possibilities of "musical" cosmology.

If, he asks, the sphere of the fixed stars were in diurnal revolution, as required by the geocentric view, then what musical pitch should be associated with it? Oresme notes that, according to many accounts, this rapid revolution of a huge and massive structure is often associated with a high pitch. Moving inward from it, the planets closer to Earth should then have successively *lower* pitches, until we reach the unmoving (and hence presumably silent) Earth itself. But this seems to him problematic, for those most exalted stellar spheres would seem more suitably associated with deep, solemn tones, not high-pitched squeaks. These musical considerations concern a cosmological decision of considerable importance, undecided on purely astronomical grounds. One wonders, too, whether he was troubled by the correlate arguments implied for the "music" of the Earth, in each case. On the contrary supposition, the Earth's immobility would be associated with the highest pitch, which also seems problematic: how can an immobile body be associated with a high degree of vibration?

Oresme does not comment on these incongruities, so it is not finally possible to assess their significance for him. If we were to take them most seriously, they would seem to impeach the geocentric view on musical grounds. As such, they might be read as forming an implicit extension of his antigeocentric arguments, amassed above, perhaps indicating to discerning readers a hidden heliocentric drift in Oresme's argument, his disclaimer notwithstanding. But nothing in the text authorizes us to take this rather conspiratorial reading as anything more than speculation. What is clear is that, for Oresme, musical arguments can address otherwise undecidable astronomical questions.

In his *Livre du ciel*, Oresme brings this approach to bear on his own inquiry into the relative status of incommensurable versus commensurable celestial movements. Here again he faces issues that are not decidable from within astronomy alone; he reminds us that nothing tells us a priori whether any given celestial sphere is or is not commensurable with another, though far more likely to be incommensurable. In his earlier *Tractatus de commensurabilitate vel incommensurabilitate motuum celi* (*Treatise on the Commensurability or Incommensurability of the Celestial Motions*, written sometime during 1340–1377), Oresme staged this problem in the form of a debate between personified figures of Arithmetic and Geometry, enacted at the command of Apollo himself. The whole dramatic scene is unique among his works, which he generally phrased in the traditional Euclidean style of geometrical propositions.

Appearing as a character in his own drama, Oresme expresses his perplexity whether incommensurability is actually present in astronomy or only a purely theoretical possibility. Then Apollo, accompanied by the Muses, Arts, and Sciences, appears to Oresme "as if in a dream." Apollo rebukes him for being "ignorant of the ratios relating the things of this world" and hence subject to "affliction of the spirit and an unending labor." Apollo phrases the problem trenchantly; "an imperceptible excess—even a part smaller than a thousandth—could destroy an equality and alter a ratio from rational to irrational." Citing the authority of al-Battani ("if you have read him"), Apollo concludes that "the ratios of these motions is unknown, and neither arithmetic nor geometry can lead you to a knowledge of it." Addressing Apollo as his "dear father," Oresme reiterates "that it is not given to human powers to discover such things" (his stated position on the geocentric controversy) yet still asks "why did you make the very nature of men such that they desire to know, and then deceive or frustrate this desire by concealing from us the most important truths?" Responding to the intensity of Oresme's pleas, Apollo, "smiling," orders the Arts and Sciences "to teach him what he asks." Thereupon Arithmetic and Geometry respectively plead their opposing positions before this highest court of knowledge. Apollo then orders them both "to defend their cause with reasoned arguments, as if they were litigants in a lawsuit," while Oresme listens "filled with wonder."[11]

As Apollo indicated, arithmetic or geometry alone cannot decide such larger issues that involve all the arts and sciences. Nor does Oresme personify astronomy as a speaker in the debate, for her status would be dependent on the result of this debate, which concerns the basis of her science. In the end, both sides invoke music as a deciding factor to break the mathematical deadlock.

Arithmetic's position is the most straightforward and traditional, as befits her claim to be "firstborn" of the quadrivium, on whose concepts all the others depend. Her biblical allusion that "the architect" built everything according to "number, weight, and measure" still does not quite resolve these more detailed mathematical issues. Arithmetic argues that "the greatest prince of all, himself one and three everywhere," the triune God, disclosed the primacy of number when he "arranged all things pleasantly, that is, harmonically," using *rational* quantities because each "irrational proportion is discordant and strange in harmony, and, consequently, foreign to every consonance, so that it seems more appropriate to the wild lamentations of miserable hell than to celestial motions that unite, with marvelous control, the musical melodies soothing a great world." Arithmetic then cites a host of ancient authorities from Hermes Trismegistus to Cicero attesting to the sublime pleasantness of the celestial concords, and hence their consonance. She also notes the consequent deduction of the Platonic Great Year and other recurrent astronomical cycles noted by the ancients. But her deepest argument seems to be that irrational proportions sound terrible and thus cannot be allowed in a harmonious cosmos.[12]

In response, Geometry does not deny "a certain eternal beauty and perfection in her [sister's] rational ratios" but wants to subsume them in a larger and less consonant musical whole. Her argument moves boldly toward a praise of artistic innovation: "The heavens would glitter with even greater splendor" if some motions were incommensurable than if all were purely commensurable. Though she disputes Arithmetic's claim of precedence as the "firstborn," Geometry does not try to argue that irrational ratios are more pleasant than rational ratios. Rather, she considers that Arithmetic's reliance on the criterion of pleasure is artistically inadequate to grasp the full complexity of cosmic music, for which diversity is Geometry's touchstone: "What song would please that is frequently or oft repeated? Would not such uniformity produce disgust? It surely would, for novelty is more delightful." Geometry asserts that purely rational music would be like the sound of a cuckoo, annoyingly repetitive not only in its uniformity of elements but in its endless recurrences.[13]

Though Geometry does not really address the troubling sensual displeasure produced by irrational proportions, she implicitly embraces it as the price of really interesting cosmic music that will not repeat itself ad infinitum. Where Arithmetic had spoken only indirectly of "the architect" (notably restrained in her reference to revelation), Geometry explicitly describes God the creator and his artistic alternatives. In contrast to the pleasures offered by Arithmetic, Geometry considers her own ideal student to be "a subtle man [who] perceives the beauty in much diversity, while an ignorant man, who fails to consider the whole, thinks that the sequence in this diversity is confused, just as he who does not realize that what we call an irrational ratio is part of our order and plan. And yet the infinite plan of God distinctly realizes this diversity which, put in its proper place, is pleasing to the divine sight and makes the celestial revolutions more beautiful."[14]

Geometry then reinterprets Arithmetic's biblical reference to "number, order, and *measure*" to mean that measured *magnitudes* are just as necessary as pure numbers. Geometry also discerns a wider horizon of mathematical possibilities for harmony, not just in planetary velocities or periods (as Arithmetic had implied) but also in the "magnitude" of the spheres, their weight or size, and hence by implication their spatial dimensions. Consistent with her specific subject matter, Geometry emphasizes the full spatial reality of the celestial spheres as structures with determined radii, not just a mathematical model (as Ptolemy had argued) but a measurable geometric edifice.

The musical implications of these questions lead us back to the issue of audibility. Arithmetic had treated celestial music as audible, hence excluding perceptibly inharmonious irrationals; Geometry, in contrast, cites the ancient authorities who treat cosmic "harmony" as inaudible, in which case sensory disharmony would be an irrelevant criterion. Even so, she keeps using *musical* language to describe the cosmic harmonies. Oresme seems to want to hang on to the imaginative and artistic possibilities of musical discourse even as he questions its sensory basis.

The ending of this singular debate leaves us hanging. After Geometry finishes speaking, "Apollo, believing himself adequately informed, ordered silence." But Oresme feels "astonished and confounded by the novelty of so many things," especially by the manifest contradictions between the arguments of Arithmetic and Geometry. Perceiving this, Apollo reassures him not to believe "that there is genuine disagreement between these most illustrious mothers of evident truth. For they amuse themselves and mock the stylistic mode of an inferior science." Apollo announces that he will now announce the truth in the form of his judgment, so that "with the most ardent desire did I await his determination, but, alas, the dream vanishes, the conclusion is left in doubt, and I am ignorant of what Apollo, the judge, has decreed on this matter."[15]

This enigmatic interruption could be interpreted as a wry expression that we cannot, after all, know the truth of such exalted matters, on the lines of Oresme's skeptical account of geocentrism. But a number of clues allow us to conclude that Oresme's own opinion lies finally with Geometry. We know, from several of his writings (including the *Livre du ciel*) that he strongly held to the inaudibility of cosmic harmonies, which put him on Geometry's side on this issue, at least. Even stronger evidence comes from Oresme's *Livre du ciel*, particularly the musical judgments he expresses there, which align strongly with Geometry's position on cosmic inaudibility and the impossibility of astronomical recurrence.

Oresme makes a point of connecting cosmic music with the essential irreversibility of events. Referring to his own earlier arguments on this subject, Oresme proceeds on the assumption that celestial motions are incommensurable, again showing his agreement with the position of Geometry. If so, "the heavenly bodies are continually and always in new positional relationships with one another so that it is naturally impossible that these positions ever repeat themselves again."[16] He immediately interprets this musically: the heavenly bodies "are continuously producing new but imperceptible music: *canticum novum*, a new song, such as never existed before."[17] He goes on to clarify his scriptural reference: "And Holy Scripture often speaks of the divine music of the angels and blessed souls caused by God Himself: They were singing a new canticle [*canticum novum*] before the throne," citing a phrase from the Book of Revelation that also figures in several psalms.[18] Oresme specifically praises the *newness* of the song, its continual novelty; though an opponent of astrology, he saw Heaven and Earth as connected, causally and musically:

Since the bodies of our world are governed by heavenly bodies and by their natural movements, as Aristotle says in the first book of *Meteors*, it follows therefore that terrestrial bodies are continuous in new and different arrangements such as never previously existed and that human affairs, except those that depend upon the will as opposed to natural inclination, are continuously different and such as they never were before in any way at all. Just as change cannot exist unless it is for better or worse—although both better and worse are sometimes for the best—and just as choral singing [*chant de pluseurs voiez*] by excellent voices is not so good if the voices always sing in absolute harmony, in the same way things here below are sometimes in better state than at other times, depending upon the variations in the imperceptible music of the spheres; accordingly, sometimes we have peace, sometimes war, as the Scripture says: *A time for war and a time for peace*; one time sterility, another time fertility, and so on with all the other changes.[19]

This striking passage opens many doors. His reference to "choral singing" is one of the rare contemporary mentions of the novel practice of polyphonic music, so significant a musical development that it deserves treatment elsewhere in its own context.[20] Here we stress its novelty: the prevalent practice of monophonic music, such as the single melodic line of Gregorian chant or troubadour song, in the centuries before Oresme had been joined by variegated and exuberant experiments in many-voiced music, from organum and the School of Notre Dame (in the twelfth century) to the ever more complex motets of *ars antiqua* and *ars nova* in Oresme's own time.

Indeed, Oresme's celebration of the "new song" is arguably an indirect reference to the musical *ars nova* not only because of the common theme of "newness" but because Oresme was directly connected with the most important master of this new style, his elder contemporary Phillipe de Vitry.[21] A renowned scholar and friend of Petrarch, de Vitry had written his *Ars nova notandi* in 1322 or 1323, setting forth the novel rhythmic procedures of the "new art." De Vitry's own musical compositions are exemplars of this new idiom (♪ sound example 2.1). Oresme dedicated his mathematical work *Algorismus proportionum* to de Vitry, "whom I would call Pythagoras if it were possible to believe in the return of souls ... so that if it is agreeable to your Excellency you may correct that which I put before you. For should it be approved by the authority of so great a man and corrected after his examination [of it], everything that has been revised by your correction would be an improvement. Then, if a disparager should open his mouth and set his teeth to rend [my work] into pieces, he would not find [what he seeks]."[22] Oresme's *Algorismus proportionum* makes no overt reference to music; it treats the addition and subtraction of "rational ratios" and "irrational ratios," evidently as part of his larger project to understand their relation. Oresme's prologue considers de Vitry not only *au courant* with this advanced mathematical investigation but capable of judging and correcting it, probably also of approving and applauding it.

Thus, Oresme's praise of the *canticum novum* accords with his dedication to the prince of the new musical art. Given his acquaintance with de Vitry, Oresme surely knew his treatise *Ars nova notandi*, which contains a considerable amount of mathematical detail as part of its exposition of the new notational possibilities he exploits in his motets.[23] Both men were part of larger currents of mathematical and musical speculation. At several points, Oresme acknowledges the prior work of de Muris, de Vitry's peer among the older generation of music theorists. De Muris's writings were also sources for the new musical practice, especially his *Ars novae musicae* (1319), whose title also registers the sense of musical innovation. As noted above, Oresme drew on de Muris's work on commensurable and incommensurable quantities.[24] These matters occupied Jewish as well as Christian scholars. De Vitry asked Levi ben Gershon (Gersonides) to help him resolve a mathematical question bearing directly on music; in turn, Oresme used this musical result to make an astronomical argument. This interchange illustrates the interweaving of mathematical, musical, and astronomical issues in the works of these men.

According to Boethius, the basic musical ratios are *superparticular*, meaning that they have the form $n:(n + 1)$, such as 2:3 or 3:4. The more complex intervals derived from this primal set involve only powers of 2 and 3. Besides these, de Vitry suspected that no other compound superparticular intervals could exist. Having learned of Gersonides probably from his *Maaseh Hoshev (Work of Calculation,* 1341), a Euclidean compilation of results in arithmetic, de Vitry asked Gersonides whether he could prove his conjecture, which he did in his brief *De numeris harmonices* (1342).[25] De Vitry probably was interested in this result less for its application to the ratios governing musical intervals, which were not really under controversy at the time, than for its implications for *rhythmic* notation, the subject of much controversy between the practitioners of *ars antiqua* and *ars nova*.[26] De Muris had already set out the complicated rhythmic issue at stake; Gersonides' result confirmed the superiority of the *ars nova* notation over its older rival. Thus, *musical* questions had led to a question in *arithmetic*, whose result then bolstered one side in the antecedent *musical* controversy. Oresme used the implications of this result to argue for a change in the fundamental concept of the harmony of the spheres, both musically and astronomically.

In his *Tractatus de commensurabilitate*, before the concluding dialogue between Arithmetic and Geometry, Oresme investigates the precise relation between the relative ratios of revolution of heavenly bodies and their conjunctions, the occasions at which they would occupy the same apparent position in the sky. His Proposition 11 demonstrated that the number of such conjunctions in any one revolution is given by the difference between the two terms of the ratio between the velocities of the two bodies, here assumed to be rational. He then notices the deep astronomical and musical problems this result implies:

If the ratio of velocities of any two celestial mobiles were in any of the principal harmonic ratios in music, namely the diapason [octave, 1:2], diapente [fifth, 2:3], diatesseron [fourth, 3:4], and tone [8:9], which make a concord or harmony, the mobiles will never conjunct except in one place only, since the least numbers of such a ratio differ only by a unit. As an example, if the mean motion of Mars were exactly twice the speed of the sun's mean motion, there would never be a middling conjunction of these two bodies for [they would conjunct] in only one place, or point. ... Since no configuration consisting of two motions is found to occur in only one point of the sky, [it follows] as a consequence that no two celestial motions have velocities related in a principal harmonic ratio. Therefore, if celestial bodies in motion produce a harmony, it is not necessary [to assume] that such a harmony arises from the velocities of their motions, but perhaps it stems from some other source for other reasons, as will be seen later.[27]

Here Oresme relies on a long-known finding of observational astronomy that the conjunctions of the planets occur at different points in the sky, not just one. Where the theorem of Gersonides proved that the "principal" harmonic ratios were all superparticular, Oresme's corollary now showed that *any putative celestial music based on those intervals was incompatible with the visible evidence of astronomy*, based on the common assumption that the "harmonies" related celestial velocities. This argument showed that the musical program of harmonizing the cosmos mathematically contradicted its own starting point, as it had been commonly understood. Oresme then generalized these results to several moving heavenly bodies or to a single body moving in several ways at once.

This, then, is the framing question that surrounds the debate between Arithmetic and Geometry: what is to become of the music of the spheres in light of these results? On the face of it, they seem to contradict the common assumption of rational (commensurable) relations between planetary orbits. As the patroness of this assumption, Arithmetic protests that "a position contrary to ours would destroy the beauty of the universe and detract from the goodness of the gods."[28] Though she also mentions Oresme's result that "if the motions are incommensurable, it is impossible that they all return [to the same place] on their circles," she seems to ignore Oresme's corollary to Proposition 11, which effectively blocks her advocacy of the alternative possibility "that the celestial *velocities* are proportioned by numbers."[29]

In rebuttal, Geometry notes that no one really supposes "that any celestial motions are related as any one of the principal concords," which seems explicitly to recognize the problem that Arithmetic had ignored. Despite this, Geometry does not abandon the concept of celestial harmony in her alternative: "However, should the celestial spheres produce some concord while moving, this ought not be measured in terms of the velocities of the motions, but rather by the volumes of the spheres, or the quantities of the orbs."[30] Both these statements accord with Oresme's own stated positions earlier in the work, thus giving further evidence that Geometry finally wins the debate, even though we never get to hear Apollo's verdict. In the wake of his Proposition 11, Oresme had offered the same response to the problem he disclosed that now appears in the mouth of Geometry: the harmonies relate *volumes* and their correlate "quantities of the orbs" (probably meaning their masses, proportional to their volumes). Neither Oresme nor Geometry abandons the concept of celestial harmony, even in the face of its inaudibility and the incompatibility of the simplest ancient versions based on the principal concords.

Geometry's resolution also comes with an important corollary: to resolve Oresme's problem, celestial harmonies must involve some aspect of incommensurability. To that end, Geometry suggests harmonizing celestial volumes, rather than velocities. The logic of her statement bears close examination: by using volumes as a way to introduce incommensurability, she seems to presume that the volumes of spheres have some kind of incommensurability. This in turn seems to reflect Oresme's knowledge of the work of Archimedes, who had proved that the volumes of spheres are proportional to the cube of their radii. As a consequence, commensurable spherical volumes have incommensurable radii.[31]

Oresme's arguments about the irreversible and never-repeating course of the universe are inextricably joined with arguments about mathematical incommensurability and with the "new song" that results in ever-novel cosmic harmonies. In this intricate tapestry of ideas, music plays an essential role mediating between astronomical, arithmetical, and geometric ideas. Oresme celebrates this "new song" in his climactic image that "the heavens are like a man who sings a melody and at the same time dances, thus making music in both ways—*cantu et gestu*—in song and in action."[32] Pure song is the melodic and harmonic ideal, incarnating mathematical relationships in the visible dance of stars and planets. The combination of new music with ancient results about incommensurability led Oresme to reject the simplest versions of the music of the spheres even as he indicated new possibilities for celestial harmony. Two centuries later, Music moved her sisters to reconsider their relation entirely.

3 Moving the Immovable

In the century after Oresme's imaginary debate between Arithmetic and Geometry, their sisters Music and Astronomy returned to the question whether a seemingly immovable center could somehow move. That center could be the Earth or the mode of a musical composition, both generally assumed to be unchanging. Because each celestial sphere was associated with a mode, a change of mode suggests motion between spheres. As innovative musical compositions used unprecedented changes of mode, the immovable musical center began to move. In the following decades, the new astronomy put forward the theory of a moving Earth. Other musical considerations moved Vincenzo Galilei to prefer heliocentrism decades before the controversy came to a head. More generally, *harmony* became a crucial term in the debates about Copernican astronomy.

Through the fifteenth century, writers on music struggled with disturbing contradictions between the ancient authorities about the exact ordering of the celestial spheres. For instance, in his *Liber de arte contrapuncti* (*Book on the Art of Counterpoint*, ca. 1476), Johannes Tinctoris expressed considerable frustration:

> I cannot pass over in silence the opinion of numerous philosophers among them Plato and Pythagoras and their successors Cicero, Macrobius, Boethius, and our Isidore [of Seville], that the spheres of the stars revolve under the guidance of harmonic modulation, that is, by the consonance of various concords. But when, as Boethius relates, some declare that Saturn moves with the deepest sound and that, as we pass by stages through the remaining planets, the moon moves with the highest, while others, conversely, ascribe the deepest sound to the moon and the highest to the sphere of the fixed stars, I put faith in neither opinion. Rather I unshakably credit Aristotle and his commentator, along with our more recent philosophers, who most manifestly prove that in the heavens there is neither actual nor potential sound.[1]

This fundamental disagreement led Tinctoris to join Aristotle in rejecting celestial sounds, but other writers generally opted for one or the other of the ancient alternatives. Thus, Giorgio Valla in 1501 followed Boethius and assigned the sun the *mese*, the "middle" note of the Greek gamut (figure 3.1a), whereas Franchino Gaffurius in 1596 followed Cicero and assigned it the *lychanos hypaton*, the "highest" note (figure 3.1b).[2] Yet both versions associated the sun with the Dorian mode (see figure 3.4), considered the first or primal

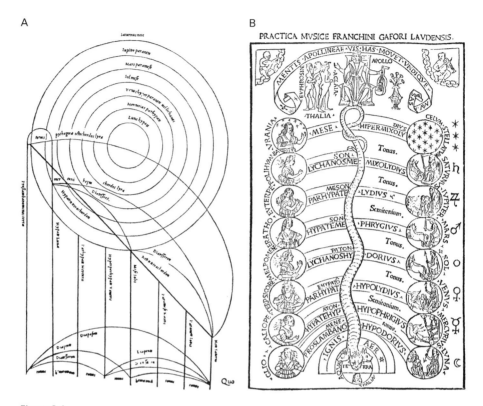

Figure 3.1
Cosmological diagrams from (a) Giorgio Valla, *De expetendis et fugiendis rebus opus*, "De musica" (1501); (b) Franchino Gaffurius, *Practica Musicae* (1596).

mode, and assigned it the middle position in their diagrams. Though practical music was still notated according to hexachords (spanning six notes), the cosmos spanned a full octave from Earth to the fixed stars, according to the ancient diatonic pattern S T T (each semitone followed by two whole tones).

This same solar primality appears in Henrich Glarean's *Dodecachordon* (1547), which summarized and reformulated the theory of the modes and their usage in contemporary composition. To the eight standard church modes Glarean added four more, thereby including the modern major (Ionian) and natural minor (Aeolian) scales on an equal footing with the older modes (see figure 3.4). His cosmological diagram displays the two rival celestial pitch-orderings side by side, but again the sun appears in a central position among the planetary notes, whichever alternative one chooses to order the spheres of the outer planets (figure 3.2a). Glarean was an innovative geographer who drew one of the first polar projections of the northern hemisphere, but his views were completely geocentric, as shown in his cosmological diagram centering the universe on his hometown (figure 3.2b), nor did

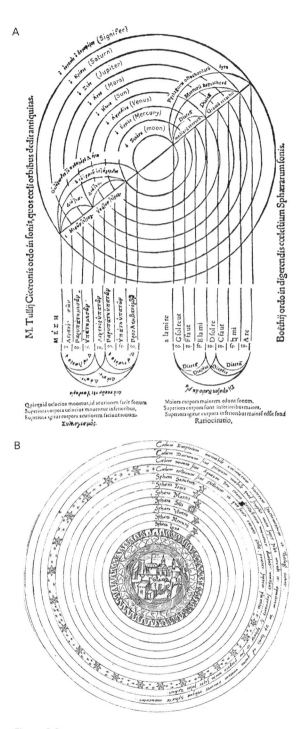

Figure 3.2
(a) Glarean's diagram comparing the two received orderings of the celestial spheres, from *Dodecachordon* (1547); (b) his manuscript drawing of cosmology, centering on the Earth and his hometown, from his *De geographia* (ca. 1510–20).

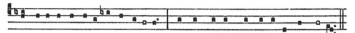

1. In éxi-tu Isra-*el de* Aegý-pto, * dómus Jácob de pópu-*lo* bárbaro :

Figure 3.3
Gregorian chant *In exitu Israel* in *tonus peregrinus*, the "wandering tone," so named because it "wanders" from the first reciting tone on A (first measure) to another on G (second measure) (♪ sound example 3.1). Text: "When Israel came out of Egypt and the house of Jacob from a strange country."

he show any awareness of Copernican astronomy. Both doctrinally and cosmologically, Glarean was conservative; he remained loyal to the Roman church and strongly distanced himself from the Protestant reformers, yet was proud to be a friend of Erasmus.

Though he was no heliocentrist, Glarean's musical work opens a new perspective on whether a seemingly immovable center can move. In compositions throughout the Middle Ages, whether chant or polyphonic works, the mode remained as unmoved as the Earth in Aristotelian physics. Only in the *Tonus peregrinus* ("wandering" or "foreign" mode) of chant was there the possibility of moving between modal centers, as in the chant sung by the pilgrims at the beginning of Dante's *Purgatorio*, the aptly chosen *In exitu Israel* ("When Israel Came Out of Egypt"; figure 3.3; ♪ sound example 3.1). But here too the chant eventually reaches a final pitch that situates it within a regular mode.[3]

In the course of presenting his novel modal ideas, Glarean also discusses ways of changing the modal center. He notices that such possibilities are beginning to be used by contemporary composers, whose practices confirm his theories by showing that, by habitually adding a B♭ to the Lydian mode, those composers essentially are writing in the Ionian (modern major) mode starting on F (figure 3.4; ♪ sound example 3.2). Such a shift from Lydian to Ionian he considers "scarcely clear even to a perceptive ear, indeed often with great pleasure to the listener," indicating the possibility of other, more radical changes: "But in other cases the changing seems rough, and scarcely ever without a grave offense to the ears, as changing from the Dorian to the Phrygian. From this I believe the adage arose: from Dorian to Phrygian, from natural to less natural, or from well-ordered to irrational, or from mild to harsh; briefly, from whatever, as they commonly say, does not keep to its course and falls from this into a different one."[4] The adage questions the status of any such fundamental change, whether in the divine order, the human polity, or music.

Aristotle considered the "manly" Dorian and "emotional" Phrygian so opposed that when Philoxenus (an avant-garde practitioner of the Greek New Music discussed in chapter 1) tried to sing a Dionsyiac dithyramb in the Dorian mode, he "fell back by the very nature of things into the more appropriate Phrygian."[5] One would expect as conservative a thinker as Glarean to reject change of mode, yet he notes that new compositions force a reconsideration: "But enough now of philosophizing. Josquin des Prez in the psalm *De profundis*, has undertaken successfully to go from Dorian to Phrygian, skillfully and

Moving the Immovable 39

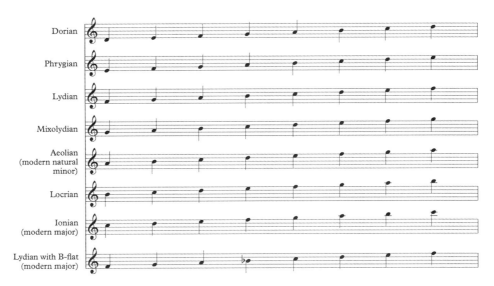

Figure 3.4
The modes according to Glarean (♪ sound example 3.2); note that the Ionian (corresponding to the modern major scale) and the Aeolian (modern minor scale) were not included among the traditional modes. Note that adding a B♭ to the Lydian mode turns it into Ionian based on F (F major).

without offending the ears." What traditional practice and proverbial wisdom would not allow, a contemporary master has accomplished, an innovation Glarean acknowledges and explores.[6]

Josquin's remarkable change of mode sets a penitential text from the Psalms: "Out of the depths have I cried unto Thee; O Lord, hear my prayer." Another dark biblical text on the lamentation of King Saul over the death of his son Absalom drew Josquin to set his motet *Abasalon, fili mi* with a shifting modal center and an extremely low tessitura of male voices. The change of modal center seems to symbolize or even reenact the felt experience of turning from the depths to the heights of divine understanding and forgiveness. As part of the general musical understanding of his time, Josquin probably knew the astronomical associations of the modes, as in figure 3.1, in which a motion from Dorian to Phrygian would correspond to going from the sun to Mars. For him and his learned hearers, this shift in mode would suggest a huge cosmic displacement without precedent in ordinary experience, musical or astronomical. Even more, Glarean notes that Josquin accomplishes this miracle without startling the ear. He bridges Dorian and Phrygian by using intermediates between them, a process that may illuminate the general problem of how the immovable might be moved.

Following well-established compositional precedents, Josquin organized his phrases through their cadences, musical formulas that serve as punctuation. The "final" or "perfect" cadence affirms the mode at the end of the composition. Intermediate cadences affirm important structural pitches within the mode. But in *De profundis*, Josquin organized his intermediate cadences so that they lead from Dorian to Phrygian structural pitches. He begins by establishing the Dorian mode sung in octaves by all voices above an extraordinary low D, far below the usual range of the voices, literally signaling the "profundity" of the depths by a low note to which we will return (♪ sound example 3.3a). Box 3.1 (♪ sound examples 3.3a–e) gives the details of how he goes from the Dorian to the Aeolian mode during the first part of the motet, by means of the Ionian mode as intermediary. As Glarean explained, these two "new" modes (Ionian and Aeolian) have many points of interconnection that render their alternation easy, for they share the same "octave species," their fundamental pattern of whole and half steps, which sets them apart from the other church modes.[7] At the same time, they complete the possible patterns started by the older diatonic modes.

The second part of *De profundis* immediately begins with a strong assertion of the Ionian (C), as if to stabilize the Aeolian–Ionian axis on which the modal transformation turned (♪ sound example 3.4a). Josquin continues to move back and forth between Aeolian and Ionian, establishing them as constituting an intermediate state in his journey from Dorian to Phrygian. When he does make his first full cadence in the Phrygian, he does so above the low E, one step above the deep D we noted at the beginning of the motet. By so doing, he reminds us both of that "profound" opening, the invocation of the depths, but also of the significant distance upward the motet has traversed, from D to E, as if mapping the distance from despair to hope (see box 3.2, ♪ sound examples 3.4a–c). The text at this point, *sicut erat in principio* ("as it was in the beginning"), is a recurrent formula that concludes many verses in the Catholic liturgy but here has a surprising significance: this arrival at the Phrygian definitely is *not* where this motet began.

His concluding reaffirmation of the Phrygian mode also recapitulates in brief the entire amazing modal journey just completed. This unprecedented transformation is so subtly accomplished that the listener may not even be aware of it, for only a highly trained ear could recollect the initial Dorian D and compare it with the final Phrygian E. Perhaps this, too, can be read as Josquin's insight that the passage of the soul out of the depths happens through means that can escape ordinary awareness, however consequential their result.

Josquin and Glarean would have looked to Aristotle for the larger philosophical context for the possibility and character of such an essential change (*metabolē*), meaning a change of regime or revolution that moves between two contradictory states via an intermediate state (*metaxu*) serving as "a contrary relatively to the extremes." Aristotle gives a musical example: "A middle note [*mese*] is low relatively to the high note [*nētē*] and high relatively to the low note [*hypatē*]."[8] A melody moves from high to low note via a middle note between them, the *mese* (associated with the sun, as noted above), both opposing and connecting the other notes. Though change of mode is a more radical change, it involves the same essential structure. In Josquin's motet, the new Ionian and Aeolian modes are the intermediaries between the contrary Dorian and Phrygian.

Box 3.1
How Josquin transforms modes: Part one of *De profundis*.

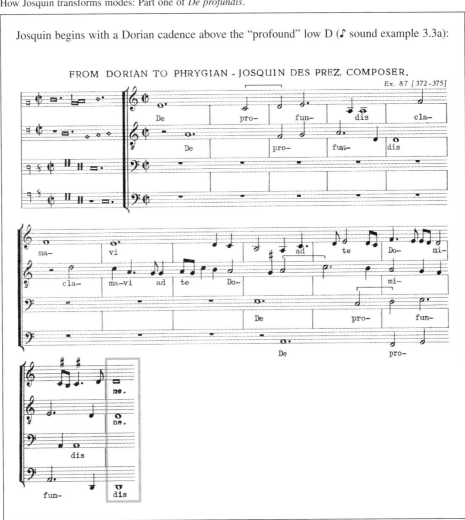

Box 3.1
(continued)

He then cadences on the Aeolian A, a fifth above Dorian D, though he only includes its fundamental (A) and fifth (E), not the third (F) (♪ sound example 3.3b):

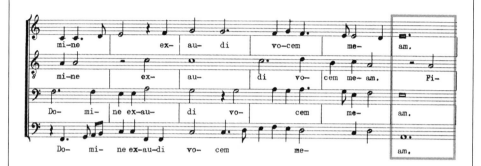

As he gradually introduces the missing F, he then leads it down a semitone to E and uses that as the third degree of C (Ionian), on which he then cadences (♪ sound example 3.3c):

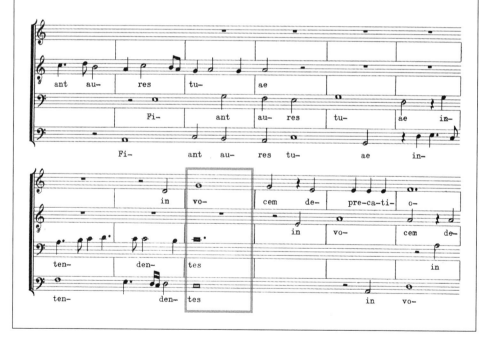

Box 3.1
(continued)

This leads to a subsequent cadence on G, the Ionian dominant (that is, the fifth above its final) (♪ sound example 3.3d):

But G is also the third degree of E; this reinterpretation of G enables the ensuing Ionian (C) harmony to be the pivot between Dorian harmonies and Phrygian. Josquin underlines the modal distance so far traversed by ending the motet's first part on a full Aeolian (A) cadence, now bringing forward in the highest voice its third (C), which also alludes to the important Ionian (C) arrivals that had prepared it (♪ sound example 3.3e):

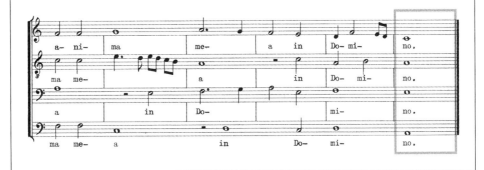

Box 3.2
Arriving at the Phrygian: Part two of Josquin's *De profundis*.

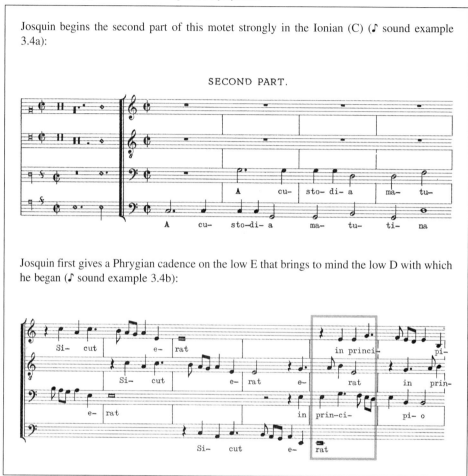

Aristotle also notes that some changes happens "by nature" while others are "against nature" and occur "by force [*bia*]," a word that came to be translated as "violence [*violentia*]." Thus, a stone thrown upward suffers a "violent" motion, compared to its "natural" motion downward, though clearly the stone is not "violated" when its natural motion is thus interrupted.⁹ Further, such "violent" change happens essentially through the action of some agent *outside* the body itself: a hand must throw the stone and likewise some external artistic force must bend the mode from Dorian to Phrygian. In Aristotle's eyes, such a modal change is violent: when Philoxenus tried to force a dithyramb into the Dorian, the song fell back into its "natural" Phrygian mode, like throwing a stone whose violent

Box 3.2
(continued)

Josquin's final Phrygian cadence recapitulates the extraordinary arc of the whole motet by moving from a Dorian harmony (D–F–A) four measures from the end via a penultimate Aeolian harmony (A–C–E) to the final, extended Phrygian chord (E–B) (♪ sound example 3.4c):

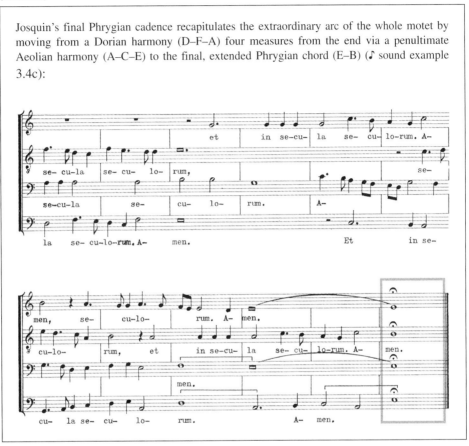

vertical rise eventually gave way to its natural fall. With this in mind, Josquin's feat exceeds the ancient example of Philoxenus, going beyond the natural (and pagan) to show the even greater force of divine grace.

The implications of Aristotle's analysis extended beyond musical examples. Oresme had used Aristotle's categories in formulating his new mathematics of change.[10] In 1624, Francis Bacon epitomized "all violent motion" by a moistened finger rubbing a glass's rim to excite its ringing, which he found similar to the chiming of a bell or the plucking of a string. This "motion of liberty, i.e. from compression to relaxation," Bacon considered "the chief root of all mechanical operations," transforming nature itself through the artful redirection of "violence."[11] Medieval alchemists theorized the possibility of elemental transmutation in terms of Aristotle's concept of an undifferentiated "prime matter" that is therefore capable of changing its outward form, so that base metals might receive the tincture of gold.[12] Even when twentieth-century physicists realized the dream of transmuting elements, they had to irradiate a stable element so that it became an unstable radioactive intermediate isotope, which then could decay into the desired product. In so doing, they used the new phenomena of radioactivity to bridge the stable elements via a common substratum of "prime matter," the electrons, protons, and neutrons they knew at the time.

Schooled in the Greek language and Aristotelian physics, Nicolaus Copernicus knew the multiple meanings of *metabolē*, which denotes change as well as revolution in the sense of cyclical, circular motion. He also realized that the heliocentric cosmology required an explanation of how the Earth could possibly move, against the evidence of our senses and Aristotle's arguments. Not until Newton's laws was there a fully worked-out replacement for Aristotle's physics; in the interim, belief in a movable Earth had to justify its seemingly paradoxical claims.[13] To do so, Copernicus gives new meaning to harmony in *De revolutionibus coelestis* (*On the Revolutions of the Celestial Orbs*, 1543).

Though he never completed his bachelor's degree at the University of Kráków, Copernicus probably studied the musical component of the quadrivium through the writings of de Muris.[14] By that time, in Paris Aristotelian natural sciences had tended to replace the quadrivial study of musical theory, but that was far less true in England and Central Europe.[15] Because of this, Copernicus (and later Kepler) had the fortune to be educated in areas where the older practice of musical-mathematical study remained in place. Though Oresme himself was not generally read in succeeding centuries, writings by others in his school, notably his teacher, Jean Buridan, and perhaps also his own, were studied in Kráków during the fifteenth century, which became a notable center of astronomical knowledge.[16] Thus, during his student days there, Copernicus may have had the opportunity to learn something of the thoughts and speculations of these Parisians of the preceding century, including their arguments that considered heliocentrism with great care.

Two passages in Copernicus indicate a significant musical connection. His early *Commentariolus* (written about 1508–1514), which first expounded his view that the Earth moves while the sun stands still, concludes that his theory suffices "to explain the entire structure of the universe and the entire ballet of the planets [*siderum chorea*]." As Oresme compared heliocentric planetary motion to a circular dance (*chorea*) he called a *carole*, Copernicus presents astronomy united with choreography and music using the same terminology. Copernicus took the Latin phrase *siderum chorea* ("dance of the stars") from Martianus Capella, whose influential musical cosmology he also cites in his *Revolutions*.[17] In that book, after addressing the objection that we see the sun rise and set but do not see the Earth move, Copernicus argues that the heliocentric theory determines the order and distances of planetary spheres, the very problem that also troubled music theorists. In the sun-centered arrangement, "we discover a marvelous symmetry [*symmetriam*] of the universe and an established harmonious linkage [*certum harmoniae nexum*] between the motion of the spheres and their size, such as cannot be found in any other way."[18]

Presenting his own cosmological assumptions at the beginning of his *Almagest*, Ptolemy made no such reference to harmony, though he refers to mathematical theories as "beautiful [*kalon*]" and praises "the contemplation of the eternal and unchanging," such as his treatise presents.[19] In contrast, Copernicus is evidently arguing against the general opinion that heliocentrism is "quite ridiculous," as Ptolemy put it, controverting common sense and plain reason.[20] Copernicus's language of *harmony* aims to reconcile his readers to these dissonances, even to help them appreciate their richness and surpassing beauty. His invocation of harmony ultimately stems from the musically formed cosmos of Plato, as do so many later invocations of harmony we will consider in science, down to the present day.

Copernicus's rhetoric relies on the implicit interconnection between astronomy and the rest of the quadrivium. Where "symmetry" tends to have visual connotations, Copernicus's word *symmetria* also has the specifically geometric meaning of *commensurability*.[21] That is, the size of each planet's sphere can be expressed in terms of the Earth–sun mean distance as an astronomical unit. By pointedly connecting the technical terms *symmetria* and *harmonia*, Copernicus signals the linkage between arithmetic and music he considers a capital feature of heliocentrism. In his dedicatory letter to Pope Paul III, Copernicus also mentions the secretive practices of the Pythagoreans; his original manuscript emphasized these Pythagorean connections even more strongly.[22] He thereby brings to mind their heliocentric cosmology, as well as their quest to understand the *harmonia* of the cosmos in musical ratios, both important precedents for his heliocentric *symmetria*.[23]

In different ways, Copernicus's early readers recognized and amplified the musical context of *harmonia*. Even before the publication of *De revolutionibus*, Copernicus's disciple Rheticus explained in his *Narratio prima* (1540) what his teacher meant by "an absolute system" in which "the order and motion of the heavenly spheres agree." Writing about earlier astronomers (such as the Arab Albategnius), Rheticus remarks that "we should have wished them, in establishing the harmony of the motions, to imitate the musicians who, when one string has either tightened or loosened, with great care and skill regulate and adjust the tones of all the other strings, until all together produce the desired harmony, and no dissonance is heard in any."[24] In favor of the heliocentric view, Rheticus notes that "all the celestial phenomena conform to the mean motion of the sun and that the entire harmony of the celestial motions is established and preserved under its control." Because Rheticus writes explicitly to explain the system of "my teacher," we might take these expressions as also having Copernicus's implicit approval. They further amplify the language of "harmony" by adding further details of its musical implications, down to the tuning of the strings to avoid dissonance.

Johannes Praetorius's *Compendiosa enarratio Hypothesium Nic. Copernici* (*Compendious Narration of the Hypothesis of Nicolaus Copernicus*, 1594) praises the heliocentric system because "this symmetry [*simmetria*] of all the orbs appears to fit together with the greatest consonance so that nothing can be inserted between them and no space remains to be filled." The explicitly musical term "consonance" expands the mathematical notion of *simmetria*. Ironically, Praetorius later turned against heliocentrism, though he had initially been captivated by its harmony.[25]

Among those who remained steadfastly attached to the new cosmology, William Gilbert emphasizes the Pythagorean connection in his *De magnete* (*On the Magnet*, 1600), listing its ancient advocates, especially Philolaus, and praising Copernicus for having discovered "the harmony [*symphoniam*]" of planetary movements.[26] Gilbert's term *symphonia* draws attention to the polyphonic fullness of the heliocentric cosmology, whose development by Kepler will occupy us in chapter 5.

But the most interesting reaction to Copernicus may be discerned in a musical text, the *Dialogo della musica antica, et della moderna* (*Dialogue on Ancient and Modern Music*, 1580–81), "surely the most influential music treatise of the late sixteenth century," by Vincenzo Galilei, a lutenist and composer who became deeply interested in the nature of ancient Greek music and its implications for the music of his own time.[27] Vincenzo had been a student of the eminent music theorist Gioseffo Zarlino and was interested in investigating the Greek sources, which had not yet been translated into the vernacular. Vincenzo was also part of the Camerata, a circle of enthusiasts around the Count Giuseppe de' Bardi, who shared an intense interest in the relation between ancient and modern music.[28]

In the course of their conversations, Bardi put Vincenzo in touch with Girolamo Mei, an older scholar who knew Greek and was engaged in the first really careful, philological examination of the ancient musical texts in the West since antiquity. Vincenzo had many questions and Mei responded at great length, often giving a very different account than what Zarlino had taught. As their exchange went on, Vincenzo became more and more excited, convinced that he was seeing entirely new vistas in this ancient music. Above all, Mei taught Vincenzo that Greek music had been strictly monophonic, a single melodic line having extraordinary powers of rhetorical persuasion and emotional effect based on its supple melody and its use of various musical modes suited to the emotions being evoked. More interested in philological sleuthing than in contemporary musical practice, Mei passed on his findings to Vincenzo, who collected his new understanding in his *Dialogue*. The Camerata doubtless discussed Vincenzo's findings and some scholars have viewed his text as a foundational document for their subsequent efforts to revive the lost powers of Greek tragic drama in a newly recreated form they called *opera*. Bardi and like-minded aristocratic patrons sponsored the first operas, beginning with *intermedii* produced at the Medici court in the 1580s. Appropriate to its ancient inspiration, early opera involved dramatizations of myths: Cavalieri's *Dialogo del anima e corpo* (*Dialogue of Body and Soul*), Peri's *Euridice*, and especially Claudio Monteverdi's *L'Orfeo* (first performed in 1607). These dramas and Vincenzo's work illuminate a crucial stage in the development of expressivity as the essential project of contemporary music.[29]

In a passing comment, Vincenzo brings forward a significant connection between these musical developments and the new cosmology. Mei himself had not been much interested in the relation of astronomy and music; he felt that ancient Greek music was more closely akin to the rhetorical and grammatical arts of the trivium than to the mathematical arts of the quadrivium. Above all, Mei was interested in how music could rejoin drama and literature, leaving behind its old connection with mathematics and astronomy. Thus, Mei's letters to Vincenzo contain scarcely any of the old planetary lore. For instance, in 1581 Mei compared the turning movements of the ancient Greek chorus to "the movement of planets from west to the east and returning to the west"; coming to a standstill, the chorus "signified the stability of the earth around which those movements are made," hence a geocentric cosmos.[30] Thus, Vincenzo's own comment about cosmology in his *Dialogue* comes as a surprise: "Like the many lines drawn from the center of a circle to the circumference, which all gaze back at the center, every musical interval in the octave sees itself as if in a mirror, like the planets [*stelle*] do in the sun, not otherwise than the way everyone, depending on individual capacity, receives from it the person's being a perfection."[31]

Vincenzo clearly indicates that he considers the *sun* to be the center of the planetary system, and his argument is based on diagrams of the musical scale such as Glarean illustrated (see figure 3.2a), in which the sun's position is surrounded symmetrically by whole tones on either side, the interval separating it from the planets on either side of the picture, and likewise of the semitones and whole tones throughout the octave, *read outward from the sun as center*, so that "each musical interval in the octave sees itself as if in a mirror, like the planets do in the sun." Vincenzo has taken a diagram based on a *geocentric* worldpicture (as Glarean and those before him had assumed) and used it to argue that the sun is the true musical and hence also cosmological center. Vincenzo uses the musical symmetries he discerns in the solar-centered octave as a way of expressing his astronomical preference for the heliocentric cosmos.

Though he does not mention Copernicus, Vincenzo is writing almost forty years after *De revolutionibus*, so that we infer that he must have known of the basic idea of the new astronomy. If so, he may have been one of the first in Italy to evince such awareness.[32] But how could he have come by it? The most likely hypothesis goes back to his teacher, Zarlino, whose own cosmology seems to have been quite geocentric, judging by the wholly traditional account of the *musica mundana*, the "music of the spheres," he included near the beginning of his seminal *Istitutioni harmoniche* (*Harmonic Institutes*, 1558) (figure 3.5). Zarlino does show an interest in symmetries of musical intervals and their correspondence with cosmology, though applied to the relations between the four physical elements (figure 3.6), rather than to the planetary system.

Zarlino, however, revealed his own interest in astronomy in two shorter writings: *Intorno il vero anno, & il vero giorno, nel qual fu crucifisso il N. S. Giesv Christo redentor del mondo* (*On the True Year of the Crucifixion of Our Lord Jesus Christ*, 1579), which addressed a long-standing chronological controversy, and *Le risoluzioni d'alcune dimande fatte intorno la correttione del calendario di Giulio Cesare* (*Resolution of All Requests Concerning the Correction of the Calendar of Julius Caesar*, 1589), which addressed the call of Pope Gregory XIII to find the best means of rectifying the growing divergence

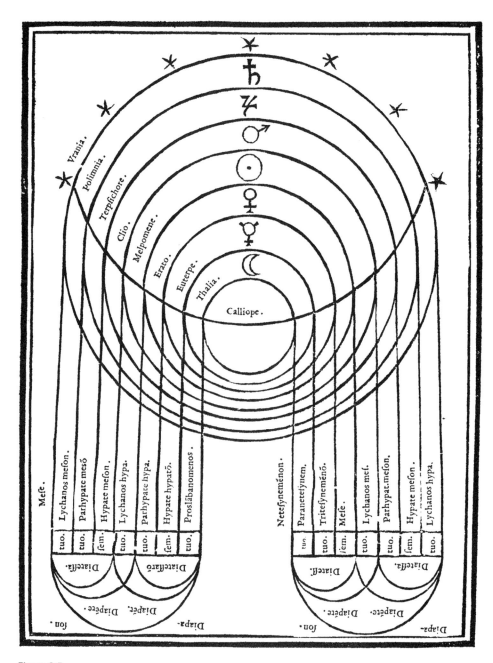

Figure 3.5
Zarlino's diagram from his *Istitutioni harmoniche* showing the orderings of the celestial spheres and their planetary designations following the traditional geocentric accounts of Cicero and Boethius.

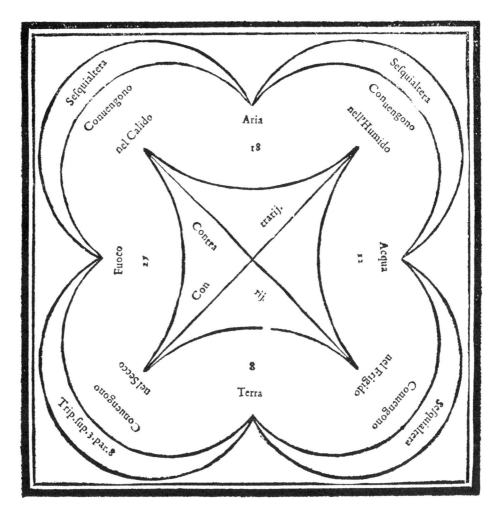

Figure 3.6
Zarlino's diagram of musical intervals, compared to the relations between the four elements.

between the Julian calendar and the observed astronomical occurrences of solstices and equinoxes.[33] Zarlino thus entered into two intricate calendrical disputes, about which his texts show his close study, evidencing his definite awareness of contemporary issues of astronomy. In his time, the Julian calendar was about two weeks in disagreement with the astronomical seasons: Julian March 21 fell two weeks shy of the observed vernal equinox. To remedy this, Zarlino proposed adjusting the calendar by twelve days (in 1582 a ten-day shift was adopted, along with the present scheme of leap years).

Even more to the point, Zarlino owned a copy of the first edition of Copernicus's *De revolutionibus*, which Zarlino signed and dated 1566.[34] This places Zarlino in an elite company of astronomers and savants who bought this rather expensive book when first it appeared. Compared to the wider distribution of the extant copies in German-speaking lands, Zarlino's copy represents a notable point of contact for Copernican ideas in Italy.

Based on the altogether geocentric descriptions in his *Istitutioni*, we infer that Zarlino was not convinced by Copernicus. If Zarlino acquired his copy of Copernicus during or before the 1560s, when Vincenzo studied with him, it seems plausible that Zarlino may have mentioned it to him, if only as a curiosity. If so, one wonders about their ensuing discussions, for at some point Vincenzo seems to have decided that Copernicus was correct. The musical justification he gives in his *Dialogue* suggests that he thrashed out the matter in the course of his studies with Zarlino, or in the subsequent years. Did they argue about Copernicus? Did Vincenzo's rebelliousness move him to sympathize with the heterodox cosmology, if only to annoy his mentor? Vincenzo certainly came to disagree with his teacher on many matters, bringing them forward both in the *Dialogue* and in his later *Discorso intorno l'opere di Messer Zarlino* (*Discourse on the Works of Mr. Zarlino*, 1589). Though at the beginning of his *Dialogue* Vincenzo paid homage to Zarlino as "one of the masters" alongside Gafurius and Glarean, his scathing criticism of what he considered his teacher's errors about Greek music was so brusque that Zarlino took offense at what seemed Vincenzo's ingratitude and disrespect.

Though these disputes had to do with musical matters, Zarlino could have been the source for Vincenzo's awareness of heliocentrism, based on his possession of Copernicus's book. There is another, not completely distinct possibility. Galileo Galilei was a teenager during the years leading up to his father's *Dialogue* and later recorded in his own *Discourses on Two New Sciences* his awareness of some of the musical issues that engaged his father. Knowing that Vincenzo had adopted the heliocentric view sometime before 1580, it seems plausible that he and Galileo discussed it. If so, Galileo might have learned of this new astronomy from his father. Writing to Kepler in 1597, Galileo remarks that he had "many years ago" adopted Copernican views, though he does not specify how long.[35] Placing his first contact with it during the 1570s would certainly fit this description, giving fully twenty years in which he could have mulled it over. Given the generally positive tenor of Galileo's references to his father, one might speculate that he took up his father's enthusiasm for the new cosmology during the period in which the *Dialogue* was being composed. But the reverse channel of influence is also possible: the teenaged Galileo may have learned from other sources about Copernicus's idea and brought it to his father, who might have also had some earlier discussions with Zarlino on this matter. Though it is beguiling to think of the intrepid teenager initiating his father into the new astronomy that would lead to so much controversy and danger for himself, no extant evidence shows how Galileo first learned about heliocentrism.

Evidence does emerge in Galileo's early notebooks, dating from around 1590 (but possibly as far back as 1584), whose section on the heavens is closely copied after a 1581 work by Christopher Clavius, the eminent Jesuit astronomer.[36] When he comes to "the order of the heavenly orbs," Galileo (citing Clavius) begins with Aristarchus and Copernicus, though the ensuing text amasses a preponderance of evidence against the Copernican view.[37] Here, it is hard to judge how far Galileo is merely copying the received view, showing his familiarity with it in order to advance his nascent academic career, or how far he himself believes the prevalent view of geocentrism presented by Clavius. At the very least, Galileo was well aware of Copernicus (and of the controversy surrounding him) by about 1590.

In any event, the evidence presented above supports Vincenzo's musical interpretation of heliocentrism, with the implication that, for him, music illuminated and underwrote a crucial astronomical innovation. As for his famous son, Galileo's arguments for heliocentrism often turn to the same musical terminology and categories that we noted in Copernicus, Praetorius, and Gilbert. For instance, in his "Considerations on the Copernican Opinion" (1615), Galileo describes how,

encouraged by the authority of so many great men, [Copernicus] examined the motion of the earth and the stability of the sun. Without their encouragement and authority, by himself either he would not have conceived the idea, or he would have considered it a very great absurdity and paradox, as he confesses to have considered it at first. But then, through long sense observations, favorable results, and very firm demonstrations, he found it so consonant with the harmony of the world that he became completely certain of its truth. Hence this position is not introduced to satisfy the pure astronomer, but to satisfy the necessity of nature. ... Who does not know that there is a most agreeable harmony among all truths of nature, and a most sharp dissonance between false positions and true effects?[38]

Here, the general language of harmony is further sharpened by specific musical distinctions between consonance and dissonance, which Galileo introduces as a higher criterion rising above the considerations of "pure astronomy," which by itself never seems to have interested him greatly. His language remains consistent in his "Reply to Ingoli" (1624) even as he turns to the more difficult issue with which we began, how the seemingly immovable Earth could possibly be movable: "Now if the nature of the earth is very similar to that of moving bodies, and the essence of the sun very different, will it not be much more probable (other things being equal) that the earth rather than the sun imitates with motion its other six consorts? Add to this another no less notable harmony, which is that in the

Copernican system all fixed stars, also intrinsically luminous bodies like the sun, are eternally at rest."[39] Galileo persuaded his readers by using familiar topics that connected astronomy and music within the quadrivium they all knew. Thus, in 1674 Robert Hooke noted that those seeking "better reasoned grounds, from the proportion and harmony of the World, cannot but imbrace the *Copernican* Arguments, as demonstrations that the Earth moves, and that the Sun and Stars stand still."[40] The language of harmony invoked a new aesthetic that would not only alleviate the dissonance of heliocentrism but invite enjoyment of its expressive power.

4 Hearing the Irrational

A century after the immovable center began to move, another seeming impossibility began to seem necessary: a new concept of number that encompassed both integers and irrational quantities. The transformation of the ancient concept of number underlies modern mathematics, and hence also much of modern science. Though a number of social, economic, cryptographic, and even legal perspectives have shed light on this mysterious shift, music (both theoretical and practical) helps illuminate the hesitation about the nascent concept of irrational number in the work of three close contemporaries: Michael Stifel, Girolamo Cardano, and Nicola Vicentino. All three worked at the frontier between mathematics and music around 1550; all three discussed the possibilities of "irrational numbers," more or less hesitantly. In the end, their different mathematical conclusions strongly reflected their different approaches to music. Only music connected Heaven and Earth, theory and experience, mathematics and feeling.

Sixteenth-century mathematicians worked in the shadow of the ancient concepts of number and magnitude, but struggled with these ancient distinctions. For instance, though Robert Recorde's *The Whetstone of Witte* (1557) notes that "*Euclide, Boetius*, and other good writers" acknowledge only "whole numbers," Recorde also includes "nombres irrationale," approximated as closely as desired by infinite series of fractions. Thus, as Katharine Neal notes, Recorde broadened his "number concept while simultaneously using labels that signaled his awareness of the unacceptability, by traditional standards, of the new numbers." Recorde observed that his number terms draw on algebra, then known as the "cossic art," whose solutions include both rational and irrational quantities. This art has many practical aspects, as Cardano and other Italian mathematicians had noted; Recorde dedicated his book to the "venturers" of the Muscovy Company, offering practical examples of military formations, bricklaying, and geography and promising a further book on navigation.[1]

Both practical and theoretical considerations moved François Viète to make crucial symbolic innovations that linked these different number concepts more closely. In his *Canon mathematicus* (1579), Viète advocated the use of decimal fractions to replace the sexagesimal calculations traditionally used for astronomy; such decimals could express both rational and irrational quantities.² Viète's reading of Diophantus and Pappus, along with his own innovative cryptanalytic work, led him to introduce alphabetic signs for unknowns as well as for coefficients, as outlined in his *In artem analyticem Isagoge* (*Introduction to the Analytical Art*, 1591).³ Because a symbol like x could now stand for an integer as well as for an irrational quantity, the new algebraic usage effectively unified these heretofore separate and opposed categories.

These innovations, however ingenious and practical, skated over a foundational abyss because they subsumed irrational and rational under a single symbol. Yet these very issues about the nature of number had emerged earlier in the context of musical theory. The nature of the musical evidence, both theoretical and practical, strongly supported the necessity and legitimacy of irrational numbers. Music was ideally situated to mediate this new understanding between her sisters arithmetic and geometry. In contrast, though the painter Piero della Francesca was an important mathematician, his writings do not show any interaction between his innovations in painting and his concept of number.⁴

The earliest explicit mention of "irrational numbers" as an intended term for these mathematical hybrids seems to have been in the *Arithmetica integra* (*Complete Arithmetic*, 1544) of Michael Stifel, a former Augustinian monk who left the order and became a friend and collaborator of Martin Luther. Alongside his work as a fervent advocate of ecclesiastical reform (he anagrammatized the name of Pope Leo X to yield 666, the Number of the Beast), Stifel was arguably the most distinguished German mathematician of the sixteenth century; his methods were crucial sources for Recorde. In *Arithmetica integra*, Stifel introduced the term "exponent" and used the signs +, −, and $\sqrt{}$.⁵

Stifel begins by reviewing "the nature and species of abstract numbers [*numerorum abstractorum*]." From the beginning, he embeds his novel term "irrational numbers" (*numerici irrationales*) in an extensive discussion of music.⁶ In Book 1 of *Arithmetica integra*, Stifel treats musical intervals in terms of the ratios of string lengths, beginning with the ancient definitions of the basic intervals as whole-number ratios. Because the octave cannot be divided into an integral number of whole tones, the construction of scales requires dividing tones in half, as Boethius had recounted.⁷ But dividing a semitone exactly in half would involve a *geometric* mean that is necessarily irrational (figure 4.1), and hence impossible in the context of the pure *arithmetic* ratios of Greek musical theory. Boethius, following his Greek sources, avoided this problem by dividing the tone unequally into a "major semitone" and a "minor semitone," which differ by the tiny interval called the "comma."⁸

But Stifel notes that "musicians speak of certain irrational proportions," implying that these proportions were already in current musical use and hence should be mathematically acceptable. In contrast, earlier theorists had held that "music does not consider irrational proportions."⁹ Stifel's statement acknowledges the new *musical* desirability of such equal

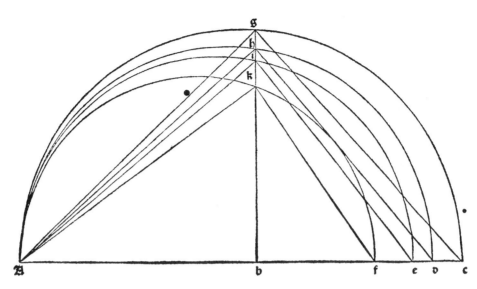

Figure 4.1
Jacques Lefèvre d'Étaples's diagram from *Elementa musicalia* (1496), demonstrating that the interval *Ab*:*bc* can be divided geometrically exactly in half (*bg*). If the two collinear segments *Ab* and *bc* are in the ratio *Ab*:*bc*:: 8:9, then (Euclid, *Elements* 6:3) erecting the perpendicular bisector *bg* on *Ac* gives the mean proportional *Ab*:*bg*:: *bg*:*bc*. Thus, a string of length *bg* would sound an exact semitone higher than string *Ab*; because 8:*bg*:: *bg*:9, in modern notation, $bg = \sqrt{72}$, hence the "ratio" of a semitone is $8:\sqrt{72}$.

division, despite its *mathematical* irrationality, in all senses. To be sure, Euclid's *Elements* used "irrational proportion" to denote "incommensurable quantities."[10] Though Oresme referred to rational and irrational proportions, he never connected number with irrationality because "no irrational ratio is found in numbers."[11] Yet neither Euclid nor Oresme ever overstepped the boundary between rational and irrational, as Stifel does. For instance, Stifel divides the tone into equal semitones following a Euclidean construction (figure 4.1).[12] Likewise, Stifel applies various arithmetic operations to musical proportions, noting that "in these ways irrational proportions of irrational terms can be computed by rational numbers through this beautiful reckoning," including his explicit halving of the tone (figure 4.2).[13] To my knowledge, this is the earliest printed statement that combines a rational (arithmetic) proportion with its irrational (geometric) mean. Acknowledging the controversy about them, Stifel still asserts that "these halvings are so certain that no one can deny them. … A tone can be divided by an uncertain number and that is constituted by no assembly of units, that is, by an irrational number."[14] Thus, in this musical context, Stifel treats these irrational "halved ratios" as though they are as valid as rational proportions.

Yet when Stifel returns to the larger question of "the essence of irrational numbers," his attitude shifts:

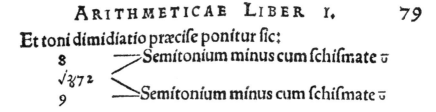

Figure 4.2
Michael Stifel's diagram showing the equal division of a whole tone, from *Arithmetica integra* (1544). He uses $\sqrt{\mathfrak{z}}$ to denote a square root and places 8, $\sqrt{72}$, and 9 in proportion.

It is properly debated whether irrational numbers are true numbers or fictions. For if we lack rational numbers in geometrical figures, their place is taken by irrationals, which prove precisely those things that rational numbers could not; certainly from the demonstrations they show us we are moved and compelled to admit that they [irrational numbers] really exist from their effects, which we perceive to be real, sure, and constant.

On the other hand, other things move us to a different assertion, namely that we are forced to deny that irrational numbers are numbers. Namely, where we might try to subject them to numeration and to make them proportional to rational numbers, we find that they flee perpetually, so that none of them in itself can be precisely grasped: a fact that we perceive in the resolving of them, as I will show below in its place. Moreover, it is not possible to call that a true number which is such as to lack precision and which has no known proportion to true numbers. Just as an infinite number is not a number, so an irrational number is not a true number and is hidden under a sort of cloud of infinity. And thus the ratio of an irrational number to a rational number is no less uncertain than that of an infinite to a finite.[15]

Here, the "cloud of infinity" is the infinite sum of fractions needed to represent an irrational quantity; such an "actual infinite" was rejected by Aristotle. Stifel's distaste for this infinitude finally outweighs his geometrical and musical arguments that irrational quantities can "take the place" of rational numbers in every effective respect. Thus, even though his musical arguments had led him to affirm irrational numbers, his concern to avoid the infinite ultimately moved him to demote them from the class of "true numbers."

Stifel's arguments show the effect of musical considerations on mathematical concerns, indicating the possibility of shifting and surprising alliances between the various parts of the quadrivium: geometric irrationalities, excluded from arithmetic, could find a place in music. Though Stifel himself finally gave precedence to an Aristotelian rejection of the actual infinite, others would take these arguments in a different direction precisely by placing new emphasis on the musical side.

Among these, the famous mathematician, physician, and polymath Girolamo Cardano has special importance, even though his writings on music are less well known than the rest of his vast output. Though Cardano's *De musica* was published only in his *Opera omnia* (1663), among his works on arithmetic and geometry, he wrote this manuscript about 1546, during the period surrounding the appearance of his most famous mathematics book, *De arte magna* (*On the Great Art*, 1545), which announced the general solutions of the cubic and quartic equations (in the midst of notorious controversies about priority and disclosure), a landmark in the development of modern algebra. Thus Cardano's presentation of what he modestly called "this most abstruse and clearly unsurpassed treasury of the entire arithmetic" should be read next to his contemporaneous musical work, which is notable for its emphasis on practical techniques related to musical instruments as well as its theoretical considerations.[16]

Cardano sang and played several instruments, including the recorder and the lyra, and was a skilled composer, as is shown by several compositions he includes in *De musica* and his careful accounts of instrumental techniques. Cardano's awareness of such modes of ornamentation as trills and vibrato draws attention to microtonal shifts that singers and instrumentalists used to decorate their melodies. He particularly emphasizes the unusual interval of a *diesis*, a quarter tone (half a semitone) that produces "such a movement [that] titillates the ear and increases its pleasure." As Clement Miller notes, "his affection for this tonal embellishment was very great and his description of the beauty and pleasantness of the effect sometimes borders on the ecstatic."[17]

Cardano's predilection for the diesis led him to put forward new opinions about its definition and that of the semitone, though he cites the mathematical problem of the exact divisibility of ratios. To divide a tone into two equal semitones (or a semitone into two equal quarter tones) "correctly and arithmetically," he acknowledges that a "true calculation" involves an irrational root, for which he accepts a rational approximation that is "closer in perception." For these "true" irrational intervals (whether of semitone or diesis), he empirically substitutes a simple rational approximation, thus conflating the geometrically irrational with the arithmetically rational. He treats the result as a "correct" system of tuning, not merely a stopgap or approximation; in fact, the application of his calculation of his approximate semitone ($\frac{18}{17}$) to fretting a lute was "the first really practical approximation of equal temperament," later (incorrectly) attributed to Vincenzo Galilei (see box 4.1, ♪ sound examples 4.1a–c).[18] Cardano treats rational and irrational intervals on the same footing primarily because of musical considerations: he chooses between rational approximations for the diesis *not* on the basis of closeness of numerical value (which would lead to $\frac{34}{35}$) but on perception as judged musically (leading to $\frac{36}{35}$). He calls "true" *both* the irrational "true diesis" and its rational, musical equivalent, which is true "arithmetically."

Though in *De arte magna* Cardano often refers to "numbers" with the sense of "integers" and never uses the term "irrational numbers," he uses the phrase "the numbers that were to be found" to refer to specifically irrational expressions. Cardano's "golden

> **Box 4.1**
> Pythagorean tuning, just intonation, and equal temperament
>
> *Pythagorean tuning* relies on the simple intervals of the smithy story (chapter 1): octave (2:1), perfect fifth (3:2), perfect fourth (4:3), whole tone (9:8). From these are derived more complex (hence less consonant) intervals, such as the major third (two whole tones, 81:64), major semitone (interval between a major third and a perfect fourth, 256:243), minor third (a whole tone and a major semitone, 32:27) (♪ sound example 4.1a).
>
> *Just intonation* is a system of related tunings that all begin with the Pythagorean octave, fifth, and fourth, and then redefine other intervals to be simple whole number ratios, such as the major third (5:4) and minor third (6:5), reflecting the growing use of those intervals as consonances after the fifteenth century (♪ sound example 4.1b).
>
> *Equal temperament* divides the octave into twelve mathematically equal semitones, each defined by the irrational expression $\sqrt[12]{2} : 1$. Except for the octave, all other intervals are slightly "out of tune" with respect to just intonation or Pythagorean tuning (♪ sound example 4.1c).

rule" shows how to find what he calls the "closest approximation" to solving an equation through finding the integers, "greater and less, which most nearly satisfy the equation," then generating a series of differences between the values those integers generate when substituted in the equation, from which a further refined estimate can be made, leading to what he takes to be a converging series of approximations. Through this procedure, "you will undoubtedly arrive at an insensible difference" compared with the true value: "This is universal reasoning and needs no other rule."[19] His procedures here seem to reflect the practical sense that comes to the fore in his musical writings: "true" geometric (irrational) and "true" arithmetic (rational) quantities *sound* the same; their differences are "insensible." Music is the sole example he has of this kind of mediation between "perceptual" and "true" quantities; he lacks any other kind of mathematical physics (as it would later come to be called) that could have confronted mathematical idealizations with physical reality. But music was sufficient for him. Musical judgments intermixed rational and irrational quantities, supporting and paralleling Cardano's working equivalence of the two in his algebraic art and providing it with crucial examples.

The grounds on which Nicola Vicentino treated irrational quantities as numbers had more to do with issues of melodic style and musical practice, as for Cardano, than with the more purely theoretical questions that concerned Stifel. Boethius had enumerated three ancient "genera" of melody, each genus designating a separate set of basic intervals on which music could be constructed. In the last chapter, we considered the *diatonic* genus, which Boethius considered "somewhat more severe and natural" (♪ sound example 4.2a). The other two genera are more unfamiliar (see box 4.2). According to Boethius, the *chromatic* genus "departs from natural inflection and becomes more sensual [*mollius*]."[20] The name "chromatic" persists even today to describe music that makes

> **Box 4.2**
> The three ancient musical genera
>
> ---
>
> The *diatonic* genus follows the pattern of a semitone followed by a tone and another tone: S T T (♪ sound example 4.2a). The *chromatic* genus has the pattern S S S^3, where S^3 stands for a "trihemitone" (an interval composed of three semitones) (♪ sound example 4.2b).
>
> The *enharmonic* genus uses a quarter tone (the diesis, abbreviated D), according to the pattern D D T^2, where T^2 denotes a "ditone," an interval composed of two whole steps (♪ sound example 4.2c).

extensive use of consecutive semitones, sometimes to evoke greater sensuality or expressivity (♪ sound example 4.2b).

In contrast, our music lacks the *enharmonic* genus, which Boethius considers "even more closely joined" than the chromatic, in the sense that the enharmonic genus uses a quarter tone (diesis). Apart from some self-conscious attempts to recreate such music that we will come to and some experimental music later still, the diesis fell out of use in Western music. Yet Boethius does not treat it as exotic but only remarks that it "is beautifully and fittingly yoked together"; indeed, its Greek name (*harmonia*) is the general word that has come down to us as "harmony": the enharmonic genus was considered harmonious *par excellence*.[21] Before the sixteenth century, only the diatonic genus seems to have evoked commentary, perhaps because it was used in musical practice. New translations brought the other genera to prominence.[22] In his book *L'antica musica ridotta alla moderna prattica* (*Ancient Music Adapted to Modern Practice*, 1555), Vicentino identified the enharmonic as the secret behind those extraordinary, lost powers of ancient music, which he decided to revive (♪ sound example 4.2c).

Vicentino was a practicing musician and composer with deep interests in the theory of music (figure 4.3). Born in Vicenza, he came under the influence of the humanist Giovanni Giorgio Trissino, who in 1524 described the enharmonic and the chromatic as "two genera that our age does not know." After studies with Adrian Willaert, the great Venetian composer, Vicentino came to Ferrara at the behest of Cardinal Ippolito II d'Este, whom he then accompanied to Rome. In 1546 Vicentino published his first book of madrigals, but around 1534 he had already begun thinking about the ancient genera.[23] In 1551 he became embroiled in a public controversy that shows the extent to which these matters provoked contention among the educated elite throughout Europe.

The argument started in June 1551 after a private performance of a motet in Rome, when those present began discussing what genera of melody were used in the composition. The Portuguese composer and theorist Vincente Lusitano maintained that the motet used only the diatonic genus, whereas Vicentino argued that it used elements of all three genera. What was at stake went beyond this single work to all of contemporary practice: what was

Figure 4.3
Portrait of Nicola Vicentino, aged forty-four, as the frontispiece of his book *Ancient Music Adapted to Modern Practice* (1555). In the outer border, his motto reads: "You have revealed to me the uncertain and hidden things of Thy science." In the inner border, he is identified as "inventor of the archicembalo and also of the practical division of the chromatic and enharmonic genera."

the true status of those ancient genera in contemporary music? The broader implications of this question concerned not only whether modern music had kept or broken faith with its ancient heritage but also the character and integrity of the cosmos, which was widely assumed to be regulated by musical intervals.

The debate began with a wager of two gold scudi and quickly became formal and public. Over a period of five days (June 2–7), Vicentino and Lusitano presented their arguments at the Vatican to "an audience of many learned men," in the presence of Cardinal Ippolito and "judges" who were singers in the chapel of Pope Julius III; the young Orlando di Lasso (then nineteen) may have been in the audience.[24] This tribunal found against Vicentino in a statement that reads (given the ecclesiastical authority of the presiding cardinal and the papal offices of the judges) like a legal anathematization, concluding that "the said Don Nicola must be condemned, as we sentence him in the wager made between them" (figure 4.4).[25] Vicentino's own account treats quite seriously what he considered a grave injustice, whether or not we take the scene as an *auto-da-fé* for "heretical pravity" that anticipates the trials of Giordano Bruno or Galileo. Perhaps Vicentino's wounded pride kept him from taking the less serious tone others may have adopted. But even a high-spirited imitation of inquisitorial proceedings presided over by an eminent cardinal seems ominous. The church, especially the Jesuits, condemned any alterations to the foundations of mathematics that would undermine its epistemic certainty and hence the unchanging rational foundations of Christian doctrine. We will shortly consider Vicentino's argument that experience was the "mistress" of musical and mathematical theory—rather than pure reason, as Boethius taught and the church insisted.[26]

The confrontation at the Vatican led Vicentino to publish his defense for a larger public interested in the case and willing to pay to read about it. By way of amplifying and illustrating his assertions, Vicentino described his newly invented archicembalo, an "archharpsichord" whose specially designed keyboard could play the complex variety of semitones and dieses necessary to execute chromatic and enharmonic compositions (figure 4.5).[27] Vicentino's keyboard mechanized the playing of quarter tones, heretofore laboriously measured and sounded one by one.[28] His new instrument enabled accurate renditions of enharmonic music, but tuning it required deciding the exact interval of a diesis. To do so, Vicentino needed to unearth the work of ancient theories who addressed these musical and mathematical questions.

In the mid-sixteenth century, the problem of dividing the tone was not yet solved uniquely; several conflicting definitions of the semitone remained in use.[29] This aggravated the problem of defining the diesis: how could one define a quarter tone if the half tone remained so contentious? The obvious approach was to define unequal major and minor dieses by dividing up major and minor semitones, but this would lead to an endless recurrence of the problem of dividing intervals.[30] New clarity was sought in the ancient sources.

LIBRO QVARTO

SENTENTIA.

Hristi nomine inuocato &c. Noi sopradetti Bartholomeo Esgobedo, & Ghisilino Dancharts, per questa nostra diffinitiua sententia & laude in presentia della detta congregatione, & delli sopra detti Don Nicola, & Don Vincentio, presenti intelligenti, audienti, & per la detta sententia instanti. Pronontiamo, sententiamo il predetto Don Nicola non hauer in uoce, ne in scritto prouato sopra che sia fondata la sua intentione della sua proposta. Immo per quanto par in uoce & in scriptis il detto Don Vincentio hà prouato, che lui per uno competentemente cognosce & intende di qual Genere sia la compositione che hoggi communamente i Compositori compongono, & si canta ogni di, come ogniuno chiaramente di sopra nelle loro informationi potrà uedere. Et per questo il detto Don Nicola douer essere condennato, come lo condenniamo nella scommessa fatta fra loro, come di sopra. Et cosi noi Bartholomeo & Ghisilino soprascritti ci sotto scriuiamo di nostra mano propria. Datum Romæ in Palatio Apostolico, et Capella prædetta, Die VII. Iunij. Anno suprascripto Pontificatus S. D. N. D. Iulij. PP. III. Anno secundo; & laudamo.

Pronuntiaui ut supra. Ego Bartholomeus Esgobedo, et de manu propria me subscripsi.

Pronuntiaui ut supra. Ego Ghisilinus Dancherts, & manu propria me subscripsi.

Io Don Iacobo Martelli faccio fede, come la sententia et le due polize sopra notate sono fidelmente impresse & copiate dalla Copia della medesima sententia de i sopra detti Giudici.

Io Vincenzo Ferro confirmo quanto di sopra.

Io Stefano Bettini detto il Fornarino, confirmo quanto di sopra.

Io Antonio Barrè confirmo quanto di sopra.

Io non uoglio dire cosa alcuna circa la sopra stampata sententia, perche questa cura lascierò giudicare al mondo, & al gran Iddio, ilquale è somma giustitia, che per suo mezzo farà cognoscere à ogniuno la ragione & il torto, come ha inspirato il sopradetto Don Vincentio, che publichi al mondo le mie raggioni con la sua opera stampata, contra de lui & delli Giudici.

Fine del Quarto Libro.

Figure 4.4
The sentence passed against Vicentino, as recorded in his book.

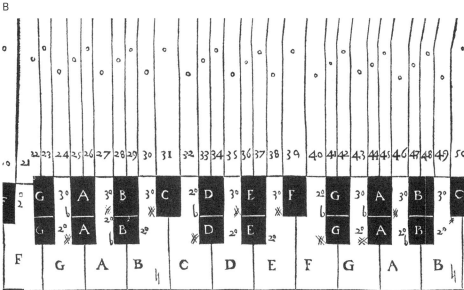

Figure 4.5
Vicentino's archicembalo. (a) Detailed view of the split keys of an archicembalo reconstructed in 1974 following Vicentino's design by Marco Tiella and made in the workshop of the organ builder Barthélemy Formentelli (Pedemonte, Verona). (b) Schematic layout of a section of a keyboard shown in a fold-out page from Vicentino's book. The split black keys allow the playing of intervals of a diesis.

What follows, then, is not a digression into antiquity but an account of how ancient problems returned to life.

Boethius had not given the enharmonic diesis a precise ratio, perhaps because of its very smallness.[31] Plato and Aristotle considered the diesis a kind of element, analogous to a vowel or consonant; Aristoxenus also judged that "the voice cannot distinctly produce an interval even smaller than the smallest diesis, nor can the hearing detect one."[32] The diesis, it was concluded, may be so small an interval that strict, secure definition is elusive. This judgment may also reflect the material circumstances and difficulties surrounding the production of this interval. Aristides Quintilianus (first century C.E.) noted that the enharmonic "has gained approval by those most distinguished in music; but for the multitude, it is impossible. On this account, some gave up melody by diesis because they assumed through their own weakness that the interval was wholly unsingable."[33] Thus, even in ancient times the diesis involved discrimination and virtuosity, as in the quarter-tonal "bending" of pitch possible on the aulos (a pipe with finger holes and a reed mouthpiece, often played in pairs). Aristotle described the aulos as "orgiastic," its shrill wails often associated with Bacchic and Corybantic rites, sending its hearers out of their minds (figure 4.6).[34] Such associations would not militate toward fussiness in intonation, if indeed the diesis were an ecstatic "bending" of a pitch not really to be measured by any ratio but only by the inspired frenzy of the Dionysian virtuoso.

Though this aural interpretation goes against Pythagorean tradition and its ratios, its ancient champion was Aristoxenus. Most of the information we have on the enharmonic diesis comes from him, suggesting that this interval may have fit particularly well into his thesis that discriminative hearing, rather than predetermined, fixed ratio, really determines musical intervals.[35] Aristoxenus stood at a critical point in the problem of subdividing intervals, which (as we have seen) involves irrational magnitudes if the divisions are to be strictly equal. In the face of this paradox, he opened the liberating possibility that we might weaken or abandon the demand that every interval be strictly rational through his reliance on the sense of hearing, rather than on reason. He himself never seems to remark that his line of argument would imply the possibility of quantities that might bridge the rational and irrational; on the contrary, he explicitly maintains a sharp "division … in respect of the differences between the rational and the irrational."[36]

By turning away from Pythagorean ratios, Aristoxenus took a crucial step toward treating an arbitrary musical quantity *as a unit unto itself, apart from whether it is or is not rational* with respect to the initial integral units of string length.[37] His demonstrations recall Euclid, who had shown in the *Elements* that magnitudes are rational or irrational only relative to other magnitudes, not in any absolute sense. The diagonal of a square is incommensurable with its side but may be perfectly commensurable with other lines (for instance,

Figure 4.6
Young man playing a pair of auloi while a courtesan dances with castanets. Red-figured cup painted by Epiktetos, ca. 500 B.C.E.

with the sides of another square built on that diagonal).³⁸ If indeed Aristoxenus built tone and semitone on the diesis as an arbitrary unit, that would ignore the inherent incommensurability of tone and semitone or quarter tone, as discussed above. At the very least, by abandoning the idealized fiction of a pure ratio underlying every note Aristoxenus was able to bring forward the practical "commensurability" of every pitch sung or played on real strings: because we can *hear* those intervals, he implies, they must *de facto* be commensurable.

At the beginning of his book, Vicentino embraces Aristoxenus and "experience as the mistress of things," paradoxically outstripping the "modern" practice of music by reviving the "ancient," specifically through the retrieval of the lost enharmonic genus he considers "more sweet and smooth than the other two genera."[39] His emphasis on experience means that musical practice is the new touchstone that can outweigh older arguments about rationality. The process of evenly dividing ever smaller musical intervals tacitly overstepped the ancient separation between geometry and arithmetic through the commonsensical identification of a physical string length with the corresponding line in a geometric diagram (such as figure 4.1), rather than with a purely numerical ratio, considered apart from any sounding body.[40]

Vicentino realizes that these quarter tones are "disproportioned and irrational. And the other parts accompanying this division cannot contain proportioned and accurate leaps because they must correspond to this irrational ratio. … Likewise, the nature of the enharmonic genus disrupts the order of both the diatonic and chromatic and permits the creation of steps and leaps beyond the rational. For this reason such a division is called an irrational ratio." In using this phrase, Vicentino is the first (as far as I can determine) to try to state in some positive (if paradoxical) way the status and character of musical intervals that are formed through *irrational*, geometric means but at the same time are incorporated into the *rational* arithmetic of music theory.[41] Vicentino's contribution may have been to state as clearly as possible, in common words, not only the inherent paradox of a new concept but its necessity and the functionality that justifies our embracing it despite and even through its paradoxicality.

Vicentino's unfolding argument presents us with both sides of the paradox: the diesis, as constructed geometrically, is irrational but, functioning as the smallest "unit" within a framework of numerical ratios, is in that sense also effectively a "proportion." Such a hybrid concept, like a centaur, needs to be grasped in its inherent duality, considered as its essence rather than as grounds to refute its existence. Vicentino's premise—that the enharmonic genus exists and is superlatively important—was attested by many ancient sources. Therefore its basis, the diesis, must also exist, and so we should take in stride whatever paradoxical qualities it may have. Vicentino anticipates that we might rightly be "astonished" at what at first seems a prodigy or monster, this "rational irrational"—rather in the way that we might consider a centaur monstrous were we not familiar with examples of wise centaurs such as Chiron, "who included music among the first arts he taught Achilles at a tender age, and who wanted him to play the harp before he dirtied his hands with Trojan blood."[42]

Vicentino's argument also builds cunningly on the successive examples of the three genera, as if in a kind of rhetorical crescendo.[43] The diatonic sets the point of departure, the purely rational. From our earlier discussion, we know that the semitone of the diatonic genus already contains the latent problem of its relation to the tone: namely, that no rational semitone can be exactly half of a tone. The stratagem of devising major and minor semitones merely conceals this problem without solving it, giving a stopgap solution that serves to make the diatonic appear wholly rational. The two successive semitones in the chromatic genus reiterate the problem: *which* semitones are they to be, major or minor? In the chromatic genus, the latent problems hidden in the diatonic come forward sufficiently to cause "disruption": incipient instability and growing theoretical uncertainty. Only with the enharmonic genus does this simmering instability disrupt *both* the diatonic and chromatic genera and "permit the creation of steps and leaps beyond all reason."[44] This, Vicentino tells us, is the cause that should move us not only to call them "irrational ratios" but also, by so naming them, to install them in mathematics and music jointly as having equal existential force with the "rational ratios" we learned from arithmetic and the "irrational magnitudes" from geometry. For Vicentino, music is the intermediate ground on which arithmetic and geometry meet in such hybrid concepts as "irrational ratios," shared between mathematics and music.

The attitudes of Stifel, Cardano, and Vicentino about these mathematical issues reflect their respective musical projects. As we saw, Stifel's closest approach to affirming that "irrational numbers" were "real" or "true" came in the context of his musical theorizing. Yet this did not prove sufficient for him to maintain this position in the face of the actual infinitude of fractional sums, the "cloud of infinity," perhaps because his involvement with music remained largely theoretical and restricted. Stifel's main foray into practical music was his thirty-two-strophe song to propagate Luther's teachings, "Johannes thüt uns schreiben" (1522), based on the popular tune "Bruder Veyt." Stifel's composition led him into a polemical war of song and countersong with the theologian Thomas Murner, both always keeping this same melody for their new lyrics (♪ sound example 4.3).[45] Though Stifel's song was very popular and went through many printings, even serving as an important early example of the power of music that may have inspired Luther himself, its melody was simple and derivative. Stifel merely provided new words to an old tune; he had no vision of reforming the elements of music that would compare with Vicentino's ambitious project to (re)create a whole new genus of music. By comparison with Stifel's song, Cardano's extant compositions are very ambitious, including a five-voice perpetual canon and a tour de force of four simultaneous three-voice canons (twelve voices in all).[46]

Among Vicentino's much larger output, his motet "Musica prisca caput" (♪ sound example 4.4) dramatizes the emergence of the enharmonic genus to glorify his patron: its first verse is in the diatonic genus, the second in the chromatic, while the final verse is in the enharmonic, dramatically reserving the introduction of the diesis to produce a special aura around the name of Cardinal Ippolito at the end of the motet. This motet's pointed delineation of all three genera provides yet another demonstration and justification of Vicentino's views to refute his critics and contest his condemnation. He tells us, as well, that the whole d'Este family, including the cardinal and the prince of Ferrara, sang this daring new music, quarter tones and all, "with the most exceptional diligence." Vicentino had evidently persuaded them that, in contrast to the public uses of diatonic music "in communal places for the benefit of coarse ears," such enharmonic music was "reserved … to praise great personages and heroes for the benefit of refined ears amid the private diversions of lords and princes."⁴⁷ Thus Vicentino brought his polemic on behalf of enharmonic music not only to experts but also to the powerful amateurs whose princely involvement he considered capital for his cause. By so doing, he carried his case to an alternative and (in his view) superior social milieu, whose approval and validation he took as definitive. In addition, he positioned himself so that his theories would be highly visible and readily available to another aristocratic set that would take up his ideas in the next generation—the Florentine Camerata and Vincenzo Galilei, whose admiring advocacy indeed vindicated Vicentino posthumously.⁴⁸ For instance, Zarlino, Vincenzo's teacher, incorporated these irrational ratios in his representation of the tuning of a lute (figure 4.7), which showed how a geometric construction can dictate the placement of the frets for equal temperament (see box 4.1, ♪ sound examples 4.1a–c). In this way, geometry set a template that could be mechanically reproduced without having to duplicate its geometric construction.⁴⁹

Even so, the use of these irrational ratios remained controversial. Though Zarlino's student Giovanni Maria Artusi accepted his teacher's geometric construction for tuning instruments, he balked at applying such irrational ratios to vocal music. Writing in 1603, Artusi objected that Claudio Monteverdi, the great exponent of the new operatic art sponsored by the Camerata, was applying irrational ratios to generate for expressive purposes intervals Artusi considered "false for singing," particularly the diminished seventh and diminished fourth (figure 4.8, ♪ sound example 4.5). Artusi complained that the use of such "irrational" intervals showed that Monteverdi had no "rational" understanding of music, as he put it. Though it was possible to play these intervals on the fretted lute, "the natural voice is not suited to negotiate such unnatural intervals by means of natural ones, not having a preset stopping place like an artificial instrument. … It cannot justly divide the tone into two equal parts."⁵⁰ Artusi's objections blend mathematical uneasiness about "unnatural" irrational ratios with his concomitant aversion to Monteverdi's expressive use of those same intervals.

Such objections show the deep and long-lasting anxieties provoked by irrational ratios, anxieties that reflect both musical and mathematical considerations. Stifel was content to remain in the realm of conventional (and popular) music, and so his "irrational numbers" drew on no particular musical justification that might help defend them against traditional philosophical objections. In contrast, Cardano's strong interest in composition and performance and Vicentino's reliance on "irrational ratios" in his music seem to have helped them sustain their effective use *as numbers*, to the extent that they advanced them as musical and hence mathematical necessities.

The comparison of these three figures as theorists, composers, and mathematicians illuminates ways in which musical concerns, both practical and theoretical, influenced the acceptance of novel mathematical concepts.⁵¹ As Guillaume Gosselin noted in *On the Great Art or the Hidden Part of Numbers, Commonly Called Algebra or Almucabala*

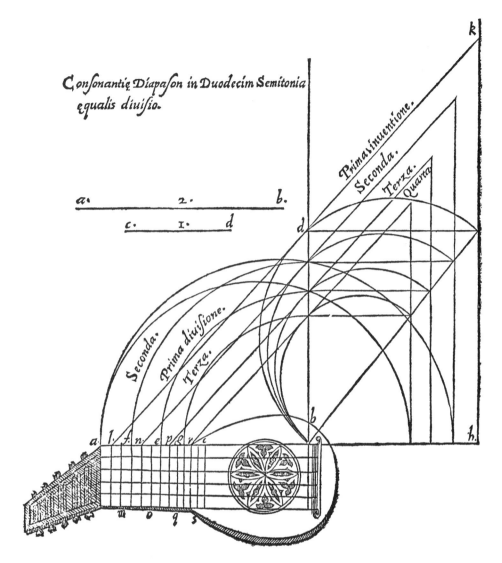

Figure 4.7
Zarlino's geometric approach (*Sopplimenti musicali*, 1588) to fretting a lute in equal temperament using the construction shown in figure 4.1. The caption reads: "The equal division of the consonance of a diapason into twelve semitones."

Figure 4.8
Claudio Monteverdi, *La favola d'Orfeo* (first performed 1607), Act II, measures 274–279 (from the edition of Venice, 1615). The messenger is recounting Euridice's dying words: "and calling on you, Orfeo, Orfeo, after a deep sigh expired in these arms." Artusi objected to the "irrational" diminished seventh between voice (B♭) and continuo (C♯) at the word *grave* (indicated by arrows), expressing the depth of her sigh (♪ sound example 4.5).

(1577), "The arithmetician sees numbers in themselves, the musician and the algebraist indeed know numbers, but in their relation to something else."[52] His implication is that musicians were essential companions on the road that led from arithmetic to algebra in that they reached beyond numbers to the "something else" manifest through music. If so, the struggle to "hear" the mathematical irrational was indeed consequential on many levels. Mediating between the realms of mathematics and felt experience, music evoked and justified new concepts of number.

5   Kepler and the Song of the Earth

For Johannes Kepler, music was crucial to the emergence of a new astronomy. Kepler's *Harmonices mundi libri V* (*Harmony of the World*, 1619) culminates in his so-called third law of planetary motion: the square of the orbital period of any planet is proportional to the cube of its mean distance from the sun.[1] This surprising connection emerged from Kepler's search for harmonic relations between planetary data and became a touchstone for Newtonian celestial mechanics. As important as this result is, Kepler presents it without fanfare, certainly not as the "law" it was later called. For him, it was only a part of his much larger project of seeking harmonious relations between planetary motions. Kepler embarked on this enterprise moved both by his Neoplatonic concern with cosmic archetypes and also by his keen interest in practical music and contemporary compositions.[2] His strong feeling for what he called the "song of the Earth" illuminated and complemented his cosmic concerns.

On the theoretical side, though he identified himself strongly with Pythagoras, Kepler reinterpreted musical consonance in terms of the ratios of the sides of polygons, instead of the pure numerology inherited from Pythagorean tradition. Indeed, *Harmonice mundi* concludes with a polemic against the English Rosicrucian Robert Fludd, from whom Kepler distinguishes himself "in the way in which a practitioner does from a theorist."[3] According to Kepler, Fludd "has advice on the composition of figured melody, an art which I do not profess" and "also digresses to various musical instruments, to which I had not even given thought."[4]

On the practical side, though the passages in *Harmonice mundi* in which Kepler turns to actual musical compositions are few and brief, they give insight into his sensibility. In those moments, he characteristically gives examples from compositions by Orlando di Lasso. Kepler does not say much or go deeply into these examples, yet they are still interesting because they figure in an astronomical work. Indeed, in the long tradition linking music and astronomy, Kepler is remarkable for citing specific musical examples, not just theoretical generalizations. These examples illuminate the context of his whole project.

Like many of the other figures considered in this book, Kepler was involved in the practice of music as well as its theory. From early childhood, Kepler was steeped in the musical traditions of Württemberg Protestantism, in school and in church.[5] From age five, he practiced German psalmody as well as the Latin sequences and hymns he later cites in *Harmonice mundi*.[6] This daily singing was supplemented by weekly lessons in theory. The standard of musical cultivation as well as of theoretical instruction in Württemberg was quite high, including also contrapuntal and figured music.[7] Kepler derived a solid theoretical knowledge and practical skill from his primary schooling, continued and deepened during his theological studies in Tübingen (1589–1594).[8] There, he, like all his fellow students, had musical instruction. The academic ordinances prescribed singing three days a week, so that the students must "always study new motets and good songs, and thus keep the exercise of music in practice."[9] Kepler participated in performances of church music and private festivities; he also encountered Glarean's musical theories.[10]

Kepler's first job was teaching mathematics in Graz (1594–1600), where he also taught Virgil and rhetoric. His abiding interest in the practice of rhetoric (and use of Virgilian images) informed his mature writings.[11] The musical life of the school where Kepler taught was many sided. The school organist, Erasmus Widmann, so favored dance styles in his sacred music that sarcastic critics wondered whether they were in a church or a beer hall.[12] During Kepler's stay, the Italian organist Annibale Perini brought Venetian musical practice to Graz. Indeed, Andrea Gabrieli had dedicated his *Primus Liber Missarum 6 V* (1572) to Karl, the Habsburg archduke resident in Graz. Karl had close links to Venice; his wife Maria had a personal bond to Lasso's family and a strong interest in his music.[13] It is not clear what part Kepler took in all this, though it seems likely that he would have been aware of these musical cross-currents. One of his letters mentions "the excellent music that Italy abounds in"; Kepler's acquaintance with Lasso's music definitely began in Graz, if not before.[14] In 1599, Kepler wrote a friend that he wished that "Orlando, if he lived," could teach him how to tune a clavichord properly, singling out the great composer as the ultimate authority on tuning.[15]

After Kepler moved to Prague in 1600, he entered the service of Emperor Rudolf II, famous for his patronage of occult arts. Unmarried, distancing himself ever further from political realities, Rudolf fostered "exact science next to the deepest superstition, religious freedom next to zealotry, a tendency to display the utmost pomp next to diseased manifestations of self-love and eccentricity, refined taste next to brutal sensuality."[16] In Rudolf's court, both practical and theoretical music were important, including some novel developments.[17] For instance, the court alchemist Michael Maier wrote fifty canons in *Atalanta fugiens* (1618), whose settings of alchemical texts would complement the manipulations of the "great art."[18] Such a synthesis would have deeply interested the alchemist-emperor. So also did the "perspective lute," which tried to relate musical tones to colors, or the court composer and organist Hans Leo Hassler's experiments with new automatic instruments.[19] As R. J. W. Evans points out, in such activities music was "practical, yet offered immediate contact with cosmic forces," mobilizing magic powers through the influence of sound.[20]

Kepler did not record his precise reaction to these developments but in a private letter wrote a stark disclaimer: "I hate all kabbalists."[21] To be sure, Kepler gave voice to mystic sentiments of his own: "For there is nothing which I examine with more scruple, and which I desire to know so much as this: whether perhaps the God whom I as it were touch by hands when I contemplate the whole universe, can also be found by me inside myself."[22] Yet Kepler noted in *Harmonice mundi* that "whoever wants to nourish his mind on the mystical philosophy ... will not find in my book what he is looking for."[23] Though he detested esotericism, Kepler was deeply interested in the larger question of how the practice and theory of music might impinge on cosmic structure. His antipathy to the occultism in Rudolf's court may have indicated his anger at what he considered the bungling of his own favorite idea that music shapes the cosmos.

Kepler's letters of the period turned to more practical concerns. Given his interest in tuning, he may well have noticed the "Clavicymbalum Vniversale, seu perfectum," a keyboard instrument much admired at court, whose octaves were divided into nineteen steps.[24] He likely attended the services of the court chapel, in which one hundred musicians (including sixty-five singers) performed music by court composers, as well as Venetian polychoral music and early monody. He could scarcely have missed the six "violinists or musicians" or the eighteen trumpeters and timpanists that were part of the imperial household.

Kepler also recorded a fragment of the prayers sung by the "Turkish priest," as he calls him, who accompanied the Turkish ambassador to court.[25] Kepler was fascinated by what he described as the priest's "practiced and fluent manner, for he did not hesitate at all; but he used remarkable, unusual, truncated, abhorrent intervals, so that it seems that nobody could with proper guidance from nature and voluntarily of his own accord ever regularly contemplate anything like it. I shall try to express something close to it by our musical notation" (figure 5.1).[26] We shall shortly return to the significance of Kepler's attempt to notate the exotic strains of Muslim cantillation. For now, it is an apt image of his alert curiosity about the possibilities of music not only in theory, but in practice.

Finally, the archduke Matthias seized power from his Prospero-like brother Rudolf, who died not long after, in 1612. Kepler did not remain in Prague but spent his last years in Linz (1612–1626) as a teacher, though retaining the title of Imperial Mathematician. There, he completed the *Harmonice mundi*, the apex of his theoretical activities, in a school reputed as "the undisputed center of musical cultivation to support the renewal of faith" and which gave the highest priority to *musica practica*.[27] Lasso had pride of place in their library, followed by other masters of the Renaissance. Following the customary academic regulations, Kepler probably would have taken an active part in the choir and the house music of the regional nobility, among whom he had many friends and patrons.[28]

Thus, though claiming no skill as composer or performer, Kepler had been surrounded with musical performance all his life and had been personally involved in musical activities on many occasions. Here, more recent distinctions of professionalism are misleading. The

Concinnus igitur & humanarum aurium judicio aptus cantus est, qui exorsus à certo quodam sono; ab eo per intervalla concinna tendit ad sonos consonos & primo illi, & plerumq; etiam inter se mutuò; dissona cursim pervolitans intervalla, in consonis verò immorans, seu mensurâ temporis, Syllabarumq; longitudine, seu crebro ad illos reditu, veluti duarum vocum inter se consonantiam affectans, unicæ vocis traductione à loco uno Systematis ad alium.   Exemplum.

Victimæ paschali laudes immolent Christiani Agnus redemit oves Cristo innocens patri reconciliavit peccatores.

Hic sonus initialis est in clavi *G*, cum quâ in cantu molli concordant *b. c. d. g.* Excurrit igitur Cantus (primùm flexu deorsum facto) ad clavem *c.* consonam, & transilit planè dissonum locum *A*; fuisset autem idem, si attigisset ipsum, sed brevi mora; tota verò series reliqua potissimùm in locis *b. d. g.* intonat, sceletô octavæ tale exprimens, in *d.* creberrimè rediens, post in *b.* in superius vero *g.* se interdum efferens, in hæc omnia loca signanter: non sic in *a.* vel in *f*, loca primo dissona: tandemq; redit ad *G.* ibiq; finit.

Figure 5.1
Kepler's *Harmonice mundi* (1619), showing the Turkish chant (top), compared to the Gregorian chant *Victimae paschali* (middle) and its melodic skeleton (bottom) (♪ sound examples 5.1–5.3).

musical experience of an amateur can be no less deep than that of a professional and, in Kepler's time, amateurs did a great deal of serious music making. Beyond the traditional school readings in the quadrivium, Kepler was largely self-educated, but with the gusto that characterized his idiosyncratic genius. One thinks of him traveling in October 1617 to save his aged mother from prosecution as a witch, taking Vincenzo Galilei's *Dialogo della musica antica et della moderna* along and reading it "with the greatest pleasure [*summa cum voluptate*]," though he disagreed with the book on many musical issues.[29] This shows that only two years before Kepler published his own treatise, he needed to catch up with contemporary theory.[30] Evidently unaware of Vicentino and his Italian sources, Kepler was able to acquire a Greek text of Ptolemy's *Harmonics* only in 1607.[31] Thus, Kepler rediscovered this important ancient source in the course of pursuing his own vision.

Though he was engaged in reviving the Platonic vision of cosmic harmony, Kepler's awareness of contemporary music informed crucial departures from the ancients. By his time, musicians needed consonant thirds and sixths for their polyphonic music, though

Boethius had assigned complex (hence less consonant) ratios to those intervals, according to the original Pythagorean tuning. Contemporary practice and theory (including Zarlino) used simpler intervals for these intervals, known as just intonation (see box 4.1).[32] Kepler sided with these contemporary musicians, critiquing the Pythagoreans for judging "from their numbers alone, doing violence to the natural prompting of hearing."[33] For his part, Kepler combines musical practice with geometric arguments about the ratios between sides of regular polygons.[34] For instance, though he used a pentagon to justify the major sixth as 3:5, he refused to use a heptagon to allow such discordant ratios as 3:7.[35] Euclidean geometry could disqualify the heptagon, which, unlike a pentagon, cannot be constructed with ruler and compass, but Kepler notes that the nascent art of algebra would allow a calculation of the heptagon's side that, if accepted, would give the heptagon as much validity as the pentagon.

Though he expressed admiration for the calculational powers of algebra, Kepler finally did not allow it full legitimacy because it implies infinite processes (akin to Stifel's "cloud of infinity") and also for musical reasons. Algebra would allow intervals like 3:7, which Kepler finds "utterly abhorrent to the ears of all men and the usages of singing, even though it may be possible for strings to be tuned in that way, seeing that as they are inanimate they do not interpose their own judgment but follow the hand of the foolish theorist without the least resistance."[36]

Going past basic issues of tuning, Kepler discusses what constitutes "naturally tuneful and suitable melody."[37] He attempts a rhetorical analysis that encompasses fine details of the melodic skeletons of two very different melodies, beginning with the Turkish chant mentioned earlier (figure 5.1, top; ♪ sound example 5.1). He treats this as a kind of anti-music, "that grating [*stridulo*] style of song which the Turks and Hungarians customarily use as their signal for battle, imitating the uncouth voices of brute beasts rather than human nature."[38] As nearby fellow-subjects of the emperor, perhaps the Hungarians' use of such signals might make their rudeness more intelligible or at least more familiar. Kepler speculates that such songs arose because their "original author absorbed uncouth melody of this kind from an instrument which was rather unsuitably shaped, and from long familiarity with the construction of the instrument transmitted such melody to his descendants and to his whole nation." The problem is not a barbaric soul but the instrument's disproportionate body, whose physical shape gives rise to its sound.

Kepler's transcription may attempt to capture the ululation of Muslim cantillation, as of a muezzin's call to prayer. Here he confronted the complex melody and pitch slides that are an essential part of Middle Eastern music. Kepler took some pains to be faithful to what he heard, though his notation and musical preconceptions were of little help. For comparison, Kepler cites a famous Gregorian chant, the Easter sequence *Victimae paschali laudes* (figure 5.1, middle; ♪ sound example 5.2). Perhaps not coincidentally, it too begins and ends on G, its highest note the g an octave higher; Christians and Muslims both acknowledge the overarching octave G as they worship the same God. In his commentary on the Gregorian chant, Kepler notes that its underlying melodic structure outlines a triad as "the skeleton of the octave." Kepler's application of the term "skeleton" (used earlier by Pietro Aron, Glarean, and Zarlino) shows his effort to understand the inner construction of melody, not just its constituent intervals or temperament. He goes so far as to write down this skeleton explicitly (figure 5.1, bottom; ♪ sound example 5.3), emphasizing its triadic shape, while leaving the Gregorian melody far behind. Seeing it rewritten thus, the reader is immediately reminded of the Turkish chant, written on the same page in the same clef (figure 5.1, top), as if to suggest that, in skeletal form, the Turkish and Gregorian chants have some relation. Still, Kepler's text mainly points to what he considers their differences: where the Turkish chant jumbles dissonance and consonance, *Victimae paschali* carefully observes their skeletal relations.

Yet Kepler never disclaims the odd resemblance between them, at least at the skeletal level. This implicit relation remains open because Kepler continues to discuss the melodic structures of both the Turkish and the Christian chant simultaneously. Ancient musical theorists gave the terms that Kepler takes up: *agogē* (literally "approach," passage from one consonance directly to another), *tomē* ("emphasis," dwelling on a consonance), *petteia* ("gaming," approach via playful "tiny motions"), and *plokē* ("twisting" that "wanders in its passage around the *agogē*, as a dog does around a passerby").[39] In the absence of any examples of ancient Greek music, Kepler interpreted these terms in light of the music he knew. He applies the same vocabulary to the Gregorian chant as he does to the Turkish.[40] Throughout, he reinterprets the ancient terminology to fit the musical realities of his examples.

Kepler emphasizes the polyphonic character of contemporary music as the model for the polyphony of planetary music, in contrast to the ancients, whose "music of the spheres" (*musica mundana*) and "instrumental music" (*musica instrumentalis*) he considers to have been restricted to a single melodic line.[41] Here, Kepler invokes no mathematical argument, only his profound feeling for polyphonic music, specified in the musical examples he instances, especially Lasso's motet *In me transierunt* (1562).[42] This particular motet was already famous as an example of the Phrygian mode, whose prominent semitone (E-F) makes it "sound plaintive, broken, and in a sense lamentable," as Kepler puts it.[43] Given the scope of his reading in contemporary German theorists, Kepler may well have known Joachim Burmeister's analysis of the musical rhetoric of this motet (1606; ♪ sound example 5.4).[44] Probably citing its opening by heart, Kepler notes the "rather rare" rising minor sixth that then descends by steps (figure 5.2, ♪ sound examples 5.5a,b).[45] He comments on the melodic shape of this opening passage using terms he had applied to the Gregorian and Turkish chants: "a single ascending leap over a minor sixth [E–C], with a downward *agogē* [approach] following, expresses the magnitude of grief, and is suitable for wailing," as the minor sixth sinks a semitone, down to a fifth.[46]

Kepler does not go further into the details of the motet, recognizing ruefully that he is not up to the task. He calls inquiry into the relation between sounds and affects "various

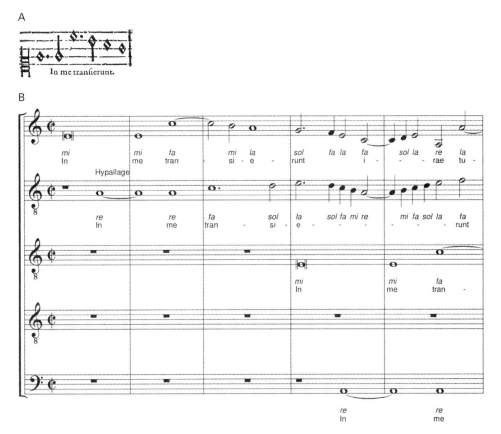

Figure 5.2
Kepler's citation of the beginning of Lasso's motet, *In me transierunt*, and a modern transcription (♪ sound examples 5.5a,b).

and manifold, and very nearly infinite. Since it is too much for my muscles, it would be more correctly passed on completely to the practical men, that is, to practicing musicians, seeing that without teaching, guided solely by nature, they emerge time and again as the authors of wonderful tunes."[47] As Kepler acknowledges the limits of his ability, he also confirms that he considers most important the testimony of "practicing musicians."

Though his own harvest of insights is limited to this one small observation, it will turn out to be pregnant, leading to the climax of Kepler's work, his description of the songs of the planets. At this point, he pauses to intone a solemn prelude:

Now there is need, Urania, of a grander sound, while I ascend by the harmonic stair of the celestial motions to higher things, where the true archetype of the fabric of the world is laid up and preserved. Follow me, modern musicians, and attribute it to your arts, unknown to antiquity: in these last

Figure 5.3
Kepler's transcription of the songs of the planets and the moon (♪ sound example 5.6).

centuries, Nature, always prodigal of herself, has at last brought forth, after an incubation of twice a thousand years, you, the first true offprints of the universal whole. By your harmonizing of various voices, and through your ears, she has whispered of herself, as she is in her innermost bosom, to the human mind, most beloved daughter of God the Creator.[48]

In fact, the planets are "singing" a polyphonic motet à la Lasso; Keper directs us to "modern musicians" in order to hear Nature's secret whispering.

In this cosmic motet, Kepler identifies the particular vocal part of each planet: soprano (Mercury), alto (Earth and Venus), tenor (Mars), and bass (Saturn and Jupiter).[49] He also notes that the motions of each planet suit its particular part: Mercury as "the treble is most free," Earth and Venus with "very narrow distances between their motions … as the alto which is nearly the highest is in a narrow space," Mars as tenor "is free yet proceeds moderately," while Saturn and Jupiter "as the bass make harmonic leaps" (figure 5.3; ♪ sound example 5.6). The interweaving of their six individual "songs" leads to a complex work of practical polyphony, in which Kepler anticipates "certain syncopations and cadences" and all sorts of passing dissonances as planets pass between rare moments of cosmic consonance, particularly when they reach perihelion or aphelion, their closest or furthest points from the sun. We shall return to the problem of reaching such cosmic cadences, moments of complete resolution and consonance.

If "the planets in combination match modern figured music," we return to the modern masters with renewed attention.[50] Kepler does not simply identify this celestial music with any existing composition; "in fact, no sounds exist in the heaven, and the motion [of the planets] is not so turbulent that a whistling is produced by friction with the heavenly light."[51] His cosmic harmony reflects the relative minimum and maximum angular velocities of the planets, as measured from the sun.

Curiously, this harmony involves certain elements that emerged when considering the Turkish chant, whose complex glissandi are not really expressible in discrete notation. Indeed, in Western music the glissando as such was not explicitly used until the cat imitations in Carlo Farina's *Capriccio stravagante* (1627; ♪ sound example 5.7). Yet glissando is a central feature of the planetary music itself. Since the planets move continuously in their orbits, their distances to the sun vary smoothly from perihelion to aphelion. As Kepler puts it, "They advance from one extreme to the opposite one not by leaps and intervals, but with a continually changing note, pervading all between (potentially infinite) in reality. I could not express that in any other way but by a continuous series of intermediate notes."[52] Accordingly, his cosmic music is really a complex interweaving of glissandi, each confined within certain limits, which D. P. Walker compares to the wailing of air-raid sirens.[53] Ironically, the same sliding Kepler found so strange and difficult to notate in the Turkish chant turned out to be an all-pervasive feature of the heavenly music. Here, the Turks and Hungarians, with their "grating," "uncouth" singing, were in touch with a dimension of musical practice that Kepler discovers in his cosmic music.

The very soundlessness of the spheres directs him all the more insistently to the modern polyphonic masters, as if their harmonies will guide him in this silent realm. In a playful marginal note, Kepler clarifies his meaning: "Shall I be committing a crime if I demand some ingenious motet from individual composers of this age for this declaration: The royal psalter and the other sacred books will be able to supply a suitable text for it. Yet take note that no more than six parts are in harmony in the heaven. ... If anyone expresses more closely the heavenly music described in this work, to him Clio pledges a wreath, Urania pledges Venus as his bride."[54] By challenging composers "more closely" to incorporate the harmonies that he has discovered in planetary data, Kepler seems to suggest that some motet already expresses the heavenly sounds "closely." Given his several preceding mentions of this work, *In me transierunt* is the obvious candidate whose text is the "royal psalter," though falling short of the challenge by having five voices, not the requisite six.[55] Earlier, Kepler had drawn attention to the prominent semitone c–b at the beginning of *In me transierunt*, which characterizes its "wailing" Phrygian modality and threads through the whole motet.

Further, this motet has a special significance in the light of Kepler's planetary melodies, in which "the Earth sings *MI FA MI*, so that even from the syllable you may guess that in this home of ours MIsery and FAmine [*MIseria et FAmes*] hold sway."[56] Kepler here uses an older system of note names than the present *do re mi* syllables. In that older system, the song of the Earth was spelled *mi fa mi*, exactly the same syllables as would have been used to spell the opening of *In me transierunt*.[57] Thus, this motet may well have struck Kepler as a powerful treatment of the song of the Earth, embedding the earthly semitone in a rich constellation of sonorities that suggest the full universal harmony.[58]

Now the Earth has a voice, no longer consigned to voiceless immobility at the center of the Aristotelian cosmos. The Earth moves and sings, and its song is not neutral and divinely impassive, like the ancient celestial monophony, but prays with the royal psalmist David, expressing desolation and seeking divine mercy. How appropriate, then, and how moving must Kepler have found Lasso's text for this motet: "Thy wrath has swept over me; thy terrors destroy me. My heart throbs; my strength fails me; my sorrow is ever before me. Forsake me not, O Lord; O my God, be not far from me." As a devout Lutheran, he viewed the semitone of human suffering as an essential part of the quest for divine grace, so that the song of the Earth resounds within the larger scheme of suffering and redemption. The semitones in Lasso's motet and in Kepler's song are signs of terrestrial dissonance that can be reconciled in celestial harmony.

This music is alive in every sense, not just lifeless intervals and ratios; Kepler takes Plato's concept of a world-soul animating the cosmos further by describing its activity in the most vivid physical terms. Emphasizing the primacy of experience and felt response, Kepler connects music with sex, both mirroring the soul's yearning to reunite with the primal archetypes that shine through visible, palpable reality. Though novel in its graphic sexuality, his idiosyncratic (and not much noticed) ideas draw on ancient connections between *erōs* and *musikē*.[59] Kepler argues that cadences, essential to musical syntax, are fundamentally sexual in character because of the underlying sexuality of numbers themselves. In a 1608 letter, Kepler identifies 2 and 10 as male, 3 and 24 as female, noting that "I do not think I can more clearly and explicitly explain this than by saying that you are to see the images here of phalluses, there vulvas."[60] A slightly less explicit figure appears in *Harmonice mundi* (figure 5.4), where he draws attention to the geometric figures he considers the source of musical ratios: "What is surprising then if the progeny of the

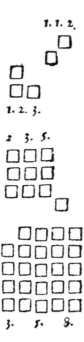

Figure 5.4
Kepler's illustration of the sexual relations between male numbers (2 and 10) and female (3 and 24).

pentagon, the hard third of 4:5 and the soft 5:6, moves minds, which are the images of God, to emotions which are comparable with the business of generation?"[61]

Accordingly, Kepler specifies that "the major third will turn out manly, the minor feminine." Between these intervals, the semitone difference is crucial, "for a semitone following after always invites the voice to climb over it, on account of its small size; for it is like a crest on a slope which gets more gentle."[62] Here Kepler specifies the position of their intercourse and the successive stages of sexual excitement expressed in melodic motion: "And every time a semitone occurs towards the upper part, it is taken as a sort of boundary to the melody, towards which it tends, and then as if the crest has been passed, and when the effort is complete it often begins to turn back to the lower part. Certainly if we sing *RE MI*, the hearing is not satisfied, but expects that *FA* should also be added."

The waves of melody parallel an increasingly urgent desire for satisfaction, which Kepler describes using Greek words to veil his meaning, perhaps to avoid the censure of the prurient, connecting musical technicalities with their explicit erotic correlates. The "hard" major third G–B strives upward, "having force which is *gonimos* [productive], and *akhmē aokhetos* [irrepressible vigor], seeking its own end," the perfect fourth G–C, "of which the semitone [B–C] is like an *ekphusis* [bursting out] for it, sought with its whole effort." In contrast, the "soft" minor third D–F "falls back" over the semitone E–F it has "climbed over …, as if content with itself, and made by nature to be overcome and to be passive, always like a hen prostrates itself on the ground, ready for the cock to tread it."[63]

No one before (and perhaps since) has described the structure of musical modes in such erotic detail. Kepler's climactic sentence describes *akhmē aokhetos*, the "highest culmination" of orgasm and *ekphusis*, bursting out as ejaculation or begetting.[64] The Greek words emphasize the union of the generative and the sexual, but are not exclusively masculine in character, for *ekphusis* can also mean bearing and generation.

The copulation of numbers is always fruitful: "For just as a father begets a son, and his son another, each like himself, so also in that division, when the larger part is added to the whole, the proportion is continued: the combined sum takes the place of the whole, and what was previously the whole takes the place of the larger part." Kepler here describes the formation of the Fibonacci series, 1, 1, 2, 3, 5, 8, 13, 21, … , remarking on its close relation to the pentagon, which is governed by the "golden proportion" that is the limit of the ratio of successive terms in the Fibonacci series. "God the Creator has shaped the laws of generation in accordance with [this series]," such as "the logic of the seeding of plants" that yields successive generations of 2, 3, 5, 8, … seeds. Kepler also describes the "weddings" by which male and female geometric figures marry and produce progeny, remarking that "the study of the sky and music … must originate from the same fatherland of geometry."[65] As Walker puts it, "Polyphonic music, with its thirds and sixths, excites and moves us deeply as does sexual intercourse because God has modeled both on the same geometric archetype."[66]

Yet though geometry is archetypal, it depends on a sexual response that is mutual between partners. Thus, describing the interaction between the sun and the Earth to make the weather, Kepler, quoting Virgil's *Georgics*, "compares the bosom of the Earth to *the thighs of a wife*, and indeed *a joyful wife*, that is, one who perceives what is happening to her with pleasure and helps her husband with suitable motions. All these things are signs of life, and suppose a soul in the body which experiences them. For it would not be easy for the Sun, destitute of suitable troops, to invade this citadel of the bowels of the earth, without the co-operation of some kind of soul, seated within, to collude with the enemy and open the gates to him."[67] This mixed image of military invasion and sexual conquest recalls Kepler's extended erotic "battle" with the planet Mars in his *Astronomia nova* and opens new possibilities in understanding the cosmic harmonies.[68] Altogether, Kepler considers sexuality an essential aspect of soul, perhaps relying on the biblical notion that knowledge is fundamentally carnal, as when "Adam knew Eve, his wife." Such knowledge encompasses both male and female sexual experience: Kepler includes both "hard" (major) and "soft" (minor) harmonies of the planets, indicating that both "masculine" and "feminine" must be given equal scope as he describes the different musico-erotic climaxes of each sex.

Kepler calculated that the universal harmonies "of the hard kind" and "of the soft kind" for six planets are notably dissonant (as one can hear in ♪ sound example 5.6) because of their prominent fourths.[69] Kepler ascribed these dissonances to deep-seated marital difficulties between Earth and Venus, whom he considers man and wife but whose music frequently conflicts because the Earth sings within a semitone (16:15), while Venus sings scarcely within a diesis (25:24). Thus, "the Earth, on the contrary, and Venus much more, on account of the narrowness of their own intervals, restrict their harmonies not only with the other planets, but most of all their mutual harmonies with each other, to a remarkably small number."[70]

In erotic terms, male and female planets battle for supremacy, the Earth "pressing on with tasks which are worthy of a man, pushing aside and banishing Venus to her perihelion as if to her distaff," or Venus beguiling the Earth "to make love, laying aside for a little while his shield and arms, and those tasks which are proper for a man; for then the harmony is soft."[71] In either case, the "harmony" of the heavens is shot through with erotic dissonance, for Kepler notes that if "this antagonistic lady, Venus," were silent, the other planets would sound more consonant chords, though this neglects further harmonic conflicts with Mars. As Turkish and Hungarian songs were appropriate to war, with all its grating harshness, the planetary music, in the throes of cosmic sex and battle, groans unspeakably.

Nevertheless, Kepler does not consider this a failure of his reasoning or an indictment of the cosmos, because "the motions of the heavens are nothing but a kind of perennial harmony (in thought not in sound) through dissonant tunings, like certain syncopations or cadences (by which men imitate those natural dissonances), and tending toward definite

Figure 5.5
The deceptive cadence in measures 6–7 of *In me transierunt* (♪ sound example 5.8). A final (perfect) cadence on the downbeat of measure 7 would have involved a G in the fifth (lowest) voice and a B in the second voice to give the "hard" major third G–B; the fifth (D) would then be implied. Instead, this deceptive cadence substitutes the "soft" minor third E–G; the fifth (B) is then given by the second voice as it resolves on the next beat.

and prescribed resolutions, individual to the six terms (as with vocal parts) and marking and distinguishing by those notes the immensity of time."[72] Here he refers to the practice of "evading the cadence" (*fuggir la cadenza*) or deceptive cadence, which Zarlino describes as "useful when a composer in the midst of a beautiful passage feels the need for a cadence but cannot write one because the period of the text does not coincide, and it would not be honest to insert one."[73] For example, Lasso's *In me transierunt* has a beautiful deceptive cadence in measures 6–7 (figure 5.5; ♪ sound example 5.8). As Claude Palisca remarks, "Lassus and other masters of the new music depended greatly on the evaded cadence, which permitted them to break up their texts into short phrases for descriptive and affective emphasis, while maintaining harmonic continuity."[74] Though Kepler did not use this precise term, he clearly understands that the cosmic harmony can immensely delay its final cadence through specific musical artifice.

Here again Kepler looks to compositional practice, in which "Man, aping his Creator, has at last found a method of singing in harmony which was unknown to the ancients, so that he might play, that is to say, the perpetuity of the whole of cosmic time in some brief fraction of an hour, by the artificial concert of several voices, and taste up to a point the satisfaction of God his Maker in His works by a most delightful sense of pleasure felt in this imitator of God, Music."[75] Kepler claims these cosmic dissonances and deceptive cadences are really pleasure made excruciating through delayed gratification. Compared to God, we experience the cosmic harmonies immensely dilated and slowed almost beyond intelligibility, but a work like Lasso's motet allows us to taste the ecstasy of deceptive cadence that is the divine pleasure.

Of course, this opens the question of whether and when a full resolution might occur. Kepler considers several possibilities. Harmonies between three planets happen rather often, but "harmonies of four planets now begin to be scattered over centuries, and those of five planets over myriads of years. However, an agreement together of all six is hedged about by very long gaps of ages; and I do not know whether it is altogether impossible for it to occur twice, by a precise rotation, and it rather demonstrates that there was some beginning of time, from which every age of the world has descended."[76] He seems to concede that "if there could occur one single six-fold harmony, or one outstanding one among several, that undoubtedly could be taken as characterizing the Creation." Yet his initial *if* marks this as hypothetical, allowing some doubt whether that initial concord really took place. If so, the uniqueness of the instant of creation is somewhat shadowed, opening the unorthodox possibility that there was no such moment.

Without fully resolving this problem, Kepler seems rather to follow the imitative texture used in *In me transierunt*, common to Lasso and his contemporaries. As the voices enter one by one in Lasso's motet, without an initial concord of all, Kepler considers the planets successively by pairs and then in larger groups (as you can do for yourself in ♪ sound example 5.6). Though there is nothing in astronomy that compels him to do so, "for some unknown reason this wonderful agreement with human melody forces me so that I am compelled" to identify planets with soprano, alto, tenor, and bass parts.[77] Throughout, Kepler relies on musical practice to guide his steps. He considers skeletons of the planetary melodies, recalling the skeleton he constructed for *Victimae paschali*, and considers how those skeletons could align to form harmonies of all six planets together. After a long series of propositions, Kepler concludes that musical and geometric constraints dictate the spacing of the planets as we find them.

In all this, the issue of the final cadence remains open. Just after stating what we now call his third law, Kepler had noted that "if we suppose an infinity of time, all the states of the orbit of one planet can coincide at the same moment of time with all the states of the orbit of another planet."[78] This is close to what is presently called the ergodic hypothesis, that eventually the planets will occupy all possible positions vis à vis each other and the fixed stars. Yet this still does not imply that the initial chord will be repeated even after an infinite time has elapsed. Already in the final lines of his first book, *Mysterium cosmographicum* (1596), Kepler had concluded, as had Oresme long before him, that "the motions [of the planets] are in irrational proportions to each other, and thus they will never return to the same starting point, even if they were to last for infinite ages."[79] Kepler reaffirms this conclusion in his notes added to the second edition of this work (1621), written after the *Harmonice mundi* and referring to it directly: "Therefore no exact return of the motions to their starting point is to be found, which can be taken as an end to the motions in accordance with form and reason."[80] Kepler's discovery of the third law reinforced this, for the relation between planetary periods and mean distances is *irrational*, proportional as cubes are to squares, and hence not expressible as any ratio of integers.

If so, there will be no final cadence to the cosmic music. Kepler's 1596 formulation excludes the repetition of any original sonority, while his 1621 addendum goes further to exclude "an end to the motions in accordance with form and reason." Did Kepler not already realize this as he wrote the *Harmonice mundi* in 1619, only reaching the more radical conclusion in 1621? This seems quite unlikely, given that he himself had established the basic result in 1596 and discovered the third law in 1618. If, then, he realized that there was no final cadence, he decided to veil this in the *Harmonice mundi*, for whatever reason. Such a suggestion of the endlessness of the world could have appeared to be dangerously heretical because it contradicts the orthodox dogma of the finitude of the cosmos, whose duration is limited by the divine creation and Last Judgment. It is not clear how this might have moved Kepler, who had already been excommunicated by his fellow Württemberg Protestants and driven out of Graz by Catholic edict.[81]

In the end, Kepler hesitated before matters lying beyond human ken. He brought his own book to its final cadence still aware that he had fallen short of comprehending the divine music. In general, "the human voice in figured melody is almost perpetually out of tune" and hence unequal to grasping the archetypal harmonies.[82] The example of figured music again comes to his aid as he contemplates what God nevertheless found "very good" in the harmonies of his creation. Kepler notes that the abstract proportions "must have given way to the harmonies" so that "the geometrical proportions in the figures strive for harmonies," not the other way around, because "life completes the bodies of animate beings," taking them beyond lifeless, static ratios to something that moves and breathes.[83] In his late *Epitome of Copernican Astronomy* (1618–1621), Kepler likewise emphasized that "the celestial movements are not the work of mind but of nature; that is, of the natural power of the bodies," which sway their souls away from uniformity and circularity to the elliptical orbits he discovered.[84] These deviations are the very signs of life by which the planetary image reflects its cosmic creator.

As in Kepler's sexual imagery, this complex motion is not a flaw but the central beauty of the design, imaging divine potency in cosmic intercourse, in cadence postponed, and in the endless ebb and flow of human desire. For Kepler, the cosmic music is an ongoing cadence that never ceases, "finite and yet similar to the infinite,"[85] signing the finite cosmos with the hidden signature of the infinite.

6 Descartes's Musical Apprenticeship

René Descartes pioneered the "new philosophy" through his achievements in mathematics, natural science, and metaphysics, yet his work in music has remained relatively unknown. He himself was diffident about his musical knowledge and accomplishments, though he first found his voice addressing musical questions. His work on music has many connections with his new physics and his view of the cosmos as a fluid continuum, whose vortices and motions explain light and celestial mechanics.

At age twenty-one, Descartes wrote his earliest essay, *Compendium musicae* (*Compendium of Music*, 1618), when he had just begun his career as a gentleman-soldier, supporting the Netherlands in its rebellion against Spanish overlordship. Stationed in Brabant, where peace then prevailed, Descartes had some leisure to think, though he wrote hastily, "in the midst of turmoil and uneducated soldiers." He formed an important friendship with Isaac Beeckman, to whom he offered the *Compendium* as a New Year's gift, seven weeks after they first met. Descartes asked that he keep his essay to himself, "forever hidden in the privacy of your desk or your library; it should not be submitted to the judgment of others," because it was written "for your sake only," in the turbulent circumstances of an army camp "by a man without occupation or office, busy with entirely different thoughts and activities."[1]

Eight years older than Descartes, Beeckman had studied theology, but doctrinal disputes prevented him from teaching in that field. He then worked making candles and water-conduits, eventually becoming a school administrator. Though he graduated in medicine the year he met Descartes, he was self-taught in science and mathematics; at times, he rediscovered results that had long been known, but he also took a fresh view of those fields, not having been steeped in Aristotelian natural philosophy.[2] Though Beeckman worked in a tremendous variety of fields, including engineering, mechanics, astronomy, logic, medicine, and music, he never published anything. Nonetheless, already in 1613 he had stated a principle of inertia and had accounted for gravity as a "force pulling by little jerks"; in their boldness, these insights parallel and even surpass Galileo's work, though Beeckman's notebooks remained private communications until their rediscovery and publication in the twentieth century.[3]

According to Beeckman, Descartes told him, "Really you are the only one who has reawakened me from idleness."[4] Beeckman noted that Descartes "says he had never found someone, except for me, who is accustomed to study in the way I prefer and accurately joins mathematics and physics. And for my part, I have never spoken with anyone apart from him who studies in this way."[5] Aristotle had argued that mathematics applies to the regular motions of the heavenly bodies, not to *physis,* the earthly realm of growth and change, because mathematical concepts were inherently unchanging. Around 1600, Francis Bacon coined the word "mixed mathematics" to describe mathematics applied to physical problems; a similar term was used by Marin Mersenne, with whom Descartes also formed an important relationship through discussion and correspondence.[6] Though Beeckman and Descartes did not use the term "mixed mathematics," they thought that understanding nature required bringing mathematics together with natural philosophy in new ways, first of all in the realm of music.

Beeckman was not really familiar with practical music, though interested in theoretical questions, and he relied on Descartes for mathematical results beyond the elementary level.[7] Descartes's exposition in the *Compendium* is a fascinating mixture of old and new; though Cohen has noted that it "adheres so closely to the Renaissance style of music theorizing," it also casts music in a new light by reconsidering its relation to mathematics and to physical sound.[8]

Descartes's terse opening words signal a shift in thinking about music, "whose object is sound," as he puts it. As Suzannah Clark and Alexander Rehding observe, "with only slight exaggeration these four words sum up the impact of the scientific revolution on music—the change from music as a divine force to music as a material phenomenon."[9] While agreeing with their general point, I think its causal force also works in reverse. Because Descartes's musical essay *precedes* the rest of his work, it is more coherent to read it as a musical argument that contributed to the formation of the new natural philosophy.

Descartes takes music as a perfect exemplar of his nascent project to "accurately join mathematics and physics," mediating between arithmetic and geometry and their physical manifestations. As sound (rather than divine afflatus), music aims "to please and to arouse various emotions," which are empirical, sensual states, rather than idealized types. Their intensity gives an observable *magnitude* of excitation, analyzable into "differences of duration or time, and its differences in tension from high to low." "The quality of tone" of the sounding body he assigns to "the domain of the physicists [*Physici*]," one of the first connections between *physici* and music. Until then, music's place was next to astronomy, the changeless heavens, rather than *physis,* the sublunary realm of change.

Pleasure is essentially a geometrical magnitude, "a proportional relation [*proportio*] of some kind between the object and the sense itself." Within the limits set by our sensory capacities, Descartes analyzes this pleasure-magnitude into sensations that do "not fall on the sense in too complicated or confused a fashion." Looking at an astrolabe (figure 6.1),

Descartes's Musical Apprenticeship

Figure 6.1
An astrolabe from the workshop of Jean Fusoris (Paris, ca. 1400) in the Collection of Historical Scientific Instruments, Harvard University. The *rete* is the inset metal structure with curved lines, showing the *mater* and *plate* underneath; the straight *rule* rotates to measure distances on the instrument.

he finds the simpler lines of the *rete* more "satisfying to the sense" than the complex design of the *mater* because the *rete* is *more distinctly perceivable*. The man who would go on to advocate "clear and distinct ideas" uses an astronomical instrument to illustrate the aesthetics of *melodies*.[10]

Sense can perceive an object "more easily … when the difference of the parts is smaller" and "when there is greater proportion between them." These proportions "must be arithmetic, not geometric, the reason being that in the former there is less to perceive, as all the differences are the same throughout," such as in figure 6.2. By comparison with the simple progression of an arithmetic proportion (in his example, 2, 3, 4), a geometric proportion involves a middle term whose relation to its neighbors is harder to discern, such as dividing an octave 2:4 in half at its geometric mean $\sqrt{8}$. Though Descartes accepts that

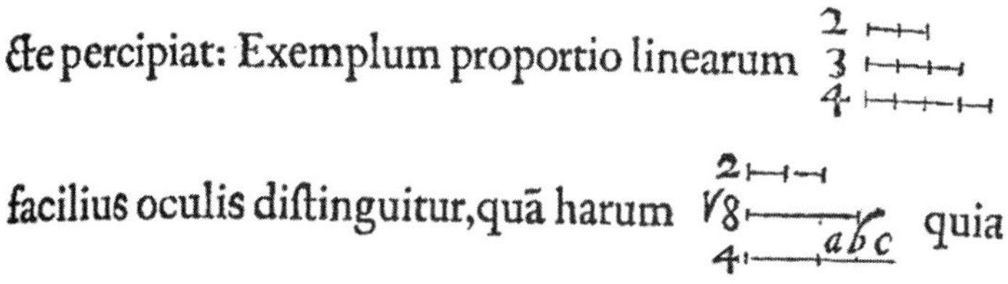

&ct;e percipiat: Exemplum proportio linearum 2,3,4

facilius oculis distinguitur, quā harum $2:\sqrt{8}:4$ quia

Figure 6.2
Descartes's comparison of arithmetic and geometric proportions from his *Compendium*: "The example of a proportion of lines [2:3:4] is more easily distinguished by the eyes than this one [2:√8:4]."

a quantity like $\sqrt{8}$ is comparable to integers, still "the mind is in this case constantly perplexed" by its irrationality. Music should rely on arithmetic rather than geometric ratios because "the most agreeable to the soul is neither that which is perceived most easily nor that which is perceived with the greatest difficulty; it is that which does not quite gratify the natural desire by which the senses are carried to the objects, yet is not so complicated that it tires the senses." Because "variety is in all things most pleasing," music relies on variation.[11]

Where Plato assumed the priority of arithmetic, Descartes allows different mathematical objects to generate various pleasures. Though he began his innovative world-project by recasting the ancient theory of music, as Cohen observed, his little treatise is perhaps the most conservative such work of its century.[12] His *Compendium* combines the ancient topics of mathematical music with the nascent mathematical physics: the numerical patterns of music theory lead directly to the perceptible, observable world. He begins with rhythm, showing how relative simplicity of ratio governs our experience, how easily we hear two even notes against one, whereas five notes against one are "almost impossible to sing." Exactly similar processes govern our awareness of the ordering of a melody as we count up its phrases, until "our imagination proceeds to the end, when the whole melody is finally understood as the sum of many equal parts."[13] Each beat physically makes us "dance and sway," accompanying "each beat of the music by a corresponding motion of our body." The downbeat of a measure, when "the sound is emitted more strongly and clearly … has a greater impact on our spirits" by sheer visceral force, so that "even animals can dance to rhythm if they are taught and trained, for it takes only a physical stimulus to achieve this reaction."[14]

Music arouses various affects precisely through various meters: "A slower pace arouses in us quieter feelings such as languor, sadness, fear, pride, etc. A faster pace arouses faster emotions such as joy, etc."[15] Triple meters (like 3/4) "occupy the senses more" than duple ones (like 2/4), "since there are more things to be noticed in them," namely three rhythmic units, rather than two. Even a solitary military drum can affect us by its beat, demonstrating the felt reality of pure rhythm. By treating "number or time in sound" before pitch (which previous theorists treated first), Descartes shows that we feel number with the full force of a martial drum.

Descartes observed that when a lute string is plucked, "the force of its sound will set in vibration all the strings which are higher by any type of fifth or major third, but nothing will happen to those strings which are at the distance of a fourth or any other consonance."[16] Here he made an important step toward the discovery of what later were called *overtones*, showing the *physical* connection between the octave, fifth, and third, rather than what the ancient theorists considered a purely numerical relation. Descartes correlated the mathematical division of the octave with physical experience, giving the first example of his new mathematical physics through the intermediacy of music.

Descartes continued to think about music long after his little *Compendium*, as shown in his correspondence.[17] He often included music among other topics in physics and mathematics. Ten years later, in his first letter to Mersenne from Holland in September 1629, Descartes begins by asking how one consonance can pass into another, such as "might offer all the diversity of music," then turns abruptly to a new "part of mathematics I call the science of miracles" that could produce astonishing illusions.[18] His next letter describes working through "all the *Meteores* [celestial phenomena]" such as parhelia (mock suns, figure 6.3), pointing to what would become his work of that name, published in 1637 as an appendix to his *Discourse on Method*, along with *La Geometrie* (on his algebraic approach to geometry) and *Dioptrique* (optical phenomena and vision). Descartes tells Mersenne that he is preparing "a little treatise that will contain the explanation of the colors of the rainbow," that archetypal miracle, but begs him "not to speak of this to anyone in the world, for I have decided to show this in public as a sample of my philosophy and to hide myself behind the picture to hear what they will say of it."[19] To his friend, Descartes discloses the masked persona he will don so that he can present his work, gauge its response, yet remain safely hidden.

This letter goes on to discuss questions of rarefaction (which might be able to explain the ancient concept of ether) and thence to a certain "book of cameos [here meaning monochrome painting on jewelry or enamel] and talismans," which he judges "only contains chimaeras," showing his disdain for the ordinary run of *trompe-l'oeil* or occult "wonders."[20] He then returns to the ways in which a unison could pass to a major versus a minor third; Descartes shows a sensitive awareness of the relation of this particular issue to the whole context of the composition in which it might occur, including the relative motions of treble and bass. He seems familiar with contemporary musical practice, not just theory in the abstract.

After this paragraph on music, Descartes turns to another question Mersenne posed: How does a pendulum move in "empty space," where the air gives no hindrance? Galileo's response was not published until 1632 and hence was unknown to these men. Descartes makes a calculation dependent on the length of the string and arrives at fractions among

Figure 6.3
Parhelia (mock suns or sundogs) in Fargo, North Dakota, 2009.

which are several well known to him from musical theory, such as $\frac{4}{3}, \frac{256}{81}$. Perhaps there is no connection in his mind between these identical results for different musical and mechanical problems, but they remind us (as may also have struck him) that both sets of questions play out on the same mathematical terrain, between pure number and sensuous reality.[21] In this letter, as in his work as a whole, musical theory had set the mathematical-empirical stage on which the new natural philosophy would perform. In his final lines, Descartes mentions a new problem straddling mechanics and music, which Mersenne had posed: what are the motions of a lute string when plucked?

Descartes enlarges the scope of such investigations in his next letter (November 13, 1629): "In place of explaining only one phenomenon, I have resolved to explain all the phenomena of nature, that is to say, all of physics."[22] Questions, first about music, then pendulums and parhelia, have mushroomed into a project to understand "all the phenomena of nature" as mathematical physics, a vast synoptic enterprise. Descartes returns to the pendulum in a vacuum and gives a more detailed account in which again the music-theoretic numbers 2, 4, 8, 9, 12, 16 figure prominently. He then returns to the problem of a plucked string, which he clearly considers a direct continuation of the problem of the pendulum swinging in a vacuum. What follows gives a window into his emergent thought process, especially its twists and turns in the face of difficulty and paradox. Descartes treats the string as making "turns and returns" away from its equilibrium position in a vacuum, each point along the string behaving essentially like the pendulum he had just considered. He concludes that the string's vibrations would damp down geometrically: if its first vibration had amplitude 4, the next would have amplitude 2; if it began with magnitude 9, then would follow 6, 4, …—again those musical numbers, here illustrating a quintessentially musical phenomenon. Then he hesitates: "I said *in vacuo*, but in air I believe that [the successive vibrations] will be a little slower at the end than at the beginning because, the movement having less force, it will not overcome the resistance of the air so easily."[23] Air is evidently more resistive than empty space, but suddenly he seems to recall the Aristotelian paradoxes of motion in a void, as he continues: "However, I am not sure of this and perhaps also the air, on the contrary, may help [the string] at the end because the movement is circular."

This moment of confusion, especially in this proudly lucid mind, gives invaluable evidence of his struggle to make physical *and* mathematical sense of a phenomenon that, so far, had purely been treated arithmetically by music theory. In the face of this hornet's nest of problems, Descartes's response is telling: "But you can experience [*experimenter*] it with the ear by examining whether the sound of a string thus plucked is higher or lower at the end than at the beginning, for if it is lower, that means that the air slowed it; if it is higher, then the air made it move faster." His purely mathematical arguments rely on a test by experiment, and a musical one at that. In the course of this letter, the problem of the plucked lute string that had begun as a purely musical phenomenon passes through a middle stage of mathematization (the analogy with a chain of tiny linked pendula), and finally returns to the realm of musical experience: the ear can test the exact influence of the resistive air.

Descartes's letter continued to discuss the vibrating string, but the remainder of his text has not survived. His next letter (December 18, 1629) returns to these matters in even greater detail. Their interchange about optical phenomena now includes Descartes's doubts about Mersenne's claim to have seen a colored "crown" around a candle flame, as if it were a miniature mock sun. Descartes also asks Mersenne, as a cleric, about the possible danger in speculating about natural philosophy in directions contrary to Aristotle, "for it is almost impossible to express another philosophy without it immediately seeming against the faith."[24] He is worried about whether there might be anything "determined by religion regarding the extension of created things, namely whether they are finite or rather infinite, and whether in all those lands one calls imaginary spaces there might be true and created bodies, for as yet I have not wanted to touch this question."[25] This hypothetical question gives a preview of Descartes's nascent project, to propose a new (and distinctly post-Aristotelian) natural philosophy disguised as the description of a purely imaginary world, which developed into his *Traité du Monde*. Galileo's *Dialogues on the Two Chief World Systems* (1629) had just appeared in Italy, though it had not reached France. Yet even before Galileo's ecclesiastical troubles, Descartes was already apprehensive.

Descartes's cautious inquiry alerts us that the mass of related questions we have been tracking were forming, in his mind, a new approach to natural philosophy that would make good on his sweeping claim to understand "all phenomena of nature." Beginning in 1619–1620 (just after completing his *Compendium*), he had begun drafting his *Rules for the Direction of the Mind*, but abandoned that work after 1628; his new project seemed to carry forward the systematic, mathematical thrust of those rules. In his letter, we seem to see his new world beginning to be synthesized as he struggles with atmospheric and light phenomena, musical questions, pendulums, and lute strings.

Immediately after asking Mersenne's theological advice, Descartes returns to musical questions, which now have a dynamic aspect: the use of various intervals depends on how the melodic lines *move*, whether they ascend or descend. For guidance, "I hold to what [musical] practitioners say."[26] Descartes notices that if, "in a musical concert, the voices move always equally or become lower and slower gradually, that puts the listeners to sleep; but if on the contrary one raises the voice suddenly, that wakes them up."[27] These musical issues merge into questions about bodies: "One can say that a deep sound is more sound than a high one because it is made by a more extended body, it can be heard from further away, etc." Here the primal quality of *extension*, so important in Descartes's later philosophy, enters the realm of sound. Returning to whether strings vibrate differently in air than in a vacuum leads him to questions about balls of different weights and materials falling from different heights, fundamental for basic mechanics (and for which Galileo had not yet published his investigations).

Throughout this letter, Descartes moves freely between musical questions, vibrating strings, and falling bodies, showing how closely these topics are related in his mind and how helpful he finds passing between them: the coincidence of the pattern of two vibrating strings can explain consonances between them.[28] Descartes also considers the relation of contemporary music to that of ancient Greece, the touchstone of music's fabled powers. His approach is not cerebral but flies free of rules:

Regarding the music of the ancients, I believe that it had something more powerful than ours not because it was more learned, but because it was less, from which it comes about that those who have a great natural talent for music, not being subject to the rules of diatonic music, do more by the sole force of imagination, which those cannot do who have corrupted that force by knowledge of theory. Further, the ears of hearers not accustomed to a music as ordered as ours are much more easy to surprise.[29]

Breaking off abruptly, he announces "I want to start studying anatomy," indicating his nascent interest in human physiology; his next sentence turns to sunspots.[30]

This amazing letter ends with a larger consideration of the nature of sound that seems to look from the contemporary theory of "pulses" of air set in motion by a vibrating body toward a more general theory of waves. Descartes notes that the vibrations that strike the ear are not those that engender the sound: he compares the process to the horizontally spreading circles caused by a stone dropped in water, "though the stone goes straight down." This was not, to be sure, an original observation; Cohen judges that it remained within the "pulse" theory of sound, which did not truly become a wave theory until a century later when it developed into a mathematical theory of longitudinal waves of compression and rarefaction.[31] But the question of the nature of sound was, for Descartes and Mersenne, alive and connected to the musical and physical issues they were struggling to resolve.

This intense exchange of letters at the end of 1629 was by no means the end of their correspondence on these matters. Thirteen other letters remain from 1630 to 1634, spaced more widely, bearing witness to Descartes's interest in music in connection with other issues in natural philosophy, especially the vibrating string and falling weights in and out of a vacuum. Yet on April 15, 1630, Descartes wrote that "in fact I cannot distinguish between a fifth and an octave," though it is hard to take seriously his assertion that he is really so tone deaf as not to know a fifth from an octave, when he has written so clearly on just that difference.[32] He goes on to make an astute contrast between the recognition of intervals heard out of context with the perception of intervals "when they are placed in a concert of music." His further remarks seem to admit knowing the very intervals he had just claimed he could not distinguish.

In a letter of October 1631, Descartes theorizes that

sound is nothing else than a certain vibration [*tremblement*] of the air, that comes to tickle our ears and that the turns and returns of this vibration are more sudden as the sound is higher [in pitch]; so that, two sounds being an octave from each other, the deeper only vibrates the air one time while the higher vibrates just twice, and likewise with the other consonances. Thus one must suppose that when two sounds strike the air at the same time, they are that much more concordant when their vibrations recommence more often with each other and when they cause less inequality in the whole body of the air.[33]

The standard theory of "pulses" of two pitches that coincide more or less completely when received at the ear here seems to shift toward a concept of *frequency* (as we now call it), which considers the element of *time* with respect to the vibration itself. At this point, Descartes wrote in the margin that "I abuse here the word 'vibration' that I take for each of the blows or little shakes that move the body that vibrates." His marginal note shows his hesitation about how exactly to describe the state of the vibrating *air*, as opposed to the vibrating body that caused the sound or the vibrating ear receiving it. This problem continued to be the central issue on which rested the mathematical theory of waves and vibrations, and hence the whole mechanical theory of continuous media. The vibrating string is the simplest example, reduced to a single dimension, of vibrating membranes and solids. Thus, this problem emergent from musical experience was foundational for the generalization of mechanics from point bodies to continuous media.

In this nexus of questions, the issue of the vacuum takes an important place; we have seen, in the moment-to-moment byplay of Descartes's thoughts, as he set them down to Mersenne, his alternate consideration of and doubt about the nature and possibility of motion in a vacuum. At first, he seems to resist Aristotle's famous arguments against the possibility of a void. Yet Descartes seems increasingly troubled by the status of a void vis à vis the resistance of air or other media. The unfolding dialectic of this letter offers evidence that *Descartes's rejection of the vacuum emerged in the context of this musical-physical problem*, one at least of the issues that moved him.

We can best see where this process took him by examining the text of *Le Monde de Mr. Descartes, ou le Traité de la lumière et des autres principaux objects des Sens* (*The World of Mr. Descartes, or the Treatise on Light and on the Other Principal Objects of the Senses*), as he assembled it during 1629–1633. His subtitle signals that, though *light* is his subject, his design encompasses a sketch of the whole visible universe and all the "objects of the senses." By labeling his treatise merely an imaginary vision, he seeks a safe way to present a new cosmology that could escape controversy or censure. Throughout this work, *sound* is the hidden thread that helps him find and state his new worldview. At the very beginning, Descartes distinguishes between "our sensation of light" and "what is in the objects that produces that sensation." For him, sound is essentially like touch, the sense "thought least misleading and most certain," yet "even touch causes us to conceive many ideas that in no way resemble the objects that produce them." Choosing an example that makes one wonder about his military experiences, Descartes notes that in the heat of battle a soldier might think he had been wounded, though "what he felt was nothing but a buckler or a strap."[34] Perhaps this was as close as he came to being wounded in action.

Descartes uses sound, understood as "a certain vibration of air striking against our ears," as a template to form his understanding of light and the cosmos, which he conceives as an infinitely divisible fluid continuum. His reluctance to accept a void, which emerged in his letters, eventuates in his view (expressed in *Le Monde*) that the continuous world-fluid is not finally atomic and hence does not admit of any void spaces, however small. As critical as he was, Descartes finally found himself on the same side as Aristotle not only because of the metaphysical problems of ascribing any positive properties to pockets of emptiness but because he considered that sound must travel in a continuous medium (as in his image of the spreading circles in a pond) that will not tolerate interruption by voids.[35]

Thus, in his figures Descartes represents the minute "parts" of his fluid cosmos by little balls, which model the interaction of the fluid on the small scale, not really atoms in any physical sense. In figure 6.4, the fluid can move by the transmission of motion from one layer of balls to the next, each relaying the "touch" to the next by a process that is essentially the same as his understanding of sound. He uses exactly the same pictures to explain

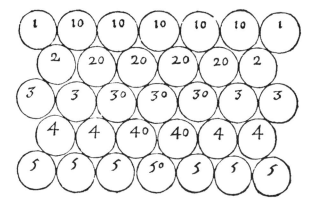

Figure 6.4
Descartes's illustration from *Le Monde*, showing how horizontal pressure to the right on the "parts" of a fluid marked 1, 2, 3, 4, 5 will in turn move those marked 10, 20, 30, 40, 50 to the right.

our sensation of light as an experience of *pressure* registered by the eyes, essentially *sonic* in its dynamics. For instance, he depicts the refraction of light rays by the reciprocal pushing and impeding of the little balls of world-fluid (figure 6.5). In his later developments of this idea, he ascribed the phenomenon of color to the different states of spin of these interacting balls, which he compares to spinning tennis balls.

On the largest scale, Descartes uses his sonic world-fluid to explain the motions of the planets in terms of a series of vortices, which carry around the visible planetary bodies (figure 6.6). His picture had the intuitive merit of explaining how the planets could move in seemingly empty space by showing the motions of the invisible fluid that guide them, those closer to the center of the vortex moving more swiftly about that center than those farther away, as is true of the planets with respect to the sun. In Descartes's picture, the sun must necessarily stand at the center of our vortex, so that his world is necessarily Copernican, probably the crucial fact he wished to conceal by presenting a purely imaginary world.

The impending controversy brought him back to this all-too-human Earth. Though in July 1633 he had written Mersenne that his treatise "is just about finished," in November of that year he learned the news of Galileo's condemnation, which "has so astonished me that I am almost resolved to burn all my papers, or at least not to let anyone see them."[36]

Descartes's new physics required the motion of the Earth and offered an alternative to Aristotelian physics. Even Galileo, convinced Copernican though he was, had no account of an alternative physics that might support the new picture of the universe that shattered the Aristotelian separation between *physis* and the perfection of the heavens (*ouranos*). Though Kepler had offered speculations about *anima motrix*, a "moving spirit" analogous to magnetic power emanating from the sun capable of moving the planets around it, he

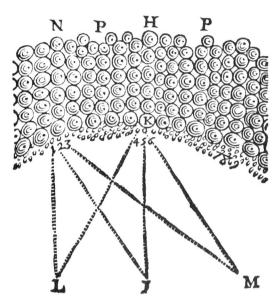

Figure 6.5
Descartes's depiction of the mechanism of light refraction, from *Le Monde*; the whole space should be considered as filled with balls representing parts of the world-fluid; the straight lines from points *L, I, M* illustrate the refraction of rays from those points via the rebound of representative balls at 1, 2, 3 and 4, 5, 6.

was not able to embed that speculation into a coherent picture, as Descartes had. Descartes's decentering of the Earth went a step beyond Copernicus and Kepler because it grants our sun no special cosmic status; our solar system is merely one vortex among many others in an endless, eddying expanse with no center or end. Bold as he was, Kepler was unable to accept such cosmic vastness, to which we shall shortly return.

Though daring in his *Monde*, Descartes was cautious in this-worldly affairs. He decided to suppress his treatise rather "than to have it appear mutilated" through politic omissions, as though it were possible to hide its central Copernican argument. Much later, in 1644, he considered the time ripe to publish his vortex-cosmos in his *Principles of Philosophy*, which had great influence in the succeeding centuries, especially on continental natural philosophy.[37]

After the 1633–1634 crisis, Descartes's correspondence contains very little reference to music. Yet, as Walker pointed out, even in 1640 Descartes was involved in musical discussions, prompted by Mersenne's commission of a vocal setting of a French poem by Joan Albert Ban, who did not know that it had also been set by a distinguished contemporary, Antoine Boësset. Mersenne sent both settings to Descartes, who wrote an extensive response that preferred Boësset; as Walker notes, "he had certainly examined Boësset's air with great attention, and his defense of it shows remarkable insight and subtlety."[38]

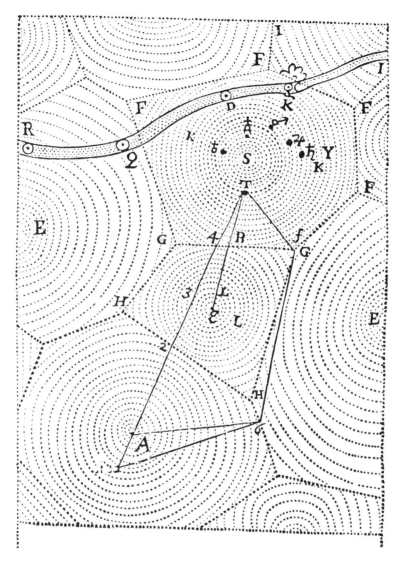

Figure 6.6
Descartes's diagram of the cosmos from *Le Monde*, showing our solar system (within *FGGF*), centering on our sun *S*, with the various planets labeled by their astronomical symbols, including the Earth. The nearby solar system within *GHHG* centers around the star *Ɛ*.

Though Descartes himself may have continued to have such interests, at least as solicited by Mersenne, the music may have drained from his imagined cosmos. If the sun is only one of a vast number of stars, there may be nothing special about the "harmony" of our particular solar system; the relative distances and periods between its planets might be utterly different in other solar systems. Kepler's cosmic harmonies sought *musical* interrelations of orbital radii and periods, on the model of understanding a motet by Lasso through awareness of its salient melodic and modal qualities. The significance of the cosmic "motet"—the harmonic ordering of the motions of the planets—is diluted, even negated, if our solar system is only one among many, seemingly random, possibilities.[39] The story of Descartes's vortex-universe sets the stage for a recurring dilemma: the quest to understand the harmony of the cosmos can lead to a plurality of worlds whose very multiplicity challenges any overarching universal music.

7 Mersenne's Universal Harmony

Advances in mathematics and natural philosophy owe a great deal to conversation, whether in person or via correspondence, contrary to the misapprehension that such work emerges in isolation. Descartes's interest in music could have remained undeveloped, had not his dialogue with Beeckman initially stimulated him to assemble his thoughts on the subject. During the critical years from 1629 to 1634 and thereafter, Mersenne's questioning sustained Descartes's continuing response.

So far, Mersenne himself has remained in the background, as if he were merely a sounding board for Descartes. To some extent, this reflects the disparity in what remains of their correspondence: only five letters from Mersenne, compared to 146 from Descartes.[1] Whether Mersenne's letters have simply been lost over time or Descartes just discarded them, we tend to read their dialogue through the one-sided perspective of Descartes's responses. But any answer must be judged in light of the precise inquiry that prompted it and, as Blaise Pascal observed, Mersenne "had a very special talent for posing beautiful questions, for which there was no one else comparable."[2] To understand the implications of any scientific theory, of any *answer*, we need always to ask: *but what was the question?* A beautiful question tends to go beyond any response it provokes, which does not close or end it but gives it further life.

In the case of Mersenne, this inquiry will lead us to further aspects of their dialogue (and of the significance of music) than were apparent in Descartes's side of the correspondence. Though for Descartes music was one among many interests, for Mersenne music was at the very center of his work, especially before 1637. He was, if anything, even more of a polymath than Descartes, as befitted his special role as the self-appointed (but universally recognized) secretary of the Republic of Letters, that cosmopolitan and self-organized network of scholars so important for the development of learning in their time. Often writing in Latin, the international language shared by the learned, these savants exchanged and transmitted enormous amounts not just of information (in the contemporary, neutral sense of "data") but of thought, observation, and (above all) salient questions.[3] Among them, no one was more prominent or more important than Mersenne in shaping the intense discourse that formed the new natural philosophy.

The last of six children born to a provincial laboring family, Mersenne did not come from the privileged world of Descartes, yet he came to share its intellectual and cultural milieu. Though they overlapped at the same Jesuit school, there remains no record of any contact between them then; Mersenne was eight years older and the Jesuits took care to separate boys of different ages. After further studies in Paris, at age twenty-three Mersenne joined the Franciscan Order of Friars Minor, known as the Minims, then considered the most severe monastic order in Western Christendom. He spent most of his life in that order's monastery in Paris, but through friendship and an immense correspondence, he reached out to a vast array of scholars across Europe. Descartes was only one among a group of correspondents with whom Mersenne maintained a particularly strong contact. In that era before academic journals, Mersenne's letters served to disseminate and exchange views so well that his correspondents were effectively publishing their letters to him. He did not scruple to hide what he learned because, for him, dissemination and discussion of important new findings far outweighed issues of priority. His activities constituted a veritable "Académie Mersenne" he organized and conducted through his correspondence, which led to the formation of the French Académie des Sciences in 1666.[4]

Mersenne's own intellectual journey took him from conventional adherence to geocentric cosmology to gradual acceptance and advocacy of the new Copernican views, even against the opposition of the Roman hierarchy. In his voluminous commentary on Genesis, *Quaestiones in Genesim* (1623), his first published work, he cited the ecclesiastical condemnations of 1605 and 1616 against Copernicus and concluded that he "could not demonstrate that the center of the universe is not our earth … whatever the explanation of Aristarchus of Samos and Copernicus after him."[5] His phrasing suggests that, though he had some sympathy for the Copernican view, he could not demonstrate it to his own satisfaction, at least to the point of holding it publicly in the face of ecclesiastical opposition. But as he assembled the materials for his work on universal harmony, Mersenne gradually moved closer and closer to the Copernican position, moved by musical arguments he discusses in his *Traité de l'harmonie universelle* (1627), a trial run for the *magnum opus* he produced a decade later. He locates the technical astronomical issues within the context of the relation between musical consonances, the heavens, and the planets.

Mersenne begins with the Platonic account and places himself with those who believe in the auditory reality of the heavenly music. We cannot hear this heavenly music, "for we are accustomed to it from the wombs of our mothers. Sometimes the sound is too far from us, too low, too high, or too great to be heard," as with the extremely quiet sounds "which ants and other little animals make."[6] Having marshaled the classical lore, he asks "if Johann Kepler has alighted on more than Robert Fludd concerning celestial harmony." Mersenne phrases the debate between Copernicus's champion, Kepler, and the geostatic traditionalist Fludd, with Brahe providing "the most correct observations that we have."

Mersenne reviews the planetary observations expressed in terms of their consonances or dissonances; where purely Copernican arguments had earlier failed to sway him, Kepler's musical treatment finally wins the day.

After explaining the Copernican system, he praises it by saying that it will

> serve musicians for entertaining their spirits in the contemplation of celestial things while playing on the spinet, lute, viol, organ, or any other wind or stringed instrument, and for admiring the providence of God, Who has preserved such beautiful proportion in the order which He has placed in all parts of the universe, I shall clearly show that man can not imagine anything excellent which is not found therein [in the Copernican system] with a singular perfection.[7]

Mersenne's "explanation" includes a detailed account of Kepler's work, which he contrasts with that of Copernicus and Tycho, who "concur regarding the sizes of the sun and the earth" and many other interplanetary distances. Mersenne prudently refrains from flatly asserting the motion of the Earth but strongly implies his views through his high praise of Kepler's detailed harmonies, which agree with observation and hence have the kind of musical provenance that Ptolemy evidently lacks, in Mersenne's eyes. His "Table of the Harmony of the Planets" marshals the observational evidence bearing on Kepler's claims, which Mersenne examines critically. Mersenne notes a number of places in which the consonances are not perfect, reminding his readers that, according to Kepler, we should judge these harmonies "as if we were seeing them from within the sun." Viewed from that central perspective, "the consonances are perfect when one considers the two points where the planets are nearest and farthest apart and when one always places two planets together," at their conjunctions.

Despite these imperfections, Kepler's planets "approach perfection so closely that the ear would have difficulty discerning what they lack." Accordingly, Mersenne feels that Kepler's planetary intervals "may serve as musical notes, not only for a simple song, but also for the four parts. Men may be said to have imitated the apparent motion of the stars in order to represent planetary motion in their songs."[8] In summary, Mersenne judges that, though Kepler "did not find all that he desired to find and we still do not know precisely enough the distances or motions of the planets, he blazed the trail and said several things never before said or even thought." Mersenne further speculates that the hymn to the sun that Kepler ascribes to the pagan Neoplatonic philosopher Proclus might be read as worshipping the Son of God "under the name of Titan or Sun, perhaps out of fear of being punished by emperors who had the Christians killed and against whom he wrote." If so, the heliocentric view could be identified as Christian and would identify this Neoplatonic sage with the Christian cause, implicitly turning the tables on those who (like the pagan Roman emperors and perhaps their latter-day heirs in Rome) proscribed heliocentrism. Mersenne's conclusion combines a clear declaration of the immovability of the sun with careful avoidance of language that might get him in trouble: "May it please God that all the musicians of the Earth should never wish to sing or compose anything but hymns and

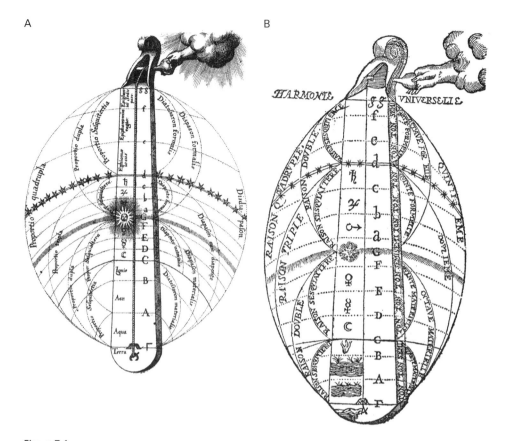

Figure 7.1
(a) Robert Fludd's image of the cosmic monochord from his *Utriusque cosmi ... historia* (Oppenheim, 1617). (b) Mersenne's similar image of "universal harmony" from the appended section "On the Utility of Harmony" from his *Harmonie Universelle*.

motets in order to dedicate them to the great Sun, which is immobile and more truly moves all creatures at will than the sun moves the planets."[9]

By contrast, his account of Fludd's arguments shows that Mersenne's devotion to cosmic harmony is tempered by his requirement of precise observation and principled argument. In some respects, Mersenne's penchant for numerology resembles Fludd's mystical numeration of the cosmos. Even in his later *Harmonie Universelle*, Mersenne included his own version of one of Fludd's famous images, the cosmos as a monochord tuned by the hand of God, "the divine Orpheus" (figure 7.1); even though he criticizes and finally rejects Fludd, Mersenne shares his fundamental premise of cosmic harmony. Mersenne presents a rather detailed account of Fludd's Neoplatonic cosmology, including a summary of his numerological design, which relies on such traditional identifications as between the

Trinity and the number three, descending into the created world as larger and larger numbers come into view. Mersenne had deep theological disagreements with Fludd's pantheism: "He mixes the Divinity with creatures, as if the latter had the divine essence for their form. ... I do not believe that he wanted to corporealize the Divinity, and I attribute it rather to his ignorance than to his malice that he considered creatures as nothing else but God."[10]

In the end, though, Mersenne judges Fludd more in terms of the empirical grounding of his claim, finding "no solidity in all this discourse [of Fludd's]"; he is rather "of the opinion of Kepler, who contends that all the harmonies of Fludd and the Platonists are but analogies and comparisons based only on imagination."[11] Mersenne dismisses Fludd's harmonies as arbitrary and ungrounded; he confronts them with what he considers proper numerology from Plato and also uses the work of Regiomontanus (the greatest astronomer in the century before Copernicus) to attack the details on which Fludd built his harmonies. Mersenne emphasizes what he considers gross physical errors in Fludd's argument, such as a triangle of strings that "can not produce the sounds of the musical scale, if they are of equal thickness and tensions," as Fludd had supposed, which Mersenne corrects in his diagram (figure 7.2).

Fludd's basic physical misunderstanding violates Mersenne's deep commitment to the facts of harmony, not just their language. After remarking in a conciliatory vein that "it is much easier to reproach others than to do better than they," Mersenne still concludes that "it is far better not to know this Harmony at all than to imagine it as something entirely other than what it is," as Fludd had. "False imaginations exercise indescribable tyranny upon our spirits, from which our spirits can disengage themselves only with difficulty."[12] Thus, Mersenne feels compelled to exorcise Fludd's false harmonies at the same time as he embraces Kepler's; both being heretics (one a pantheist, the other a Protestant), the grounds of Mersenne's choice remain musical, rather than theological.

Though himself orthodox, Mersenne's passion for universal harmony led him to espouse a teaching anathematized by his superiors. Far to the north of the Curia, from within the proudly independent French church, Mersenne was able to advocate the Copernican cause in ways not possible to Galileo, whom he attempted to befriend and defend.[13] But this was only one facet of Mersenne's amazingly diverse interests, which extended throughout the whole realm of science and engineering. This got him into trouble during his first foray abroad, to the Netherlands in 1628–29. Though himself without funds, he traveled at the expense of his order and exhibited such curiosity about everything he saw that, near Anvers, he was arrested as a spy and thrown into the tower overnight. On another occasion, a suspicious soldier shot at him, though fortunately the weapon was not loaded.[14]

During the height of the correspondence we examined in the last chapter, the Galileo affair was just becoming known in France while Mersenne was immersed in the preparations for his largest work, *Harmonie Universelle* (1636–37). Throughout its fifteen hundred pages, this book demonstrates many times over the centrality of music in his work on

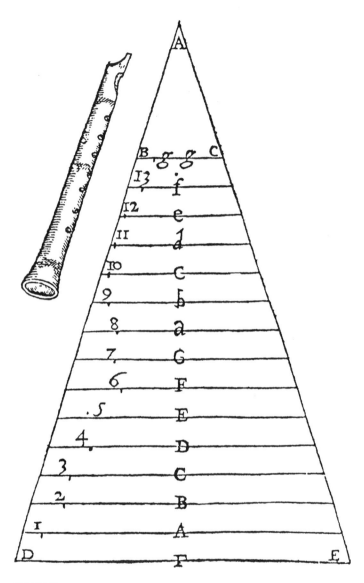

Figure 7.2
"The System of Robert Fludd," from Mersenne, *Traité de l'harmonie universelle* (1627). Correcting Fludd's mistake, Mersenne places a bridge at the number indicating each string, showing the actual length required to sound the stated note. The flute illustrates another argument by Mersenne against Fludd's pantheism.

natural philosophy. Music played a crucial role in the discoveries that Mersenne himself made and was constantly present in his correspondence.

Both on the large and the small scale, *Harmonie Universelle* shows the closest interaction between practical music, its theory, and natural philosophy. Its subtitle proclaims that it treats "the nature of sounds, and of movements," as well as of "consonance, dissonance, genres, modes, composition, voice, chants, and all sorts of harmonic instruments." Its elegant frontispiece (figure 7.3) depicts Orpheus playing his lyre to charm the animals, in which a lion lies next to a lamb, identifying the scene with the biblical vision of the end of time and the divine singer with the Redeemer. This is indeed a vision of universal harmony. Through his motto from the Psalms, Mersenne depicts his project as a confession of "thy truth with the instruments of the psaltery," in which singing and the harp itself are not only the means of divine praise but also the instruments through which divine truth is revealed. In the process, warring animal natures are subdued: the lion looks down rather fiercely on his prey, who turns demurely away. The image hints at the possibilities but also the limits of taming the passions; only the innately peaceable animals seem totally immersed in the music, like the ecstatic turtle at Orpheus's feet or the rapt sloth in the tree above his head.[15]

Mersenne begins with "the nature and properties of sounds," "the movements of all sorts of bodies," and "the movement, tension, force, weight, and other properties of harmonic strings and other bodies," followed by treatises on the voice, on chant, consonances, dissonances, the art of composition, and all sorts of instruments. He situates his encyclopedic treatment of music in relation to the new Galilean theories of motion. He does not disguise his adherence to the heliocentric view, but (in contrast to his preliminary *Traité de l'harmonie universelle*) he does not devote a section of the new book to astronomy or cosmology, standard topics though they were in treatments of *musica mundana*, thereby avoiding dangerous debates. He anticipates that, though practicing musicians may find his first books "the most laborious of all," they will still understand the necessity of connecting music with fundamental physics. In short, he imagines his readers to be, like himself, passionately interested in music as practical art and physical phenomenon. He believes that they, like he, want to confront facts established by precise experiments, not just plausible reasonings or empty words; he takes nothing simply from authority, not even Galileo's, but wants to test assertions for himself and urges his readers to do likewise. As a prime example, in his preface he emphasizes the surprising result of that archetypical Galilean experiment, the "Pisan drop": "two bodies of the same size but one weighing eight times more than the other will fall in almost the same time when dropped over one hundred feet." He takes this sheer fact as crucial in light of the long-received Aristotelian arguments that deny this. As he insists on positive proof by experiment, he also defends using the technical terms of the art of music, whose practical aspects are, for him, a direct source of experimental knowledge.

Figure 7.3
Frontispiece of Mersenne, *Harmonie Universelle* (1636). The motto is drawn from Psalm 70:22: "For I will also confess to thy truth with the instruments of the psaltery: O God, I will sing to thee with the harp, thou holy one of Israel."

Mersenne also emphasizes that the investigations of sound and light mutually clarify each other, so that the study of sound is the royal road to understanding nature as a whole. Ever since his visits with Descartes and Beeckman in the Netherlands in 1628–29, he had considered light, like sound, to be a purely corporeal phenomenon. This required a certain adjustment in reading the divine poetry of scripture so as not to conflict with these scientific insights. In that spirit, Mersenne interprets sacred scripture to present the study of universal harmony as the key to human excellence, whose understanding of nature culminates in personal and religious perfection.[16]

Writing in the aftermath of Galileo's pioneering work, and by comparison with it, one might think of Mersenne as derivative or as essentially a reporter, not an originator of discoveries. Yet Mersenne reached certain insights well before Galileo. As he presents his work in *Harmonie Universelle*, he connects these insights within his overarching musical context. Though he was rarely, if ever, truly original in the sense of initiating a new question or line of investigation, Mersenne was able to use his extraordinary persistence and awareness of the history of prior developments to extend them still further. Some of these have an obvious musical origin and importance, such as his demonstration that pitch is proportional to frequency and hence musical intervals are ratios of frequencies of vibration. Though G. B. Benedetti, Vincenzo Galilei, and Beeckman had already established the fundamental argument underlying this general proposition, Mersenne showed how to count the slow vibrations of very long strings against "a heartbeat, or a very slow and lazy pulse" he takes as measuring a second of time.[17] In so doing, he gives an experimental and observable actuality to this proposition, in accord with his principled preference for deeds over words. Mersenne's musical-physical experimentation gave the first absolute measurement of the frequency of a vibrating body, which Galileo had thought to be impossible because the rapidity of audible vibrations blurs them together so that they cannot be counted by sight.

Mersenne solved this problem by using a string 17½ feet long, "a lute or viol string of the size one mounts on racquets," alluding to the vogue of tennis that also touched Descartes, though this Gargantuan string is "made from a dozen sheep's intestines."[18] Essentially, Mersenne magnified a musical string to the point where its vibrations are commensurate with human sense capabilities. Stretched under a weight of half a pound, his string vibrates at two cycles per second, just countable because its cyclical "turns and returns" have been sufficiently slowed. Then Mersenne increases the weight on the string: under two pounds, it vibrates four cycles per second; under eight pounds, eight cycles per second. Though he does not make this explicit, these observations depend on his musical awareness, for these specific cases correspond to successively higher octaves above the lowest tone, as given by the series of ratios of frequencies 2:4:8 given by the ratio of weights ½:2:8. One infers that he adjusted the weight upward and listened for the octaves (and perhaps also noted the characteristically doubled visual wave-form of the string).

From these observations follows an empirical proportionality between the vibrational frequency of a string and the square root of its tension (here measured by the weight); in other experiments, he likewise showed that frequency varied inversely as the string's length and its cross-sectional area, results now called "Mersenne's laws."[19] He also established similar relations for wind and percussion instruments, demonstrating their general application to vibrating bodies. These findings allowed him to carry his result from the ultra-slow vibrations of his giant string to the realm of more ordinary frequencies. He observes that a section of the same string about one foot long stretched under an eight-pound weight sounds in unison with a four-foot organ pipe pitched at the *ton de chapelle*, one of the standard pitches in use at the time. From his empirical laws, he deduces that this string was vibrating at 84 cycles per second, a frequency sufficiently high that, as Galileo had surmised, it could not have been counted directly.[20] Mersenne goes on, in his methodical way, to tabulate the frequencies of notes over eight octaves.

He notes that "a string must beat at least 20 times a second in order to be heard, and only 42 times a second for its movement to be seen by the eye, nevertheless without being able to count its returns until it only makes more than ten," indicating the greater sensitivity of the ear to discern these very slow vibrations. Thus, Mersenne's experimental technique essentially depends on the ear even as it explores realms of frequency that are no longer aurally discernible.

At the same time, Mersenne became interested in aspects of sound that would not depend on the observations of a trained ear. By applying the results of his empirical laws, he was able to show that "a deaf man can tune a lute, viol, spinet, and other string instruments and find the sounds he wishes, if he knows the length and size of the strings." He provides a "harmonic tablature for the deaf" that enables them to find the visible characteristics of different notes they might be asked to produce (figure 7.4). Perhaps this was addressed to his friend Descartes, who by 1638 described himself as "almost deaf."[21] In the following generation, Joseph Sauveur made important contributions to acoustics (even providing that name to the field) though profoundly deaf and mute until age seven.[22] Conversely, Mersenne demonstrated that "one can know the size and length of strings without measuring or seeing them, through the means of sounds," so that hearing can substitute for the other senses.[23]

A similar blend of practical musical consideration and theoretical speculation characterizes Mersenne's other investigative initiatives. He may have been the first to measure the speed of sound and to show its independence of pitch and loudness, a proposition he tested by various kinds of echoes. Mersenne's experiments involve using language itself to probe the speed with which the echo is formed; "it is certain that all sorts of echo that repeat seven syllables pronounced in the time of a second must cover the distance of 485 feet," which he compares to the firing range of an arquebus.[24] He repeated the syllables *Benedicam Dominum* ("Let me bless the Lord") at higher and lower pitches, softly and loudly, in foggy and clear weather, to determine that the speed of their sound, measured by the

Tablature harmonique pour les sourds.

		Table I. La tension des chordes proportionnées selon la raison doublée des interualles.				Table II. La grosseur des chordes proportionnée selon la raison simple des interualles.		Table III. La longueur des chordes proportionnées selon la raison simple des interualles.			Table IV. La Tension des chordes proportionées selon la raison simple des interualles.			
Les 8 sons, ou notes de l'Octaue.	Les 7 degrez de l'Octaue.	liures.	onces.	gros.	grains.	parties de ligne	dixièmes.	pieds.	poulces.	lignes.	liures.	onces.	gros.	grains
1 VT		1	0	0	0	10		4	0	0	2	0	0	0
	ton mi.													
2 RE		1	4	15	54	9		3	7	2⅖	1	12	12	58
	ton mai.													
3 MI		1	10	9	0	8		3	2	4⅘	1	9	9	43
	sem.mai.													
4 FA		1	14	3	32	7½		3	0	0	1	8	0	0
	ton mai.													
5 SOL		2	6	4	0	6²		2	8	0	1	5	5	24
	ton mi.													
6 RE		2	14	3	32	6		2	4	9½	1	3	3	14
	ton mai.													
7 MI		3	11	12	18	5⅓		2	1	7⅕	1	1	1	5
	semi.maj.													
8 FA		4	4	0	0	5		2	0	0	1	0	0	0

Figure 7.4
Mersenne's "Harmonic Tablature for the Deaf," showing (leftmost column) the notes of the scale, along with the tension, size, and length of the string needed to produce these sounds.

clarity of their echo, did not depend on these factors. Though his monastic brethren would have recognized his words, his intent went beyond ordinary prayer. He also refers to an echo in the Tower of Metallus near the Aventine Hill in Rome, which (or so he recounts) can repeat eight times the recited opening of Virgil's *Aeneid*, *Arma virumque cano qui primus ab oris* ("of arms and the man I sing ..."). Mersenne used this curious lore to extend his calculations. He notes that this verse cannot be said clearly and distinctly in less than two seconds, so that the eightfold repetition would take thirty-two seconds (evidently allowing equal time between each successive echo). Based on his calculated velocity of sound, the eight echoes traverse 1,296 toises, traveling back and forth, about half a league (2.78 km). He speculates that this might give a way of measuring a large distance (such as the width of a city) by measuring how far sound could be heard across it.

Having used an instrument string to measure the frequency of sound and its laws, Mersenne points out that "if one brought a piece of music from Paris to Constantinople, to Persia, to China, or elsewhere, along with those who understood the notes," they could perform the piece "according to the intention of the composer" because they could adjust the pitch to the Parisian standard, using his laws to generate the corresponding frequency.[25] Further, the tempo could be specified in universal units, such as beats per resting pulse or per second by the clock.

These issues of musical time also are connected with Mersenne's reconsiderations of the clock itself and the means by which it might measure time more accurately. This forms part of his book on the "movements of all sorts of bodies," on which depends the problem of vibrating strings. Here, he is much influenced by Galileo, whom his own activity in this field outstripped on some occasions. In June 1634, he noted that the frequency of a pendulum is inversely proportional to the square root of its length, a full year before Galileo found this result. In his *Harmonie*, Mersenne provides a table showing this result, noting that physicians might use such a simple pendulum "to find out how much faster or slower is the pulse of their patients at different hours and days, and how much the passions of anger and other hasten or retard it." He also noted that watchmakers could also use this device to improve time-keeping; though the pendulum watch was not patented by Christiaan Huygens until 1656, with improvements that were important for it to reach sufficient accuracy for navigation and other precise uses, Mersenne's insight was an important step.[26]

Mersenne's detailed treatment of the mechanics of falling bodies, inclined planes, and pendulums clearly supports and enables his ensuing deductions about vibrating bodies, following out Descartes's insight that a vibrating string could be understood as an ensemble of pendulums, one for each point along the string. As Peter Dear puts it, "Mersenne accomplished the harmonization of mechanics through the mechanization of music."[27] But Mersenne did not only move in one direction with these deductions, from musical observations to physical theories. He also moved in the other direction, from the physical propositions he had established to their musical applications. For instance, he studied the various sounds made by falling bodies, which vary in pitch depending on the height from which they fall. From what heights, he asked, ought they be dropped so as to produce any given consonance or dissonance? He worked out an elaborate table that he mapped into a striking crisscross circular design, in which the entire musical scale is cross-referenced with the appropriate falling body (figure 7.5). This construction hovers somewhere between the conceptual and the observational; it does not seem credible that he has actually *heard* the pitches of these "singing" bodies with any degree of accuracy as they fell. Mersenne gives the authority of Aristotle for the general premise that "a sound is that much higher if it is made by a faster movement," to which he adjoins the Galilean law of freely falling bodies. From this, Mersenne extrapolates the degrees of velocity reached by two falling bodies (depending on the height from which they were dropped) and then rather arbitrarily

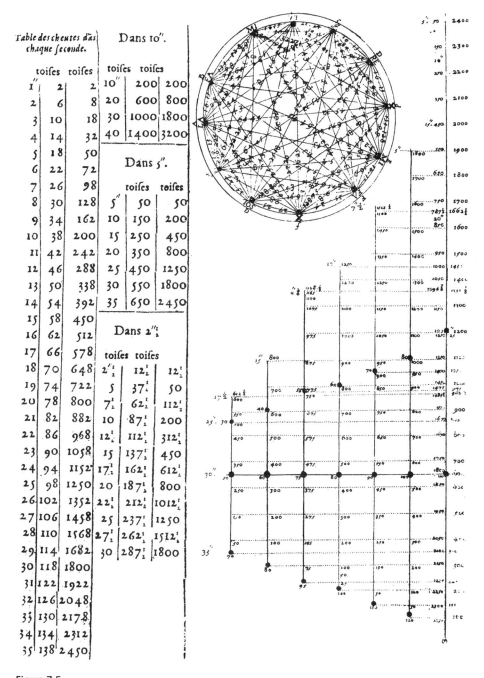

Figure 7.5
Mersenne's diagram showing the distances traversed by freely falling bodies (left), which he converts to the appropriate ratios needed to sound musical intervals (right). In his circular diagram, the pitches of the scale are shown around the circumference, while the relative distance a body needs to drop to make a given interval (as he calculates it) is shown by a chord between its two notes.

Figure 7.6
The fundamental pitch (C) and the first four overtones (*petits sons*) above it, corresponding to integral ratios, as observed by Mersenne.

takes the ratios of these speeds to represent the musical ratio that they *would* sound at that terminal velocity. He does not correct for air resistance or any other factor that would indicate the kind of experiment he elsewhere insists on; this whole proposition seems a *jeu d'esprit*, an imaginary return from physics back to the spirit of music from which it had emerged. Though allegedly staged on Earth, its pattern of deduction is much closer to that which Kepler (and Mersenne) had used to examine the harmonic relations between the planets. In that sense, Mersenne is trying to bring back to Earth what he had learned in the heavens.

Above all, Mersenne's musical motivations led him to investigate the physics of sound. Descartes had noted overtones at the octave, fifth, and third; Mersenne noticed other *petits sons*, "little sounds" that could be heard from a single string a second octave higher and, above that, the note a major third higher still (figure 7.6). Their successive ratios of string lengths fall along the series of the first five integers. He noted that "it is necessary to find complete silence to perceive them, although this is not necessary when one has a trained ear." These overtones were overlooked because musicians "are so anticipating and preoccupied with the natural tones of the string that there is (it seems) no place in their ordinary senses or imagination to receive the idea or species of these small, delicate sounds." He notes that it is easier to hear them played by a bass viol in the silence of the night, details that suggest the range of his experiments. Despite the difficulty of producing and distinguishing these sounds from the fundamental tone (or the habitual sound of the string in the player's ear), "I have had no difficulty and I have met many musicians who hear them as well as I and undoubtedly ever one hears them when they lend the necessary attention."[28] Though the *petits sons* had been there all along, Mersenne was the first to hear so many of them because of his extraordinary attention to every detail of musical experience and instrumentation. As precedents he mentions only Aristotle, not Descartes, though Mersenne knew of Descartes's *Compendium*. Even so, Mersenne went much further than Descartes precisely because of his far greater interest in the fine details of music. Where Descartes barely commented on the extra tones, Mersenne verified his observations "very exactly more than a hundred times, on the viol and on a theorbo, as well as on two monochords," in order to exclude the possibility "that these different tones do not come from other strings that are on the instruments and that tremble without being played," by sympathetic vibration, "since the single string of the monochords produce the same sounds."[29]

Mersenne goes even further when he notes that there are "*at least* five different tones at the same time," implying the possibility that there are still more. He hears "still a fifth one higher" that "produces the major twentieth with the natural tone." If so, this seventh overtone (counting the fundamental as the first) would correspond to the note A above the last overtone shown on figure 7.6; later observations placed it closer to B♭. Given the increasing faintness of each successive overtone, it is not surprising that he would have had difficulty with this even fainter one, but theoretical reasons may have made him hesitate. Up to this point, the overtones had coincided exactly with the well-known consonances: octave, fifth, major third. Theorists soon noticed that Mersenne's series of four overtones (figure 7.6) in fact sounded the ordinary triad based on the fundamental tone, laid out in exactly the way that practitioners had found most euphonious. This was, for many music theorists, the fundamental justification of the triad as the "chord of nature."[30]

Compared to that purely triadic standard, the seventh overtone is a rogue, an outlier, rather like Boethius's version of the fifth hammer in the blacksmith shop. Mersenne's hesitation about its status reflects its deviance from ordinary music theory and practice. So strong was his commitment to the conventional intervals that he felt implicit pressure to reduce it to one of the notes he did know (namely, the twentieth, two octaves and a sixth above the fundamental). When he turns to wind instruments, he confirms the series of overtones so far, which are formed by "overblowing" the airstream. The trumpet of his time, a tube with a flared horn and no valves, produces this series with great clarity and volume, including the troublesome seventh partial (♪ sound example 7.1). Players then and now know that this tone lies flat from the equal-tempered scale (and also from just intonation), so that it needs to be adjusted upward in pitch by using the tension of the lip.

Mersenne includes the trumpet in his extraordinary survey of all the instruments in the known world, a musical *tour d'horizon* that had little precedent in earlier writings and which remains an invaluable source about the practice of his times. He describes the history and construction of the trumpet and lists its range, which coincides with the series of overtones but omits the seventh, skipping from the sixth to the eighth and also omitting two other higher overtones that sound flat (figure 7.7a). When he illustrates the instrument in its military uses (figure 7.7b), though, he includes all these partials in one panel but not the other, which presumably includes only the notes used in fanfares and excludes these problematic overtones. His text is hesitant and puzzled on the whole matter; he notes "many difficulties" that begin with the seventh overtone. The sequence of pure consonances is broken at this point; he has heard a discordant pitch in the place of the seventh partial but does not want to admit its existence, calling it a "wound," even a "vice."[31] Musical practice maintained so strong a hold over him that he tried to exorcise his experimental results as numerologically impossible or devilish. Though he tried to amend the faint seventh overtone of a viol string, the trumpet's was harder to ignore.

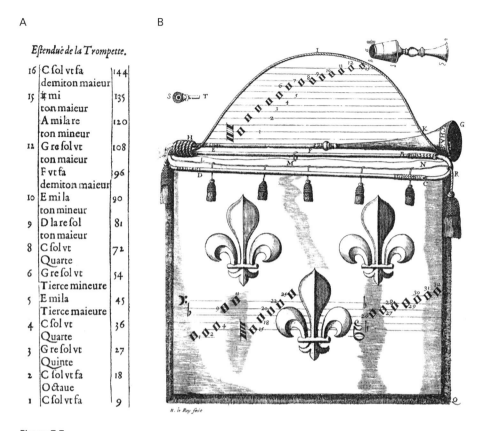

Figure 7.7
(a) Mersenne's list of the range of the trumpet, which omits the discordant seventh, eleventh, and thirteenth overtones. (b) Mersenne's illustration of the military trumpet with its range, including the problematic overtones (above), but not in their musical application (below).

The multiplicity of overtones posed an even more fundamental musical and logical problem: "How is it possible that a single string can make many sounds at the same time?"[32] Further, "why does it make no sound lower than what is natural to the string?" In a corollary, he argues that "it is more probable that these different sounds come from different movements of the exterior air rather than those of the interior [of the string]."[33] Yet he refers his readers to his discussion of bells, which he says use the "motion of their parts, which imprints a similar movement on the air with which they are surrounded."[34] His discussion of bells intensifies the problem because they so loudly exemplify this multiplicity: even to a relatively unpracticed ear, a single bell produces a complex sound, not a single pitch.

Among all the many instruments he described, Mersenne devoted special care to bells and organs, as befits one who spent his life in churches (figure 7.8).[35] Mersenne notes how

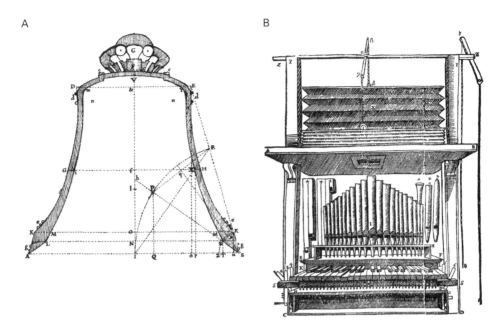

Figure 7.8
(a) Mersenne's cross-sectional diagram of a bell. (b) Mersenne's diagram of an organ.

striking a bell at different points elicits different sounds. Noticing that, in different bells, the various overtones appear with different strength, he brings forward information about the possible combinations of metals, showing that he had studied the artisanal knowledge of bell-makers, though many of them "have been wrong in all sorts of bells, making them too heavy, or too light, or too straight, or too large."[36] He lists the most successful mixtures of metals and the best proportions for bells but is not able to account for them mathematically. He compares the various overtones of bells with those of strings, the voice, and organ pipes.[37]

In his preliminary summation, the movements of all "solid and hard bodies which are made of complete vibrations, which hold some elasticity, are species of tremblings and shakings." The problem is "how the shuddering is made without the bells being burst, for if all their parts are moved, when they tremble, some must give way to the others … [and] find space for moving and vibrating." He concludes that the body must be "more or less porous," having "a great number of small empty places" as in the ancient atomic theory, which he thinks "can easily explain the vibration of bells. For when one strikes them their atoms are stirred and crackled in changing place, and in occupying the spaces of the small vacuums, and then they return many times into their ordinary place, and return into the said vacuums until they become quiet."[38] Here Mersenne is in sympathy with his close friend Pierre Gassendi, who was deeply interested in atomism, though it was ecclesiastically suspect because it seemed to leave no room for the transubstantiation of bread and wine into divine body and blood.[39]

Still, Mersenne notes the unsolved problem of why the atoms return to their place, or what force might so move them; "it is not enough to say that it is natural to them," because "atoms are indifferent to all sorts of places and movement is as natural to them as the repose is contrary." He speculates about "little hooks or crotchets of the other atoms of the bell, which draw them back into their ordinary place," or of atoms having tetrahedral or octahedral shapes (as in Plato's *Timaeus*) that might somehow explain their internal forces and movements.[40] But he could go no further in advancing this microscopic view or in formulating the macroscopic mathematics of vibration. Even so, Mersenne raised the perplexing multiplicity of overtones into a principle of harmonic pleasure: "The sound of any string is the more harmonious and agreeable, the greater the number of different sounds it makes heard at a time." If, he continues, it is permitted to "translate physics into human actions, one can say that each action is much more agreeable and harmonious to God as it is accompanied by a greater number of motives, provided that they are all good."[41] Mersenne's own vibrating plethora of investigations and speculations testify to his devotion to the God of Universal Harmony. His questions, his beautiful questions, he left for those who came after.

8 Newton and the Mystery of the Major Sixth

Though Isaac Newton considered poetry "ingenious nonsense," music had a significant if limited place in his intellectual world.[1] His youthful manuscripts demonstrate the scope of his knowledge and interest. Later, at a critical point in his optical writings, he relied on a musical analogy to compare the seven notes of the diatonic scale and the seven colors he likewise attributed to the spectrum. A close examination of his use of this analogy discloses its power and implicit limitations. Newton's case may be read as a cautionary tale about the way musical analogies can open possibilities but leave important matters provocatively undecided.

Almost all of Newton's musical writings are in a notebook he used during his undergraduate period (1664–1666), spanning his *annus mirabilis* 1665, the year in which he first grasped his celebrated insights about gravitation and light during an enforced stay at home to avoid the plague.[2] In this notebook, Newton first compiled drafts that work out, in intense and obsessive detail, fundamental definitions of musical intervals (figure 8.1). He includes circular diagrams resembling some in Descartes's *Compendium musicae*, supporting the possibility that Newton studied that work, as he studied Descartes's other writings at the time.[3] Newton's manuscript "Of Musick" clearly demonstrates his involvement in the study of music as part of the quadrivium. Though brief, the work's level of detail and methodical enumeration of possibilities show the ways music was important to him, not merely a perfunctory study.[4]

Beyond summarizing commonplaces, the ordering and minute details of Newton's text register his unfolding sequence of further reflections. He begins by analogizing the "Clef or Key" of a piece of music to the concept of "an unit" as mathematical center. Turning then to the octave and the intervals within it, Newton considers their "order of concordance" but goes beyond commonplaces about their ratios to an argument that is interestingly physical in character: "As too sudden a change from less to greater light offends the eye … so the sudden passing from grave to acute sounds is not so pleasant as if it were done by degrees, because of too great a change of motion made thereby in the auditory spirits."[5]

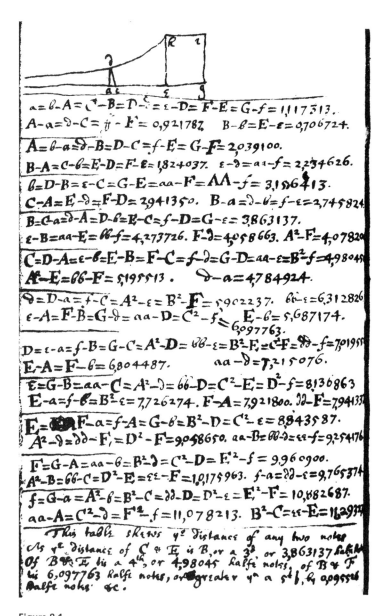

Figure 8.1
A page from Newton's undergraduate notebook, dated November 1665: "ye distances of any two notes" (f. 104r). (By permission of The Syndics of The Cambridge University Library.)

In this passage, Newton independently puts forward an analogy between optics and music years earlier than Robert Hooke, who communicated this analogy to Henry Oldenburg in a 1672 letter later forwarded to Newton.[6] In his notebook, Newton extends this analogy from sound to light to heat: "Thus a little heat is least perceptible to one newly come from a greater." In framing this analogy, Newton might well have recalled several ancient texts he had studied: a famous passage in Plato's *Republic* described the inability of the eye to deal with sudden passage from light to darkness, but there is no precedent in Plato for the comparison with sound, though Aristotle does discuss such a comparison. Thus, Newton may have synthesized Aristotle's comparison of light and sound with Plato's description of the physical (almost visceral) effect of abrupt transitions in light.[7]

Returning to enumerate the parts of an octave, Newton begins a new thought in square brackets, which he used to set off speculative or interpretative comments: whole tones might be divided into semitones and quarter tones, "but they would be of no use" because "the number of discords twixt each concord would much more bee harsh than the concord would be pleasant."[8] Here, as with Vicentino before him, quarter tones open new theoretical possibilities. Though Newton initially reverts to the conventional view that semitones and quarter tones are "unpleasant" discords, he then strikes out his closing bracket in order to open a new thought: "Yet perhaps ½ or ¼ notes passed over very hastily with a larger stay upon the concords twixt which they are, might be delightful." This shows a certain curiosity about new musical possibilities; there is no evidence that he ever read Vicentino, though he studied Kepler closely. Newton closes off this line of thought by noting that "since they are such discords, inserted as 'twere by accident only to graduate concords, & so quickly slipped over, the sense cannot perceive any error or exactness in them, & therefore be they useful yet to treat of them would be lost labor." Even the way he decides to end his bracketed digression shows his preoccupation with "the sense," here meaning the experimental judgment of the ear, rather than any prior theoretical considerations. This experiential orientation accords with his earlier optical/musical analogy, which emphasized the physical response of eye and ear.

Newton limits himself to the diatonic order in his following discussion of how the modes "much limit the parts of the tune from discord sounds of one with another, particularly because tunes framed by divers of them differ in their airs or Modes." He enumerates all the possibilities of ordering tones and semitones and reconstructs the conventional modes, presented in tabular form. Newton several times refers to "sweetness," "grace," or what is "grateful to the ear," rather than numerical theory, as his criteria in setting out "the 12 Modes in their order of elegancy," again seeming to prefer empirical, physical criteria of satisfaction to pure numerology. This represents an important divergence from Boethius's preference for rational judgment over the evidence of the ears; Newton too followed Aristoxenes' empiricism, whether knowingly or not.[9] Likewise, in his "Questiones quaedam Philosophiae" (composed in the early 1660s), Newton had noted that "the senses of divers men are diversely affected by the same objects according to the diversity of their constitution. To them of Java pepper is cold."[10]

Among the modes, Newton puts in the highest place the Mixolydian, which "excels" the Dorian he places second (see figure 3.4); all the other modes he considers even more "diminished" in "sweetness." Here Newton seems to follow contemporary trends in music theory, which gradually came to prefer what we call the major mode (which is essentially Mixolydian with a raised seventh, F♯) over the church modes, among which Dorian was the first in the usual order. In his final sections, Newton also notes that "Tis usual to pass from one mode to another in the midst of a song," the practice of modulation we considered in chapter 3: what during the sixteenth century was rare became common in Newton's time, during the same period when the formerly unimaginable motion of the Earth became widely accepted.[11]

A decade after "Of Musick," Newton's analogy between optics and music came forward in his comparison of the spectral colors to the seven notes of the diatonic scale, first presented publicly in his second letter on light and colors for the Royal Society (1675).[12] There, Newton observes that as vibrating bodies excite sounds of various tones, so does light excite the optic nerve,

much after the manner, that in the sense of hearing, nature makes use of aereal vibrations of several bignesses to generate sounds of divers tones; for the analogy of nature is to be observed. And further, as the harmony and discord of sounds proceed from the proportions of the aereal vibrations, so may the harmony of some colours, as of golden and blue, and the discord of others, as of red and blue, proceed from the proportions of the aethereal. And possibly colour may be distinguished into its principal degrees, red, orange, yellow, green, blue, indigo, and deep violet, on the same ground, that sound within an eighth [octave] is graduated into tones.[13]

Newton goes on to describe how he had projected prismatic colors in a dark room and asked "a friend to draw with a pencil lines cross the image, or pillar of colours, where every one of the several aforenamed colours was most full and brisk, and also where he judged the truest confines of them to be. And this I did partly because my own eyes are not very critical in distinguishing colours, partly because another, to whom I had not communicated my thoughts about this matter, could have nothing but his eyes to determine his fancy in making those marks." This notably solitary and secretive worker used "a friend" to check his own lack of critical judgment about colors, though Newton never made a similar acknowledgment about his sense of pitch; in his account, the tones within an octave seem a better established frame of reference against which he judges the vagaries of color perception. Though Newton acknowledges that "the just confines of the colours are hard to be assigned, because they pass into one another by insensible gradation," he notes that "this observation we repeated divers times" and that "the *differences* of the observations were but little, especially toward the red end." By "taking means between those differences" Newton judged that the length of the whole image "was divided in about the same proportion that a string is, between the end and the middle, to sound the tones in the eighth [octave]," as he illustrates in figure 8.2. He regarded the color spectrum as similar to the musical divisions of a string.

Though Newton presents the analogy as the result of careful experiment, consciously constructed to avoid prior hypothesis, he asserts more generally that "the analogy of nature is to be observed" with respect to the corresponding natures of the different human senses, rather than between the natures of light and sound as such.[14] As Newton went to apply this analogy to *different* optical phenomena, though apprehended by the *same* sense, he obtained results whose divergence cannot be due to that single sense but to the nature of light itself.

The crux of Newton's analogy was that, as the upper note in an octave stands to the lower (as d would to D an octave lower), so do the extremes of color, namely "deep violet" and red, likewise represent an "octave" in color, within which the intermediate hues should occupy the traditional seven scale degrees, thus interpreting the colors of the spectrum as corresponding to musical notes spanning an octave. From this flows his assertion that, at the appropriate points in the scale, the spectral colors orange and indigo should be inserted at the very points in which the chosen mode (Newton took this to be Dorian) has the semitones E–F and B–C. For those who came after, Newton's musical analogy is the source of the widely held opinion that orange and indigo are actually intrinsic in the spectrum, despite the great difficulty (if not impossibility) of distinguishing indigo from blue, or orange from yellow, in spectra. Thus the authority of Newton, even speaking far from his primary expertise, carries unquestioned weight even to the present day. Yet in his *Optical Lectures* (1670–1672) Newton had been rather diffident about the analogy and admitted that "I could not, however, so precisely observe and define this without being compelled to admit that it could perhaps be constituted somewhat differently."[15]

Here Newton acknowledges the difficulty of dividing the spectrum into seven "more prominent" colors "proportional to a string so divided that it would cause the individual degrees of the octave to sound." That is, he admits that he imposed the seven colors by analogy with the (Dorian) mode without being able to demonstrate that those specific colors must necessarily be placed at those scale steps—hence his admission that the color correspondence "could perhaps be constituted somewhat differently."[16]

Newton's primary assumption is that color, like sound, admits of octave (2:1) ratios. This assertion bears strongly on Newton's theories about light. Though familiar with the experiments of Francesco Maria Grimaldi that seemed to show wave effects in light (figure 8.3), Newton argued that a light wave passing an obstacle should "bend into the shadow," which he felt had not been demonstrated even by what Grimaldi called "interference."[17] Though he preferred a particulate description of light emission, Newton never presented his preference as more than an hypothesis; while denying that light itself was a wave, he put forward his idiosyncratic (and rather puzzling) "fits of easy transmission and

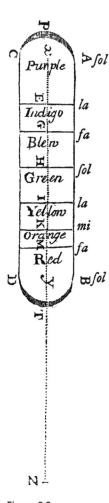

Figure 8.2
Newton's illustration (1675) of the analogy between spectral colors and the seven notes of the diatonic scale. The note names follow the older nomenclature spelling the Dorian mode.

reflection"—sudden seizures in the behavior of light—as a way to incorporate certain aspects of wave theory into a predominantly particle view.[18]

Newton's musical analogy, however, has an unexpected relation to the wave theory. Already in a 1672 manuscript, he supposed that "the vibrations causing the deepest scarlet to be to those causing the deepest violet as two to one; for so there would be all that variety in colours which within the compass of an eight [octave] is found in sounds, & the reason why the extremes of colours Purple & scarlet resemble one another would be the same that causes Octaves (the extremes of sounds) to have in some measure the nature of unisons."[19] Here Newton seems to assume that the "resemblance" of purple and scarlet parallels the "resemblance" of octaves.

In this manuscript, Newton tried to find empirical support for the 2:1 ratio of the "vibrations" of purple and scarlet in the ratios between spaces of colored rings from illuminated lenses (first described by Hooke, though usually known as "Newton's rings," figure 8.4). Yet in Newton's rings the ratio of the extreme colors was "greater than 3 to 2 & less than 5 to 3. By the most of my observations it was as 9 to 14."[20] In his *Opticks* (1704), Newton stated that rings "are to one another very nearly as the sixth lengths of a Chord which found the Notes in a sixth Major," such as from D to the b above it, compared to the octave D–d.[21] Here Newton reduces the number of his "principal colours" from seven to five, which probably stemmed from his observations of the rings, in which it is hard to observe minute color nuances. Thus, he was open to altering his musical enumeration of spectral colors.

Newton's hesitation between octave and major sixth shows the difficulty and importance of the point. He had initially assumed an octave, based on his prior ideas about the perfection and completeness of that interval, whereas a major sixth clearly comes from empirical observation and seems to indicate some quality inherent in light itself. Indeed, a wave theory can far more naturally explain this ratio than can a particle theory, which lacks a concept of wavelength (whose place Newton tried to supply with his "fits"). In terms of wavelength, visible light spans roughly only a major sixth, about a ratio of 700:400, corresponding to the modern conventions for violet at 400 nm and red at 700 nm, noticeably short of an octave. In short, the human eye has never experienced an octave relation, whereas the human ear recognizes many octaves.[22] Newton's analogy is therefore in tension with this fundamental inconsistency. Clearly troubled by the discrepancy between octave and sixth, in order to agree "something better with the Observation" Newton then reinterpreted his measurements through a rather intricate stratagem. He suggests that the rings' major sixth could be understood "as the Cube Roots of the Squares of the eight lengths of a Chord" in an octave, thus rather tortuously reinterpreting the musical interval of a *sixth* in terms of an *octave*.

This connection between cube roots and squares resembles Kepler's third law connecting the cube of the distance of a planet from the sun and the square of its period.[23] Newton considered that relation crucial to establishing the inverse square law of gravitation; in the

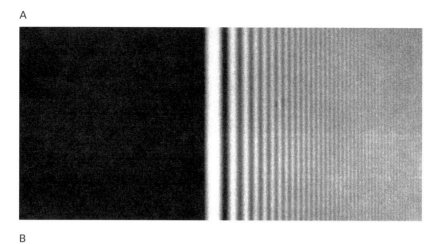

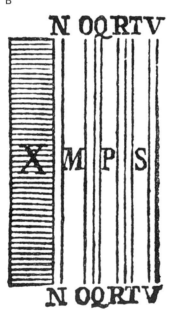

Figure 8.3
(a) A modern photograph of the light fringes seen next to a sharp edge. (b) Grimaldi's 1665 diagram showing the dark fringes *N, OQ, RT, V*.

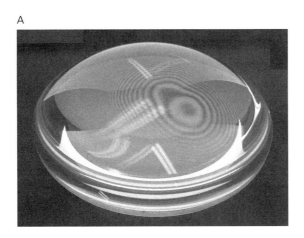
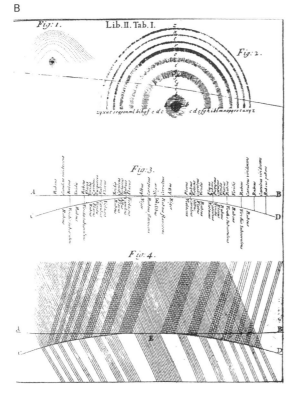

Figure 8.4
Newton's rings. (a) A modern recreation of his experiment using two plano-convex lenses pressed against each other, showing the characteristic moiré pattern. (b) Newton's diagram of the appearance of the rings, their colors in relation to the curved lenses, and their explanation in terms of his "fits of easy transmission," from his *Opticks* (1704).

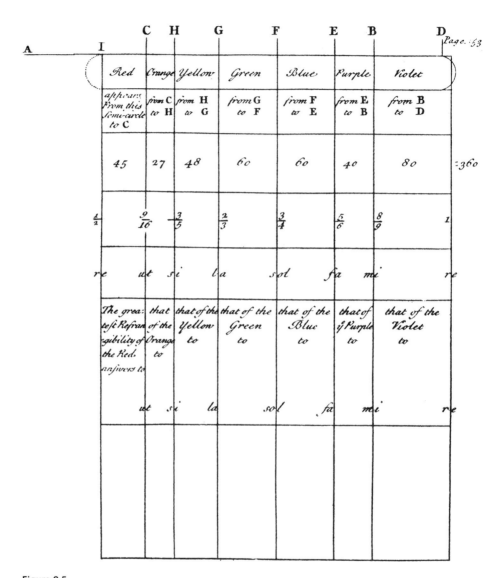

Figure 8.5
Voltaire's illustration of Newton's musical analogy between musical notes and colors, from *The Elements of Sir Isaac Newton's Philosophy* (London, 1738).

case of light, his use of a similar proportion allows him to subsume the major sixth in an overarching octave.[24] His reinterpretation sets forth a rival Keplerian "third law," here for the harmonies of the colored rings, rather than of the planets.

Having done so, he seems to have satisfied himself that the major sixth is a masked form of the octave. Thus, he was not inclined to interpret the major sixth as having some important significance of its own, much less that it could be interpreted in terms of a wave theory of light. Nor did those who followed him immediately notice this mystery. For instance, in 1712 Nicolas Malebranche argued that "*different colors* consist only in the different *frequency* of the pressure vibrations of subtle matter, as *different tones* of music result only from the different *frequency* of the vibration of gross air."[25] Thus, Newton's analogy persists in Malebranche's wave account, which incorporates Newton's octave of color without noticing the problem of the major sixth. Voltaire also featured "the Resemblance between the seven Primitive Colours and the seven Notes in Musick" in his popular treatment of Newtonian philosophy (figure 8.5).[26]

The picture was so beautiful, the analogy with the octave so fetching, that all trace of the troubling major sixth was mostly forgotten until Leonhard Euler and Thomas Young took note of it decades later, as we shall see. For Newton himself, other more weighty arguments and concerns may have relegated this musical puzzle to the sidelines. His opposition to the wave theory of light was long considered and deeply held; his optical writings are far greater in volume than the few pages he devoted to music in his youth. Yet musical theory supported his interest in the importance of ratios as applied to physical phenomena and thus was a helpful touchstone for his mathematical natural philosophy. In later life, he remarked that "Pythagoras's Musick of the Spheres was gravity."[27] Though Newton took music sufficiently seriously to register the mysterious major sixth, in the end his desire to lay the mystery to rest may have missed its surprising import because he did not take music seriously enough.

9 Euler: The Mathematics of Musical Sadness

Among Continental scholars who advanced and reconsidered Newtonian physics, Leonhard Euler was probably the greatest and surely the most prolific. Of his thirty thousand published pages, only a few hundred are devoted to music, but these have a special significance among his works. Music was one of the first topics he addressed at length, and he returned to it several times throughout his life. Moreover, musical questions led Euler to consider new mathematical topics and devise new approaches that then characterized several of his most important initiatives in mathematics and physics. Indeed, Euler's individual mathematical discoveries, great as they are, need to be placed in the context of his larger role in the beginnings of modern number theory and topology. As familiar as these mathematical disciplines have become, we cannot take them for granted but should try to understand how they came into being in Euler's hands. In this story, his musical writings open surprising perspectives.

At age thirteen (1720), Euler matriculated at the University of Basel, which included musical studies in its curriculum and was an important center of musical thought. His father, a Calvinist pastor, introduced him to Johann Bernoulli, whom Euler visited on Saturday afternoons to discuss mathematics. Bernoulli noted his extraordinary talents and persuaded Euler's father to allow his son to follow his mathematical interests; thereafter, Bernoulli continued to correspond with Euler about mathematical, scientific, and musical questions, as did his son Johann II.

Indeed, Euler was much occupied with music throughout his life. Nicholas Fuss, his student, son-in-law, and secretary, recorded that "Euler's chief relaxation was music, but even here his mathematical spirit was active. Yielding to the pleasant sensation of consonance, he immersed himself in the search for its cause and during musical performances would calculate the proportion of tones."[1] This quest for a new mathematics of music began in his earliest works and persisted throughout his productive life.

Euler's early scientific notebooks include an outline he prepared at age nineteen (1726) for a projected work he entitled "Theoretical Systems of Music," an ambitious survey he intended to include sections on composition in one and many voices, treating both melodic and harmonic writing.[2] His outline also envisaged chapters on various dances, as well as larger musical forms. Euler's interest in music encompassed many aspects of contemporary composition and practical music-making, not only its mathematical elements. In his early manuscripts, notes on musical theory precede any material referring to his second printed work, "Physical Dissertation on Sound" (1726), indicating the path that led him from music to the mathematical physics of sound.[3]

In this work, Euler addresses all kinds of sources, from musical instruments to thunder and snapping twigs, whose sounds all arise "from the sudden restitution of compressed air, and as a stronger percussion of the air." Especially, he addresses wind instruments such as the flute, "since no one up to the present has given anything of substance concerning these instruments."[4] He extends Newtonian methods by considering an air column to vibrate "following the amplitude of expansions and contractions in the same manner as strings, and thus I can consider that same air column as a bundle of air strings with the tension given by the weight of the atmosphere." Here he faces the inherent mathematical difficulties of the vibrations of a cylindrical pipe, for which his treatment is only a beginning.[5] Still, he expresses satisfaction in his general result that "the sounds of flutes will be sharpest in pitch with the maximum heat, and the air the least dense, but to be lowest pitch with the maximum cold and the most dense [air]. This difference of sounds is especially observed by musicians and organists. But since all flutes have the same change in place equally, the melody is not changed."

Euler puts forward his work "to be examined along with the distinguished candidates," showing that he considered his treatment of sound a calling card demonstrating his skill as he searched for a position. Euler's work on sound led to further problems in mechanics and thence to questions regarding the relative stability of ships with varying heights of masts and sizes of sails. His first foray into nautical science forms the pendant to his work on sound, which informs the mechanics of masts and winds. The juxtaposition of these diverse topics shows their interconnection in his mind: in the study of sound, practical and theoretical concerns unite mechanics, metaphysics, and nautical engineering.[6]

During the same period as he was preparing his work on sound in music and physics, Euler was also working on a more speculative, larger work, his *Tentamen novae theorae musicae ex certissimis harmoniae principiis dilucide expositae* (*Essay on a New Theory of Music Based on the Most Certain Principles of Harmony and Clearly Expounded*, written 1730, published 1739).[7] Not able to find a job in his native city, Euler left Basel in 1727 to move to St. Petersburg, where he obtained the chair of natural philosophy in 1730, the year he completed his *Tentamen*. By devoting so much of his attention to this work during the crucial period in which he needed to establish himself in a permanent position, Euler showed how integral he considered music to be to mathematics and natural philosophy. Writing in 1731 to Daniel Bernoulli, a fellow pioneer of the mathematical study of vibrating strings, Euler clarified his larger intent: "My main purpose was that I should study music as a part of mathematics and deduce, in an orderly manner, from correct principles, everything that can make a fitting together and mingling of tones pleasing. In the whole discussion, I have necessarily had a metaphysical basis, wherein the cause is contained why a piece of music can give one pleasure and the basis for it is to be located, and why a thing to us pleasing is to another displeasing."[8] By Euler's time, music was well on its way to its present status as a fine, rather than liberal, art, grouped with painting and architecture rather than with mathematics. Not satisfied with the classical accounts, Euler wanted to find new principles connecting mathematics with music and pleasure.

Euler begins his *Tentamen* by reviewing his earlier work on the physical basis of sound, understood as "the perception of successive pulses which occur in the air particles situated around the ear."[9] He reviews the mathematics of strings, vibrating bodies in general, and his own "entirely new theory of sounds provided by wind instruments."[10] Though he takes note of the Pythagorean teachings about musical ratios, he seeks to put them on a new mathematical basis. As he noted in a 1752 letter to the great composer and theorist Jean-Philipe Rameau, "the Pythagoreans were early misled in their numbers and treated them capriciously, as when they maintained that only superparticular ratios furnished consonances, a principle devoid of all foundation, and in this regard the Aristoxenians were right to mock their false theory."[11] Where Boethius had assumed that simple ratios like 1:2 (octave) were *more perfect* than complex ones like 243:256 (semitone), Euler wished to demonstrate that they were *more pleasurable* and to calculate the exact degrees of pleasure involved.

The difference in these fundamental categories reveals the profound shift toward an aesthetics premised on sentiment and pleasure, rather than pure order and its concomitant goal of moral perfection. In his reply to Euler, Daniel Bernoulli expressed some puzzlement:

I cannot readily divine wherein that principle should exist, however metaphysical, whereby the reason could be given why one could take pleasure in a piece of music, and why a thing pleasant to us, may for another be unpleasant. One has indeed a general idea of harmony that it is charming if it is well arranged and the consonances are well managed, but, as it is well known, dissonances in music also have their use since by means of them the charm of the immediately following consonances is brought out the better, according to the common saying *opposita juxta se posita magis elucescunt* [opposites placed together shine brighter]; also in the art of painting, shadows must be relieved by light.[12]

Bernoulli shares the common presupposition that pleasure is fundamental, but he does not see how it could be mathematically or metaphysically grounded beyond itself. As if expressing the widely shared admiration of sentiment, Bernoulli relies on purely intuitive notions of pleasure through contrast, which his example from painting underlines; if so, the fine arts all share this fundamentally nonmathematical reliance on contrast. His invocation of charm sustains this aesthetics of ineffable sentiment.

Euler seeks to combine what his friend considered two rather antithetical approaches. To find a mathematics of sentiment, Euler had to make a new construct, for the traditional accounts did not seek to bridge these two; with few exceptions to this day, he was a lone pioneer of mathematical aesthetics.[13] He founds his new theory on "the exact knowledge of sound," understood from the mechanics of waves, and on "metaphysics": "Led by reason as well as experience, we attacked that problem and drew the conclusion that two or more sounds are pleasing when the ratio, which exists between the numbers of vibrations produced at the same time, is understood; on the other hand, dissatisfaction is present when either no order is felt or that order which it seems to have is suddenly confused." To make this quantitative, "we graded this perceptive ability in certain degrees, which are of greatest importance in music and also may be found to be of great value in other arts and sciences of which beauty is a part. Those degrees are arranged in accordance with the ease of perceiving the ratios, and all those ratios that can be perceived with equal facility are related to the same degree." He calls this their *degree of agreeableness* (*gradus suavitatis*), using a Latin word that might also be translated as *sweetness, charm,* or *tunefulness*.[14]

Though Euler's exposition of these degrees may remind us of the ancient ordering based on the *perfection* of intervals, his definition reminds us that he ranks "how much agreeableness each consonance has in itself or, what amounts to the same thing, how much facility is required for perceiving it." Where the ancients had placed the priority on the intervals and ratios themselves, Euler now places it in the perceiving human subject. Still, his prior mathematical sense of the relative simplicity of various ratios informs his ensuing definitions (see box 9.1 for details).

Box 9.1
How Euler constructed his degree of agreeableness

> First, he assigned degree 1 to 1:1 and degree 2 to 1:2, which sets the basic pattern: "by the simple operation of halving or doubling, the degree of agreeableness is changed by unity." Then to ratios of the form $1:2^n$ he assigned the degree $(n + 1)$ because "the degrees progress equally in ease of perception. Thus, the fifth degree is perceived with more difficulty than the fourth," and so on. For ratios of the form $1:p$, where p is prime, he assigns the degree p; thus, both 1:3 and 1:4 have degree 3, to accord with both principles he used thus far. He then argues that $1:pq$ (where both p and q are prime) has degree $p + q - 1$. A few more steps led him to the general conclusion that for any composite number m composed of n prime factors whose sum is s, the ratio $1:m$ has the degree of agreeableness $s - n + 1$. Based on this, he then argued that the degree of a series of proportions such as $p:q$ or $p:q:r$ (where p, q, r are primes) is the same as that of $1:pq$ or $1:pqr$ respectively, where Euler calls the least common multiple of these primes the *exponent* of that ratio. Hence, he assigned to $1:pqr$ and to $p:q:r$ the same degree, $p + q + r - 2$. Thus, the fifth (2:3) has degree $(5 - 2) + 1 = 4$, the same as the degree of 1:6. A major triad C–E–G (4:5:6) has the same degree as $1:4\cdot5\cdot6 = 1:120$, whose exponent $60 = 2^2\cdot3\cdot5$, $s = 12$ and $n = 4$, thus the degree is $s - n + 1 = 9$, the same as the dissonant major seventh C–E–G–B, 8:10:12:15, whose exponent is also 60. A minor triad like E–G–B (5:6:7) has exponent $210 = 5\cdot6\cdot7$, so $s = 18$, $n = 3$, and degree $s - n + 1 = 16$.

Euler illustrates his reasoning with a diagram (figure 9.1) showing "the pulses in the air as dots placed in a straight line. The distances between the dots correspond to the intervals of the pulses." He sees this diagram as visualizing their degree of understandability and hence agreeableness. At the same time, though, this diagram represents the coincidences between the "pulses" and hence represents geometrically the interrelation between sound waves. Implicitly, Euler's two different meanings converge: agreeableness correlates with the alignment of the two wave-forms, which Hermann von Helmholtz made explicit in his physical theory of consonance over a century later (with due acknowledgment to Euler).[15] In his *Tentamen*, Euler restricted himself mainly to the traditional just intonation using simple whole-number ratios, not the newer temperaments intended to allow free modulation between all keys (see box 4.2).

Within these limitations, Euler's quest for a precise degree of agreeableness informs his mathematical rankings. In light of this, he chooses the degree always to be integral, never fractional, "since in this case the ratio would be irrational and impossible to recognize," implying an underlying rationality to the felt quality of agreeableness. He sets out the result in a table that goes far beyond the traditional set of musical ratios (figure 9.2). Euler's mathematical schema leads him to include ratios that have no precedent in traditional music theory; Zarlino, for instance, argued that only numbers up to six (the *senario*, as he called them) are allowable in musical ratios, but Euler makes a case for going past six. In so doing, Euler makes consonance and dissonance really a matter of degree, as opposed to the traditional tendency to distinguish sharply between them. He is led to this notably innovative step by his mathematics, which phrases both in the same general language of ratios, as well as by his awareness of the expressive power of dissonance.

Euler thus found a new numerical index that, to some extent, correlates with traditional (and aural) judgments of relative consonance but is far more precise: the lower the degree, the more agreeable the sound. Yet in his system an interval between *two* notes can have the same degree as a *triad*, which has a more fundamental status in harmony. Worse, Euler's scheme assigned the same degree to the most familiar triadic harmony (like C–E–G) as to the dissonant major seventh chord (C–E–G–B) (♪ sound example 9.1).[16] Still, Euler's numerical rankings illuminate a long-standing theoretical problem: the status of the minor mode. After its discovery by Mersenne and Descartes, music theorists realized that the overtone series provided a natural justification for the major mode because the first six overtones sound a major triad. Yet the minor mode had no such acoustical justification. Further, why does the major mode sound "happy," the minor mode "sad"? From where, exactly, does the minor mode derive its origin and its emotive power?

Euler argues that "everything pleases us in which we perceive perfection to exist, and so we are pleased more when we observe more perfection. On the other hand, we are displeased by those things in which we perceive a lack of perfection or much imperfection."[17] Hence, we should absolutely prefer music that keeps to the lowest degrees of agreeability (in his scale) and be displeased by any deviation toward the higher degrees.

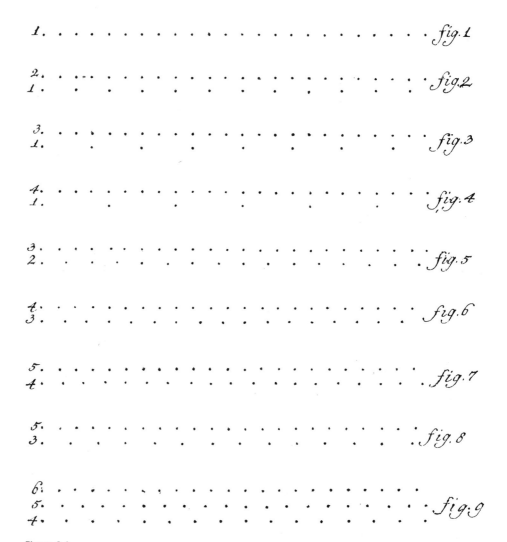

Figure 9.1
Euler's diagram visualizing the relative agreeableness of various simple ratios of sound pulsation, from his *Tentamen* (1739).

Gr. II.	2:5.	Gr. IIX.	3:7.	3:64.	1:160.
1:2.	1:18.	1:14.	1:25.	1:256.	5:32.
Gr. III.	2:9.	2:7.	1:28.	Gr. X.	1:162.
1:3.	1:24.	1:30.	4:7.	1:42.	2:81.
1:4.	3:8.	2:15.	1:45.	3:14.	1:216.
Gr. IV.	1:32.	3:10.	5:9.	6:7.	8:27.
1:6.	Gr. VII.	5:6.	1:60.	1:50.	1:288.
2:3.	1:7.	1:40.	3:20.	2:25.	9:32.
1:8.	1:15.	5:8.	4:15.	1:56.	1:384.
Gr. V.	3:5.	1:54.	5:12.	7:8.	3:128.
1:5.	1:20.	2:27.	1:80.	1:90.	1:512.
1:9.	4:5.	1:72.	5:16.	2:45.	
1:12.	1:27.	8:9	1:81.	5:18.	
3:4.	1:36.	1:96.	1:108.	9:10.	
1:16.	4:9.	3:32.	4:27.	1:120.	
Gr. VI.	1:48.	1:128.	1:144.	3:40.	
1:10.	3:16.	Gr. IX.	9:16.	5:24.	
	1:64.	1:21.	1:192.	8:15.	

Figure 9.2
Euler's table of the first ten degrees of agreeableness of musical intervals.

But when we calculate the degrees associated with major versus minor triads, in general and in the most common keys, the major triads are of *lower* degree (see box 9.1). If so, we should always prefer music with major triads and avoid minor ones, which "will be almost painful."[18] Yet Euler knows that music contains many minor triads, and he points out that "the more easily we observe the order in a given thing, the simpler and more perfect we consider it, and therefore we receive pleasure and delight from it. On the other hand, if the order is discerned with difficulty and seems less simple and distinct, we perceive something like sadness [*tristitia*]. In either case, as long as we sense order, the given object pleases, and we conclude that the object has agreeableness."[19] Euler thus connects the greater "difficulty" of minor intervals (and higher-degree "dissonant" intervals) with their perceived affect of *sadness*. Hence, he explains the sadness of the minor mode (for instance) as the direct correlate of its *epistemological* status: what is harder to know is felt to be sad simply because we struggle to discern its order. In that sense, sadness seems to be the felt effect of the pain we experience in the face of cognitive dissonance.

Euler interprets this sadness as part of the larger project of the pleasure conveyed by music, which includes both happiness and sadness. He connects the mathematics of sadness with the experience of drama: musical harmonies "are like comedies and tragedies all of which should be filled with agreeableness. The comedy should fill the spirit with joy and the tragedy should convey sadness. Thus it is clear that something can please and evoke joy, and something else can please and bring sadness."[20] His paradoxical context points to Aristotle's discussion of the "joy of tragedy," the tragic pleasure that purifies the soul through pity and fear.[21] In Euler's account, sadness and joy both have a precise numerical degree: mathematics is capable of rendering these seemingly incommensurable mental states literally commensurable by calculating their common measure. The sadness we experience hearing minor harmonies is not simply dismay at "contemplation of the imperfect," as we would feel hearing a blatantly wrong note. Instead, we locate the minor harmonies (and dissonances in general) as parts of the larger perfection of the whole musical edifice. "For music, since it tries to please, neither intends nor is capable of much sadness. Thus sadness simply involves more difficult perception of perfection or order and differs from joy only in degree."[22] By providing commensurate degrees for both joy and sadness, Euler shows the basis on which the mind can integrate and reconcile them in the overall pleasure conveyed by the entire musical work. In so doing, he also illuminates the nature of *tristitia* by revealing it as the sensation of the mathematical mind laboring to understand difficult ratios.

Even as it struggles, the mind experiences the concomitant pleasure of connecting its complex labors with the relative resolution felt in simpler states, which are then perceived as joy. This does not mean that we simply suffer through the sad parts in order to enjoy the relief of their ending, as if they were a kind of toothache whose passing gives us the relative pleasure of anguish ended. Though he does not spell it out, Euler's argument clearly implies that a mind capable of contemplating complex things has a more intense response to the work of music than would be felt by a less percipient—and less intelligent—hearer.

This analysis of mental mathematical activity also informs Euler's parallel inquiry why "barbarians get little or no enjoyment from our music," whether that pleasure comes "of familiarity alone" or because "there is far more order and agreeableness in our music, of which only the least part is perceived by the barbarians."[23] Though he acknowledges the power of familiarity, ultimately his argument puts much more weight on the trained mind's ability to discern order. More complex ratios may indeed weary and sadden us, but they are the indispensable grounds for our experience of joy, which we know through our *mathematical*, even calculational, faculties. Euler thus identifies mathematical awareness as the core of our ability to experience joy and sadness, whose inherent nature in fact requires the connected experience of both. The young mathematician here seems to anticipate and to welcome his coming lifetime of struggle and triumph, with all the sadness and joy his endeavors will entail. He implicitly places these under the aegis of music by treating "music as a part of mathematics," connected just as are joy and sadness.

In the remainder of his *Tentamen*, Euler gives evidence of such mutual interactions between music and mathematics. At many points, he goes into considerable musical detail.[24] Not content only to schematize degrees of agreeableness, he gives detailed theoretical examples of increasingly complicated harmonic structures to illustrate his fundamental concepts (figure 9.3). To the more complex of these harmonies, Euler adds a sort of figured bass notation. When played in order, these complex harmonies become progressively more audacious, even weird (♪ sound example 9.2); indeed, his "figured bass" is more a kind of shorthand notation than anything conforming to the musical usage of his time.

To simplify calculations in his *Tentamen*, Euler was one of the first to treat musical ratios with logarithms, which reduce multiplication to addition and division to subtraction.[25] This *musical* application then induces Euler to take a new *mathematical* step, because expressing a logarithm's magnitude calls for the use of irrational numbers, in general.[26] For instance, using logarithms to calculate the ratio of the octave to the fifth, Euler gets decimals, which he then converts to the expression

$$1 + \cfrac{1}{1 + \cfrac{1}{2 + \cfrac{1}{2 + \cdots}}}.$$

He can then obtain approximations by truncating the denominator of this *continued fraction* at successive points downward.[27] While preparing for the publication of his *Tentamen*, Euler wrote "On Continued Fractions" (1737), the first sustained treatment of this new kind of mathematical object.[28] He realized that continued fractions, as they emerged in his musical treatment, were ideal arenas for considering irrational numbers, each of which turns out to correspond to a *unique* continued fraction, which displays the inner structure of that number in a different (and often more perspicuous) way than its decimal expansion. On the other hand, Euler demonstrated that the converse is not true, for it is possible to express any ordinary (rational) fraction as a continued fraction.[29] Among irrational quantities, the celebrated "golden ratio" φ has the beautifully simple form

$$\varphi = 1 + \cfrac{1}{1 + \cfrac{1}{1 + \cfrac{1}{1 + \cdots}}},$$

which exposes an inner structure not manifest in its decimal expansion, $\varphi = \frac{1+\sqrt{5}}{2} = 1.6180339887\ldots$. Then too, that most familiar of irrational numbers has the beguiling expression

$$\sqrt{2} = 1 + \cfrac{1}{2 + \cfrac{1}{2 + \cfrac{1}{2 + \cdots}}},$$

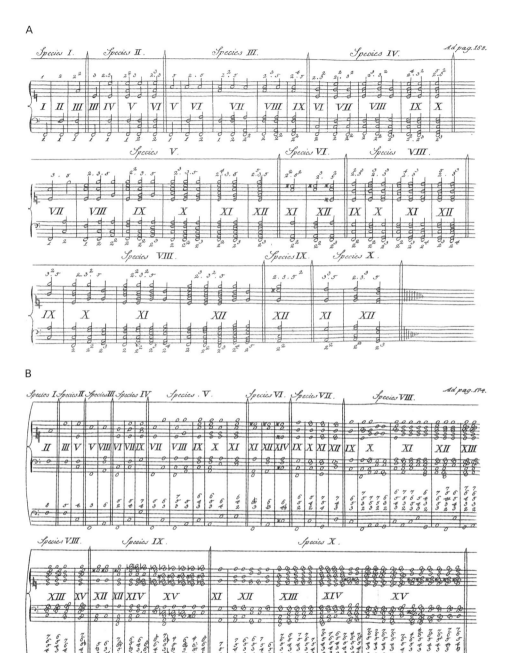

Figure 9.3
(a) Euler's musical illustration of the first eight species of harmony, according to his degrees of agreeableness.
(b) Species I–XV, with Euler's figured bass notation of the harmonies (♪ sound example 9.2).

in which the ever-recurrent 2s seem to echo the initial 2 from which $\sqrt{2}$ is drawn. These two examples show something of the visual poetry of continued fractions, which powerfully symbolize and expose the infinite processes and relations that form the inner structure of irrational numbers.

In "On continued fractions," Euler gave the first proof that $e = 2.71828182845904...$, the base of the natural logarithms (and the crucial constant describing exponential growth or decay), is in fact irrational, which had been suspected but not proved. To accomplish this important step, he showed that it could be written as a continued fraction:

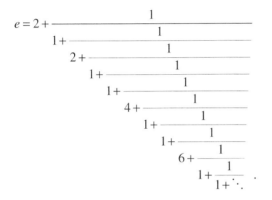

Here one wonders what the pattern of *even* integers (2, 1, 1, 4, 1, 1, 6, 1, 1, 8, 1, 1, …) spaced regularly along the diagonal has to do with this *irrational* quantity; Euler's algebraic deductions do not give an intuitive meaning for this pattern, though they implicitly generate it.[30]

Nor were the mathematical effects of his musical work restricted to this one technique. Though Euler's name later became so closely associated with number theory, his interest in this field began *after* his earliest work on music, such as his 1726 notebook entries. Only after his arrival in St. Petersburg in 1727 and his subsequent correspondence with Christian Goldbach (who moved to Moscow shortly after Euler's arrival) did Euler's interest in number theory really begin, during the period of his greatest activity preparing the *Tentamen*. For instance, in December 1729, Goldbach wrote Euler to ask him whether "Fermat's observation [is] known to you, that all numbers $2^{2^n}+1$ are prime? He said he could not prove it; nor has anyone else done so to my knowledge." Euler's rather indifferent response indicates that, even by that date, he was not greatly interested in this fundamental question. Only after a subsequent letter from Goldbach prodding him did Euler catch fire and disprove Fermat's conjecture by showing that the fifth Fermat number, $2^{2^5}+1 = 4,294,967,297$, is evenly divisible by 641.[31]

After that, Euler read Fermat ever more closely and took up number theory with particular passion. His first result already underlines his phenomenal abilities as a calculator;

only as a result of that special skill, combined with his mathematical acumen, could he have achieved such a factorization, long before computers or any other mechanical calculators. The same fascination with the pure manipulation and calculation of numbers also pervades his musical *Tentamen*, which contains many tables of numbers that have some importance in his musical scheme. Study of Euler's early notebooks (around 1726) shows that, as he prepared his *Tentamen*, he was not then aware of Leibniz's (1714) view that "music charms us, even though its beauty consists only in the harmonies of numbers and in a calculation, which we do not perceive but which the soul nevertheless carries out, a calculation concerning the beats or vibrations of sounding bodies, which are encountered at certain intervals."[32] After he met Goldbach, Euler became aware of these views, which Leibniz had described to Goldbach in a letter of April 1712.[33] Though Euler conceived his musical theories independently, Leibniz's writings supported them, for Euler had set out a precise scheme whereby the soul might accomplish its musical counting quite consciously. Given Euler's staggering calculational abilities, including lightning mental computations, one can readily imagine that he himself may have been able to compute what he was hearing, perhaps even in "real time." At the least, his *Tentamen* contains his retrospective account of musical awareness in terms of explicit arithmetic.

The juxtaposition of the musical and arithmetical concerns in Goldbach's correspondence with Leibniz helps underline the many ways in which these two themes arguably overlapped and intersected in Euler's mind through his interchanges with Goldbach. Yet even before then, Euler's absorption in the intricate arithmetic of his music theory provided the fertile ground on which his ensuing interest in number theory grew. The modern concept of "pure mathematics" should not blind us to the many ways in which, in Euler's time and before, no hard barrier separated it from the "applied" branches of what we now call physics, engineering, music theory—disciplinary names that he would neither have known nor separated absolutely. Nothing would have been more natural for Euler than to follow his intricate musical arithmetic into the further studies of the properties of numbers that only came to be called "number theory" in the aftermath of his own work.

Looking back to the *Tentamen*, many of Euler's musical arguments directly imply arithmetical problems that lead straight to the more general questions he later addressed about the properties of numbers. His definition $s - n + 1$ for the *gradus suavitatis* of a musical interval involves counting the n prime factors of the interval's exponent and their sum s (box 9.1), which are central topics in his ensuing number theoretical work. The Pythagoreans had begun to investigate perfect numbers (each equal to the sum of its proper divisors, such as $6 = 1 + 2 + 3$) and pairs of amicable numbers, for which each is the sum of the other's proper divisors, such as 220 and 284.[34] Both types of numbers became important for Euler, but he had already laid the groundwork for their study in his *Tentamen*. In 1747, Euler published thirty new pairs of amicable numbers, compared to the four pairs previously known, listing them in a format that recalls his diagrams ranking musical intervals in his *Tentamen*.[35]

> **Box 9.2**
> Euler and the harmonic series
>
> Following the ancient definition of a harmonic mean (box 1.1), Oresme proved that the *harmonic series* $\sum_{k=1}^{\infty} \frac{1}{k} = 1 + \frac{1}{2} + \frac{1}{3} + \frac{1}{4} \cdots$ diverges. Euler proved that $\sum_{k=1}^{\infty} \frac{1}{k} = \prod_{p} \frac{1}{1-(1/p)}$, where the product on the right-hand side goes over all prime numbers $p > 1$. Euler also showed that in general $\zeta(s) = \sum_{k=1}^{\infty} \frac{1}{k^s} = \prod_{p} \frac{1}{1-(1/p^s)}$, whose name $\zeta(s)$ came from Bernhard Riemann, who brought the properties of this "Riemann zeta function" to the center stage of mathematics. These expressions relate the *sum* over the reciprocal of each number 1, 2, 3, … (raised to the power $s > 1$) to the *product* only over the prime numbers, indicating the deep structure whereby the primes underlie all other numbers.

Euler went on to conduct many other inquiries into amicable and perfect numbers, among a vast variety of topics related to the abundance of prime numbers of different kinds, including his profound relation between the harmonic series and the prime numbers (box 9.2). To be sure, Euler does not make any explicit connection between the harmonic series and harmony, but he knew well that this series had its origins in music, for the Pythagoreans already defined the harmonic ratio as a way of mediating between arithmetic and geometric ratios (see box 1.1).[36] In making his arguments for this deep and surprising result, Euler used the tools of analysis—that is, differential and integral calculus—as well as those of traditional arithmetic, not only to find individual results such as this one but to open a whole new field of mathematics. André Weil observed that "one may well regard these observations as marking the birth of analytic number theory," as it came to be called.[37]

The influence of Euler's musical work can also be seen in a very different arena of his activity, his famous solution of the problem of the Königsberg bridges. In that city (now called Kaliningrad in Russia), the island Kneiphof in the river Pregel joins various parts of the city via seven bridges (figure 9.4). Euler became aware of the "quite well-known problem" whether someone could take a walk that would return to its starting point after crossing each of the seven bridges only once. His letters reveal that, even in 1736, he considered the problem "banal" because its solution "bears little relationship to mathematics, and I do not understand why you expect a mathematician to produce it, rather than any one else, for the solution is based on reason alone, and its discovery does not depend on any mathematical principle."[38] Euler's distancing of this problem from what he considered "mathematics" helps clarify the new step he made by considering it (as he puts it in his 1736 paper) an example of a branch of geometry "that has been almost unknown up to now; Leibniz spoke of it first, calling it the 'geometry of position' [*geometria situs*]. This branch of geometry deals with relations dependent on

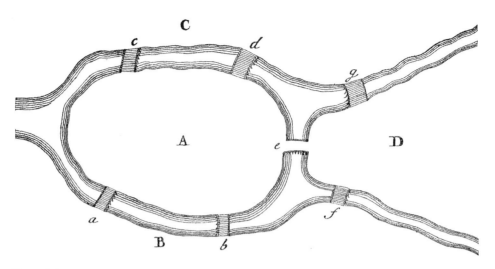

Figure 9.4
Euler's diagram of the city of Königsberg, the Kneiphof island (A) and the seven bridges over the River Pregel, $a, b, \ldots, g$.

position alone, and investigates the properties of position; it does not take magnitudes into consideration, nor does it involve calculation with quantities."[39] Euler's letter of 1736, though, shows his puzzlement as to what this *geometria situs* really means: "You have assigned this problem to the geometry of position, but I am ignorant as to what this new discipline involves, and as to what types of problems Leibniz and Wolff expected to see expressed in this way." Euler's paper was first presented in 1735; the field became known as *analysis situs* and is now called topology, of which this paper is one of its first great results.

Euler immediately generalized the Königsberg problem to "any configuration of the river and the branches into which it may divide, as well as any number of bridges, to determine whether or not it is possible to cross each bridge exactly once," which has come to be called an *Euler walk*.[40] He reduced topography to alphabetic symbolism and derived simple rules, though without defining a numerical index that would "involve calculation with quantities," as he put it.

Euler did devise such an index when he returned to the "geometry of position" in his "Elements of the Doctrines of Solids" (1752), the first of two papers in which he studied the relations between the number of vertices (V), edges (E), and faces (F) of a polyhedron (figure 9.5).[41] Euler's crucial innovation was defining the edge (*acies*) of a polyhedron, which, curiously enough, had never before been stated. Euler also identified the polyhedron's faces (*facies*) and its *angulus solidus*, by which he means not "solid angle" (as a subtended, finite angle) but the point from which such an angle emerges, only called

Euler: The Mathematics of Musical Sadness

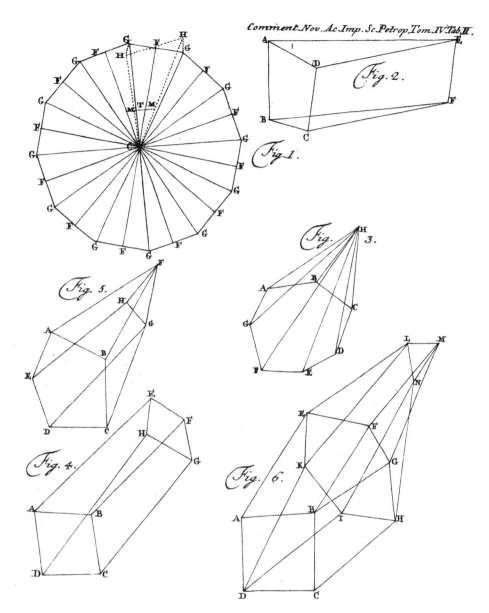

Figure 9.5
Euler's 1752 illustrations of various polyhedra.

a "vertex" by Legendre (about 1794). If a solid polyhedron (closed but not necessarily regular) is bounded by planar faces, Euler argued that "the sum of the number of solid angles plus the number of faces exceeds the number of edges by 2," or $V + F - E = 2$ (to state it algebraically, which he does not do), now widely called "Euler's formula." Though his arguments are somewhat flawed, the truth and depth of this proposition make it one of his most celebrated results. It represents a three-dimensional generalization of the Königsberg bridge problem, in which the requirement of *closure* for the solid polyhedron corresponds to the *connectedness* of an Euler walk, its return to its starting point.[42] By identifying and tabulating V, F, and E, Euler was able to calculate the index $V + F - E = 2$ that also characterizes such polyhedra.

The structure of this relation between vertices, edges, and faces is strikingly similar to the structure of the degree of agreeableness of musical intervals, $s - n + 1$. Without intending any direct connection between polyhedra and Euler's hierarchy of musical intervals, as such, both these relations ($V + F - E = 2$ and $s - n + 1$) give the kind of general categorization we now think of as *topological* and which Euler thought of in terms of *geometria situs*. To be sure, these relations are very different, and not just in the objects they describe. Euler's formula is an *equation* describing a necessary and sufficient condition for closed, convex polyhedra; his formula for musical degree defines a hierarchy between different intervals. They both pose a general schematization that categorizes a vast domain, of polyhedra or of musical intervals, respectively, subsuming many different individuals under a larger genus. Thus, polyhedra of many different shapes and numbers of sides fall under Euler's formula, which (as modern topology phrases it) describes polyhedral surfaces of *genus 0*, those having no "holes" or "handles." Later topologists generalized Euler's formula to manifolds of higher genus than zero (such as a doughnut whose hole gives it genus 1, for instance) by defining the *Euler characteristic* $\chi = V + F - E$. In this way, χ gives a "degree" of such surfaces that is analogous to the musical degree $d = s - n + 1$, which gives the "topology" of musical intervals, their general grades of classification.[43]

Thus, Euler's early work classifying musical intervals grouped different intervals under a single degree, expressing a higher commonality among them, despite their differences. He did so without any earlier precedent in mathematics, for the traditional hierarchy of musical intervals was based on fairly arbitrary numerological criteria of "simplicity."[44] Euler's degrees group together intervals by cutting across these traditional classes; his criterion for setting up his degrees is freely chosen according to his notions of what would be more "intelligible" and hence more "agreeable" (*suavis*). In his musical work, Euler first devised the general classificatory strategy that he then applied to the bridge problem and later to polyhedra. To use a later mathematical term, his approaches in these cases were *isomorphic*, that is, they had the same essential structure. Because the musical example came first, it arguably was the arena in which he first found and applied the kind of approach that he later (and perhaps without realizing it) then found appropriate to bridges and polyhedra.

Euler thus discovered not just the first important insights that later grew into the field of topology but also, more deeply, *indexing* as a crucial (and novel) tool of what became *the topological approach itself*. Music was a particularly appropriate first venue for this new topological thinking because musical intervals do not have the kind of spatial structure that seems to govern elementary geometry. The lack of visible evidence—and his judgment of the insufficiency of the traditional criterion of "simplicity" of ratio—opened the door to his definition of degree, tied to the sensual criteria of *suavitas*. After Euler took this initial step away from the traditional givens of mathematics, such as pure ratio, it was probably easier to think in essentially the same way when he came to the bridge problem and then to polyhedra. To be sure, the concept of degree was already familiar in the realm of algebra, such as the degree of a polynomial equation. But the example of music required a still bolder application of this general concept of degree in a case where it has no obvious prior meaning, unlike the algebraic degree of an equation that is manifest in the highest power of the unknown.[45] In the cases of music, bridges, and polyhedra, Euler had to devise a degree for each that would have a decisive, *invariant* significance; this required him to discern from a number of surface details those that could constitute the kind of parameter that would answer his questions. For Euler, musical questions opened the way to a new mathematics.

10 Euler: From Sound to Light

Besides his enormous achievements in mathematics, Euler was deeply involved in many areas of physics. His early work on music had a direct bearing on his study of sound, which in due course contributed to his studies of the mechanics of continuous bodies, the transmitters of sound vibrations. These important advances in continuum and fluid mechanics also moved Euler to advocate a wave theory of light, as against Newton's emission (particle) theory. Throughout, Euler used the examples of sound and music as exemplars for a new understanding of light and color.

In the century after his seminal work on optics, Newton's theories remained the locus of considerable controversy. On the Continent, his work found supporters as well as notable critics. Leibniz, for one, was impressed but advocated careful repetition of Newton's experiments. The prolific Christian Wolff became the chief popularizer of Newton's theory in German lands, but many scholars were more attracted to versions of Descartes's theory of a vibrating medium or ether that pervaded space.[1] Nor can such medium theories be too sharply distinguished from Newton's, who also argued for a "subtle ether" that would fill space but not retard the heavenly bodies, though he considered light an emission phenomenon rather than a state of the ether as such.

Others within the Cartesian tradition took the idea of a light-bearing medium in quite different directions. For instance, in 1690 Christiaan Huygens considered light to be a sequence of pulses traveling at a finite velocity within the medium. The word "pulse" here should be distinguished from "wave" because Huygens dismissed the possibility that the pulses follow each other at regular intervals, as would wave fronts. Yet some of his concepts carried over to the later wave theory, most notably the formation of collective fronts of pulses or waves through what now is called Huygens's principle (figure 10.1). Though some accepted Huygens's theory because of its account of the perplexing phenomenon of double refraction—the passage of two different light rays at two different speeds through a calcite crystal—others had trouble with his account of how simple rectilinear propagation could be reconciled with ever-spreading circular pulse fronts. Most troubling, his theory gave no account of colors; in the half century after 1700, it fell from sight in research publications, though it was noted favorably in German textbooks.

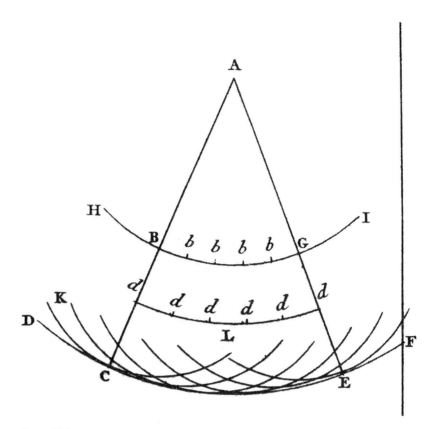

Figure 10.1
Huygens's principle, showing the constructive interference of pulse fronts along the line *DCEF*, from his *Traité de la lumière* (1690).

Still others offered various attempts to synthesize the Newtonian emission theory with some kind of medium, most notably Johann II Bernoulli, Euler's childhood friend and son of his Basel mentor. In 1736, the younger Johann submitted to the Paris Académie a physical and mathematical investigation into the propagation of light. Johann II considered the ether to be filled with infinitely many tiny vortices, interspersed with hard, small particles that are sent into longitudinal vibrations about their equilibrium positions as a result of compression transmitted through the vortices. This resulted in what he called a "light fiber" (*fiber lumineuse*) stretching along the line of particles, its luminosity propagated from one particle to the next. The first particle in the series determined the way the light would travel down the fiber, in which particles much larger or smaller than the first would be unable to follow the traveling motion of the light ray. Bernoulli compared this process to the resonance of a string instrument in sympathetic vibration, responding to vibrations of another nearby string tuned at the same pitch.[2]

Bernoulli's analogy shows the power of the example of musical sound as a formative influence in the emergent wave theory of light, in which his theory was an interesting attempt at a synthesis of Newton's concept of color-making particles with a Cartesian vortex-filled medium. Bernoulli's use of the analogy with sound followed other attempts to apply that analogy, including (as we have seen) Newton's own. Bernoulli may well have drawn on the earlier work of Dortous de Mairan, whose theory of sound had moved the other direction, by analogy with Newton's color-making particles, to propose the existence of air particles that could transmit only one specific tone.[3]

Though he rejected Huygens's principle and did not discuss interference effects, Bernoulli noted that sound can travel obliquely, compared to the rectilinear propagation of light. In that sense, he follows Newton in denying any positive evidence of wavelike behavior of light. But where Newton used the same model for the medium of light as for sound, Bernoulli considered their media to be fundamentally different. He worked out the mathematics of the displacements of his fibers, noting especially the comparison between the *longitudinal* motion of light along a fiber and the *transverse* displacements of a vibrating string (that is, its motion perpendicular to itself). Here again, he makes use of what he considers a "far-reaching similarity between the motion of the string of a musical instrument and that of a fiber."[4] These two perpendicular kinds of propagation in continuous bodies will return at several points in later developments; for Bernoulli, their differences and similarities emerged in the context of musical instruments.

When Euler brought forward his own theory of light, "the most lucid, comprehensive, and systematic medium theory" of his century, he first of all posed it on the analogy with sound, rather than in the context of the long-standing debates between emission and medium theories.[5] He first announced his theory in a 1744 lecture to the Berlin Academy, "Thoughts on Light and Colors," whose opening section announces the analogy on which he builds:

There is such a great connection [*rapport*] between light and sound that the more one studies the properties of these two objects, the more one discovers resemblances. Light and sound both come to us in straight lines if nothing impedes their movement, and if there are obstacles, the resemblance does not cease to hold. For as we often see light by reflection or refraction, these two things are found also in the perception of sound. In its echoes we hear sound by reflection, in the same way that when we see images in a mirror, the refraction of light is the passage of rays through transparent bodies, which always produce some change in the direction of the rays; the same thing is found with sounds, which often pass through walls and other bodies before reaching our ears, so that the walls and other similar bodies are in connection with sound the same as transparent bodies are for light. ... So great a resemblance does not allow us to doubt that there is such a harmony between the causes and the other properties of sound and of light, and thus the theory of sound will not fail to clarify considerably that of light.[6]

Apparently, this analogy was so coolly received by its first audience that, when Euler came to write his extended presentation of his "New Theory of Light and Color" (1746), he made substantial changes to his argument by foregrounding the well-known controversy between medium and emission theories as the ground on which his analogy, and his ensuing new theory, stood. Still, his original statement clarifies the preeminent significance of this analogy in his thinking.[7] As he clarified his argument, Euler also sharpened the contrast between medium and emission theories, which until then had often (as we have seen) been combined in various ways, rather than being considered entirely exclusive of each other. Euler used the analogy with sound to ground his polemic for his medium theory.

To do so, Euler addressed head on a difficulty that probably led his 1744 audience to doubt his argument, the very difficulty Newton had emphasized: whereas sound pulses entering a room through an opening penetrate the whole space, light rays do not seem to behave similarly. Euler's reply reasserts the power of the analogy with sound, which he contends that Newton misunderstood. Euler argues that Newton ought to have compared the optically opaque room with an acoustically "opaque" barrier because sound can pass even through normal walls, not just through an opening between rooms. Only a perfectly soundproof barrier could rightly be compared with a visually opaque wall, though Euler acknowledges the practical difficulty in constructing such an ideal acoustic barrier. His insistence on the sound/light analogy brings forward the precise limitation of Euler's views that Young will address in the next chapter. Euler asserted that sound had not been observed to spread out laterally in a room but propagates linearly, as does light. In his view, one sound pulse could not prevent or interfere with another so as to allow sidewise spreading. Indeed, the question of the spreading of sound propagated through a medium remained controversial for the next half century; Euler was relying on the absence of what he considered sufficient evidence for spreading, but he also seemed to have used the analogy in reverse, applying the straight-line propagation of light rays to sound.

This example shows the multiple possibilities that lay within the application of the sound–light analogy. Rather than imposing the later perspective that Euler simply erred, my point is more that the example of sound was so strong for him (as for Newton himself) that it remained potent even when its conclusions may seem paradoxical. Yet there was no paradox for Euler: his argument about sound-transmitting walls seemed to answer Newton's assertion neatly, nor did Euler have any positive information on sound propagation that would have differed from what was common knowledge about the rectilinear propagation of both sound and light. Euler fortified his argument with several others against Newtonian emission (such as that it would deplete the sun's matter and would preclude the possibility of transparent materials). Consistently, he found no evidence that a beam of sound or of light could interfere in any way with another such beam, which an emission theory would portray as the collision between the emitted particles. Ironically, this important advocate of wave theory seemed to overlook what, in the sequel, seemed to others its most salient feature, interference. He did so as much because of his arguments for the lack of interference between sound sources as because of his arguments for the lack of interference between light beams: above all, he maintained the *analogy* between them.

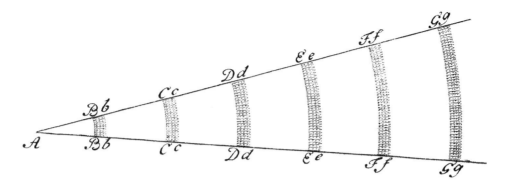

Figure 10.2
Euler's diagram of the propagation of a sequence of pulses, from his "New Theory of Light" (1746).

On the other hand, Euler's arguments brought forward other aspects of waves that are no less consequential and cannot be taken for granted. He took the pulse theory as it had developed in the work of Descartes and Huygens but changed it by adding a new element of *periodicity*, so that Casper Hakfoort cautions us to call it a "periodic pulse theory" rather than simply a "wave theory."[8] This leads to the most innovative part of Euler's theory, which connected this periodicity to the phenomenon of color. In this development, already his earliest auditors remarked that "what is special in the hypothesis of Mr. Euler is its parallel between sound and light."[9] Euler specifically applied this analogy to extend the propagation of pulses from sound to light. Beginning with his 1727 calculation of the speed of sound (in his youthful "Dissertation on Sound"), Euler now applied a similar argument to light, considered as pulses in a "subtle" ether that has a finite, if small, density and hence a finite velocity of propagation, as opposed to Descartes's and Huygens's instantaneously propagating medium. By comparing the ratio of known sound and light velocities with his calculations about the strength of materials, Euler arrived at very nearly the same ratio of the ether's elasticity to its density from both calculations, which he took as further confirmation of his fundamental analogy and its physical preconditions.

From there, Euler constructed propagating light pulses, which are not yet fully waves but still very close to sound pulses, each considered as a sequence of traveling zones of higher state of motion, whether in air or in ether (figure 10.2). Defining the distance between pulses as d, their frequency f, and velocity of propagation v, Euler gives their fundamental relation $d = v/f$, which is the basis for his account of color. Note that he does not speak of "wavelength"—for indeed these are not waves but traveling pulses—but of the frequency of the pulses. Not a whole "wave" but the individual pulses are primary for him; Euler considers that the pulses can be "isochronic," equally spaced pulses corresponding to a single musical pitch of frequency f, or "nonisochronic" pulses having no single spacing d and hence no single frequency.

Thus, Euler's concept of frequency, perhaps his most important contribution to the theory of light, comes directly not just from the general analogy with sound but from the specifically musical concept of a single pitch: as our perception of high or low pitches depends on how many times a second our ear receives sound pulses, so do our eyes distinguish more or less frequent impacts by their color. When he goes on to address Newton's seminal finding that white light is a composite of many colors, Euler took his musical analogy one step further by explicitly comparing a "composite ray" of light to a multinote chord, under the implicit premise that the eye blends the "notes" of that chord into a single perception of color (say, white), whereas the ear does not blend the chord tones but hears them as separate, though perhaps related harmonically. Euler's ingenious suggestion, however, raises the unacknowledged question: do "composite" light rays themselves really blend or do they keep separate their constituent chordal (and separately pure) "notes?" Yet this bold application of music to harmony represents, for Euler, the power of the sound–light analogy, when taken to its furthest extent.

In the remainder of his treatment of light, Euler continued to use the musical underpinnings of his theory to guide him, especially in difficult cases, such as the problem of the colors of opaque bodies. Newton had somewhat tortuously argued that their colors tended to come from iridescent layers, such as the colors in soap bubbles, a peacock's tail, or thin layers of air (as in Newton's rings; see figure 8.4). Even the blue of the sky was, for Newton, to be understood in terms of such seemingly evanescent phenomena.[10] Euler emphasizes that Newton's examples of iridescence have very different appearances and colors when seen from different angles, whereas most opaque bodies do not exhibit any such iridescence. Instead, Euler compares the vibrating particles of opaque bodies to a number of taut strings, each one resonating only at its own particular frequency.[11] He carries this comparison even further to hypothesize color overtones, analogous to those produced by sounding bodies: "Let us suppose that a ray representing a red color carries f pulses to the eye in one second; and, just as in music sounds are held similar which have vibrations, produced in the same length of time, that bear a double, quadruple, eight-fold etc. ratio [to the main tone], so simple rays containing, in one second, $2f$, $4f$, $8f$ etc. or $\frac{1}{2}f$, $\frac{1}{4}f$, $\frac{1}{8}f$ etc. vibrations, will all be considered red."[12]

Euler offers no proof of these assertions, other than the analogy with musical sounds, which he seems to treat as established fact; one wonders whether he meant to go so far as to include all the higher overtones of a given frequency, not just those corresponding to octaves. Even more striking, he also includes *under*tones, those corresponding to fractional underoctaves below the fundamental frequency f. This he could not have based simply on acoustic theory; following Mersenne, only overtones had been experimentally recognized. Because of this, we must conclude that Euler got his idea of undertones not from natural philosophy as such but from music theory, particularly that of Rameau. Euler corresponded with Rameau, who, though respectful, was critical of Euler's degrees of agreeableness. Euler was surely aware of Rameau's concept of *sous-entendre*, which held that the hearer would "sup-pose" or "hear below" the lowest notated pitch the fundamental bass needed to make sense of an inverted chord, in which that fundamental note may be buried in the middle. In Rameau's own theoretical writings, it is not clear whether he took this *sous-entendre* to be essentially an action of judgment, though perhaps almost unconscious, whereby the hearer gravitates toward the fundamental bass, "the musician's invisible guide, *which has always directed him in all his musical works without his having yet noticed it*."[13] In this reading, *sous-entendre* is an act of *implication*, of discerning a note that is not physically there but whose presence is implied by the other sounding pitches. Alternatively, the *sous-entendre* could be interpreted as responding to a subtle physical phenomenon of undertones, vibrations below the fundamental pitch of a string precisely on the analogy of overtones.

The notion of undertones went on to a long history of its own; as late as 1875, the musicologist Hugo Riemann made delicate nighttime experiments trying to hear the undertones of a piano.[14] Though this extreme view of audible undertones eventually fell into disrepute, its significance here is independent of experimental judgment. Euler was moved to assert color overtones and undertones *simply on the authority of Rameau's harmonic theory* and its agreement with Euler's sense of mathematical symmetry. The only comparable example of a natural philosopher being so swept up by the prior force of musical theorizing may have been Newton himself trying to impose the scale on the spectrum.

Though Euler proposed his color undertones already in his 1744 summary announcement, he reiterated and expanded them in his 1746 "New Theory." Just as a musical octave includes many microtonal pitches, some of which have not been named, so too the color spectrum contains many unnamed colors, along with those whose names reflect their "musical" interrelation. This may have been Euler's version of Newton's musical spectrum; Euler seems to imply, as Newton had, that the well-known colors correspond to the principal notes of the musical scale. In his 1744 "Thoughts," Euler treated the higher "octaves" of color as appearing brighter and more vivid than the lower. By his 1747 "New Theory," Euler is aware of Newton's mistake: the extreme visible frequencies in sunlight differ *less* than by a factor of two, the octave factor Newton had implicitly assumed in his color scale. Thus, Euler recognizes that the naked eye cannot discern an "over-red" that would be an "octave" $2f$ above the frequency of ordinary red, f.

Even though the naked eye could not discern these "derivative colors" that represent octaves above or below those we see, Euler considered that such undercolors or overcolors might have some experimental reality, perhaps observable through the phenomena of resonance or sympathetic vibration: a body resonating to red might also perforce resonate to over-red, which shares the same submultiple frequencies as red. Essentially, Euler's musical theorizing drew him to propose the physical existence of what we now call infrared and ultraviolet light, demonstrated experimentally around only 1800, as we shall see.

Euler's "New Theory" became the most influential work on the theory of light for the rest of the century. He himself returned to various optical topics many times over the succeeding thirty years. For instance, he changed his mind several times about whether red or violet had the higher frequency; his *voltes faces* caused some amusement among savants. But his main initiatives turned from theory to practice; he demonstrated the possibility of a lens that could be free of chromatic aberration, which would be of great importance for all optical instruments, such as eyeglasses, telescopes, and microscopes. This aspect of his work culminated in his magisterial *General Theory of Dioptrics* (1765). His work on acoustics itself continued with studies of the propagation of sound in the atmosphere (1759). During those decades, Euler also returned a number of times to musical questions, demonstrating that his interest was not merely a youthful fancy but a continuing preoccupation alongside his other work in mathematics and natural philosophy.

As with his later work on optics, Euler's interests in music became more practical, devoted more to issues closer to the composition and performance of music than to its theoretical foundations. His 1764 paper "On the True Character of Modern Music" took up a theme already broached in his 1730 *Tentamen*, the possibility of using intervals having the number 7 in their ratios.[15] Euler sets forth a contrast between "ancient" and "modern" music, which he evidently assumed would be clear to his readers. Though he does not specify the exact chronology or stylistic periods, his distinction seems very close to that of Vincenzo Galilei, contrasting the serene polyphonic practice of composers like Palestrina to the expressive, monophonic art of the early operas, reviving the fabled powers of ancient Greek music. Among composers or theorists, Euler refers only to Rameau, a preeminent "modern." Euler's own preferences emerge in his characterization of modern music as "sublime, because its character consists in a higher degree of harmony," compared to ancient music as "common [*commune*]," in the sense of adhering to common harmonic practice. Yet he never cites a single musical example that would give specific insight into his compositional tastes. Disconcertingly, his sole musical example is a formulaic cadence that violates elementary rules of voice-leading by allowing parallel octaves (figure 10.3a; ♪ sound example 10.1). Were these solecisms just typos, or did the great mathematician finally have a tin ear?[16] Or was he quoting crude hymnody he remembered from the Calvinist services of his childhood?

Euler lived in an age of vivid musical controversy, between French and Italian styles, between partisans of ancient and modern practices. A full study of the issues involved would call for another book; in our present context, Euler is offering an implicit defense of the modern practice of Rameau and others by explaining its modernity as a new freedom in the use of dissonance. Euler perceptively locates this modern dissonance in what he himself and his contemporaries were beginning to call the chord of the dominant seventh (figure 10.3b; ♪ sound example 10.2). Euler argued that the aural pleasure of the progressions enabled by this chord justifies its dissonance. Quoting Rameau, Euler notes that this dissonance serves to alert the hearers to the key they are in; in figure 10.3a, the tritone B–F in the penultimate dominant seventh chord tells us that we are in C major

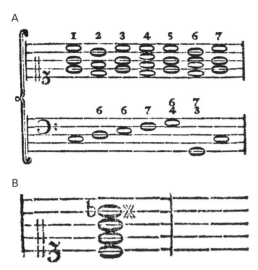

Figure 10.3
(a) A progression cited by Euler outlining the key of C especially through the penultimate dominant seventh chord (marked $\frac{7}{3}$). Note the parallel octaves between the lower voices, from the second to the third (D–E) and fourth to the fifth (F–G) chord (♪ sound example 10.1). (b) A dominant seventh chord built on the note C (identified by the C clef on the bottom line of the staff): C, E, G, B♭, as dominant seventh in the key of F (♪ sound example 10.2). From Euler, "On the True Character of Modern Music" (1764).

because B–F moves strongly to the resolution E–C (♪ sound example 10.3). The 4:5:6:7 ratios of a dominant seventh chord involve the previously forbidden sevenths, which Euler call "strangers" (*étrangers*) but invokes "a *musical license*" that allows these "foreigners" to be welcome. Thirty years after his youthful *Tentamen*, Euler was extending and continuing its argument to include contemporary harmonic practice. In 1773, Euler returned yet again to the foundations of his harmonic theory and the role of interrelated keys in harmonic practice in "On the True Principles of Harmony Represented in the Mirror of Music."[17]

Euler also put music at the center of his popular account of contemporary science for an intelligent reader, his *Letters to a German Princess* (1768–1771). Written at the behest of Catherine the Great (herself originally a German princess) after his return from the Berlin Academy to the Petersburg Academy, these letters may be the first work of popularization by a great scientist, but they are also an important document in the history and philosophy of science. Euler discusses music in far more mathematical detail than any other subject in his *Letters*, subjecting the princess to quite a bit of the argument of his *Tentamen*. After he emphasizes the expressive, dramatic side of music and its attendant sentiments, Euler reviews his arguments on the nature of light based on its analogy with sound, often using musical examples. He explains that "difference of color is to the organ of vision what sharp or flat sounds are to the ear," going even further than in his earlier optical papers to emphasize that "the parallel between sound and light is so perfect, that it holds even in the minutest circumstances." Thus, the mysterious glowing of phosphorescent substances, "which, once illuminated, preserve their light for some time, though conveyed into a dark room," are sympathetic vibrations made visible.[18] Euler's musical examples reflect not only the genesis of his theory and its original arguments, but its justification and popularization in musical terms.

11 Young's Musical Optics

The crucial evidence for the wave theory of light was the work of an amazingly multitalented individual, who, though surely unique in his constellation of abilities, manifests the fruitful breadth of scope so important in the advances made by other contemporary natural philosophers. Thomas Young used studies of sound and music to advance the theory of wave motion, especially the concept of interference, which he learned from sound and then applied to light. Sir John Herschel singled out Young's insight into sound interference as "the key to all the more abstruse and puzzling properties of light, which would alone have sufficed to place its author in the highest ranks of scientific immortality, even were his other almost innumerable claims to such a distinction disregarded."[1] Young's awareness of sonic and musical phenomena prepared the ground for his work on light, down to the precise details of the experiment that would finally satisfy Newton's stipulations.

Though born in modest circumstances to a pious Society of Friends (Quaker) family, Young's uncle was an eminent physician and member of the Royal Society. Early on, Young showed prodigious talent for languages, though basically self-taught. By age nineteen he was fluent in Latin and Greek, had a good command of the principal European living languages, could read biblical Hebrew, and had also studied Chaldean, Syriac, and Arabic.[2] He translated Shakespeare into classical Greek. Young also taught himself mathematics and developed an interest in science. He read Newton's *Principia* by himself; he ground pigments to make paint, studied drawing, and constructed scientific instruments. After leaving one of the local schools, he devoted himself "almost entirely to the study of Hebrew and to the practice of turning and telescope-making."[3] Yet despite his amazing breadth and depth of learning, Young's Quaker upbringing removed him from the ordinary activities of his contemporaries.

Whatever may have been his personal preferences, his family's finances dictated that he take up a career in medicine, following his uncle's lead. This he did without complaint, seemingly considering it a continuation of his interests in physics and mathematics, now extended to a physiological sphere. Following the practices of the time, Young first served an apprenticeship in London as a pupil in St. Bartholomew's Hospital and showed his extraordinary abilities in anatomy. At age twenty (1793), he made a major discovery about the function of the lens in accommodation, the process through which the eye adjusts its focus from near to distant objects.[4] In studying the eye of an ox, Young thought he had found evidence of fibers inside the lens that could plausibly act as focusing muscles, which earlier anatomists had conjectured but not seen definitively. Through the good offices of his uncle, Young read a paper on his discovery to the Royal Society, which led to his being elected a Fellow at age twenty-one, though this accolade was overshadowed by controversy. John Hunter, an eminent anatomist, claimed Young's discovery as his own, while another anatomist asserted he could find no such muscular structures in the lens. At that point, Young withdrew his discovery, in deference to this authority, though he later reasserted it in light of further research.

Young's medical apprenticeship led him next to Edinburgh, where many Quakers chose to study, excluded from Oxford and Cambridge on account of their faith.[5] Still, at Edinburgh Young began to play the flute and to take dancing lessons, which disobeyed Quaker precepts, as did his incipient experiments in theatergoing.[6] Not surprisingly, the experience of new places and people helped Young break away from the doctrinal limitations in which he had been raised. He further broadened his horizons in Göttingen, where he attended the lectures of Georg Christoph Lichtenberg, who presented and critiqued Euler's theory of light.[7] Young's doctoral dissertation (1796) concerned the physiology of human speech, including an alphabet of forty-seven letters intended to convey every sound of which the voice is capable.[8] In this work, his interests in sound directly address his ongoing linguistic and phonological concerns.

Young's disorientation in adjusting to foreign customs paradoxically intensified his pursuit of the social and artistic activities excluded from his Quaker upbringing. He began to take dancing lessons five or six times every week; as he wrote an English friend, nor was he "very punctual in some of the medical courses." George Peacock, Young's early biographer, noted that "it was in vain that his fellow-students, whether in banter or in earnest, told him that his musical ear was not good, and that he would fail to acquire ease and grace as a dancer. A difficulty thus presented to him as insuperable was a sufficient motive to attempt to conquer it; and though different opinions have been expressed with respect to the entire success of the experiment, there is no doubt that the mastery of those arts, which he really attained, was another triumph of his unconquerable perseverance."[9]

Precisely because they were relatively late interests that emerged in his formative years and spoke to a part of his nature that had been underdeveloped, the musical side has special importance for Young.[10] His final stage of medical apprenticeship led him to matriculate at Cambridge (1797) and to break with the Westminster Quaker meeting, which formally disowned him in 1798. Young still struck his Cambridge classmates as having "something of the stiffness of the Quakers"; he did not associate much with the other young men, who called him "Phaenomenon Young," indicating both their respect and their disdain.[11] One of them recalled that "he read little, and though he had access to the college and university libraries, he was seldom seen in them. There were no books piled on his floor, no papers scattered on his table, and his room had all the appearance of belonging to an idle man. I once found him blowing smoke through long tubes [though Young never smoked tobacco], and I afterwards saw a representation of the effect in the Transactions of the Royal Society to illustrate one of his papers upon sound; but he was not in the habit of making experiments."[12] We will shortly return to this scene. Young himself noted, shortly after arriving in Cambridge, that, starting with his Göttingen thesis on "the various sounds of all the languages that I can gain knowledge of," he had "of late been diverging a little into the physical and mathematical theory of sound in general. I fancy I have made some singular observations on vibrating strings, and I mean to pursue my experiments."[13]

In 1797, Young's uncle died, leaving generous bequests to his friends (and patients) Edmund Burke and Samuel Johnson, as well as to Young himself, who was now free to follow his own interests without financial concerns. The following year, after an accident and broken bone kept him from his usual exercise, Young devoted himself to what he called "observations of harmonics," by which he meant experimental studies of wave motion in sound.[14] During his recovery, he also read contemporary French and German mathematics and noted that "Britain is very much behind its neighbours in many branches of the mathematics; were I to apply deeply to them I would become a disciple of the French and German school; but the field is too wide and too barren for me."[15] His choice not to engage further with Continental mathematics had lasting consequences, as we shall see.

The course of Young's work in the years after his early paper on the accommodation of the eye clearly shows the interweaving of music, sound, and light in his subsequent work. Three essays he published in the year 1800 show the remarkable overlay and simultaneity of his thinking in these domains. In January 1800, while still at Cambridge, he read to the Royal Society his "Outlines of Experiments and Inquiries Respecting Sound and Light," which in essence lays out the fundamental premise of his ensuing research and whose title emphasizes the yoking of these two fields.[16] Young's experiments measured the quantity of air discharged through an aperture, the direction and velocity of the air stream, the velocity of sound, its degree of spatial divergence, the harmonic sounds of pipes, and the decay of their sounds, ending with a general discussion of the vibration of various elastic fluids. He often connects his work with those who preceded him, especially Euler, whose arguments about the wave theory in sound and light he had studied closely.[17] Here, and throughout the later works we will discuss, Young often interweaves musical references very naturally, as if he clearly expected his audience to find them familiar and congenial. Such connections between music and more general scientific topics were evidently widely shared.

Young structures his acoustical investigations first to elucidate how pipes make harmonic sounds. Using simple equipment (glass tubes, funnels, bladders), he devises ways to measure the flow of air through a pipe. He is careful and observant, often referring to common phenomena, such as still liquid disturbed by a stream of air directed toward it, or the deviations wind causes in the shape of a flame or of smoke ascending from a chimney. The physician in him notes the slight effect of every pulsation of the heart on the lungs blowing air through a glass tube. In fact, he is writing from his rooms at Cambridge, where (as we have seen) Young blew smoke through various tubes, to the mystification of his classmates. He now records more detailed experiments in a meticulous table comparing the varying pressures required to sound various harmonics from organ pipes (figure 11.1).

When Young turns to "the analogy between light and sound," he lists the evidence that light is a wave, including Newton's rings.[18] Young notes the difficulty and complexity of Newton's putative "fits of transmission and reflection" and adds that the recurrence of the same color in Newton's rings is "very nearly similar to the production of the same sound, by means of a uniform blast, from organ pipes which are different multiples of the same length."[19] For instance, four-foot and eight-foot-long organ pipes under the same pressure sound the note C an octave apart. Young notes that Euler had already noticed this analogy, "although he states the phenomena very inaccurately."[20] Though Young himself leaves his exact analogy somewhat unclear, he considers the *recurrence* of colors in Newton's rings to be precisely comparable to the recurrence of pitches produced by organ pipes, a phenomenon he finds incomprehensible to particle theory, which has nothing like a series of overtones underlying it.[21] Young also draws attention to the "tone, register, colour, or *timbre*" of the organ and other instruments as a neglected subject that should be studied by natural philosophy.[22] Thus, when Young compares Newton's rings to an organ, we realize the full appropriateness of his application of timbre or sound-color to visual color, for Newton himself had noted the importance of the recurrent pattern of the coloration in his rings. Young "hears" Newton's rings as resembling an organ's cyclical structure of pitches, overtones, and stops, in which the pressure of the air stream can excite recurrent harmonic pitches as the pressure exerted on the glass can evoke the recurrent colors of the rings. Implicitly, Young translates a *temporal* phenomenon (the frequencies of the organ pipes) to a *spatial* one (the varying lens thicknesses producing Newton's rings).

Having established this fundamental analogy between music and light, Young then turns to a musical phenomenon that will provide a crucial insight into light. His point of departure is a troubling assertion by Robert Smith, the eminent Cambridge astronomer, in his *Harmonics, or, The Philosophy of Musical Sounds* (1749), that "the vibrations constituting different sounds should be able to cross each other in all directions, without affecting the same individual particles of air by their joint forces." On the contrary, Young notes, "undoubtedly they [the vibrations] cross, without disturbing each other's progress; but this can be no otherwise effected than by each particle's partaking of both motions." As proof, he instances "the phenomena of beats" as discussed by Smith. To explain them, Young devises a kind of thought experiment, supposing "what probably never precisely happens, that the particles of air, in transmitting the pulses [of sound], proceed and return with uniform motions," drawing their motion along the horizontal axis, their displacement along the vertical (figure 11.2).[23] Young includes a number of different cases in which, "by

Young's Musical Optics

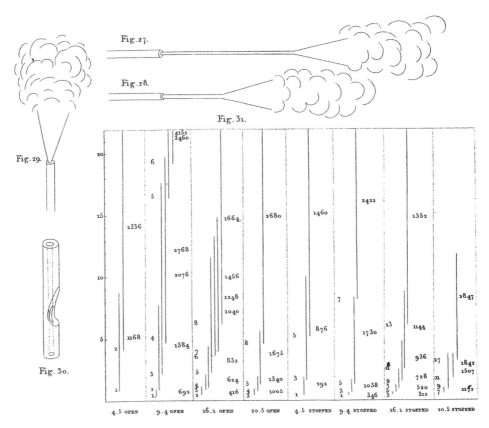

Figure 11.1
Young's illustration of his experiments in "Experiments and Inquiries Respecting Sound and Light" (1800), with his caption: "Fig. 27. The appearance of a stream of smoke forced very gently from a fine tube. Fig. 28 and 29, the same appearance when the pressure is gradually increased. Fig. 30. A mouth piece for a sonorous cavity. Fig. 31. The perpendicular lines over each division of the horizontal line show, by their length and distance from that line, the extent of pressure capable of producing, from the respective pipes, the harmonic notes indicated by the figures placed opposite the beginning of each, according to the scale of 22 inches parallel to them. The larger numbers, opposite the middle of each of these lines, show the number of vibrations of the corresponding sound in a second."

supposing any two or more vibrations in the same direction to be combined, the joint motion will be represented by the sum or difference of the ordinates." Thus, two sounds of nearly the same strength and pitch will produce a "joint sound" called a "beat" that reaches its maximum (the sum of the maximum of each component) on a slow rhythm determined by the exact difference between their respective frequencies or pitches. Young's sequence of cases show the graphic difference between the joint sounds produced by different components, noting that "the greater the difference in the pitch of two sounds, the more rapid the beats, till at last, like the distinct puffs of air in the experiments already related, they communicate the idea of a continued sound; and this is the fundamental harmonic described by Tartini."[24] His diagrams (figure 11.2) show "snapshots" of the vibrating string, translating its temporal motion into instantaneous spatial wave-forms.

At this point, Young's description breaks free from the presumption that sound is a vibrating *body* by noting that sufficiently frequent puffs of air by themselves "communicate the idea of a continued sound." Thus, the locus of the investigation of sound has been shifted to the vibrating air, away from the body no longer needed to produce it. We now realize that, in his student rooms, Young had been producing not just puffs of smoke but *a sound of very low frequency*, as if he had slowed the phenomenon of a musical pipe down to an immensely slower time scale on which it could be carefully observed and thoroughly compared with the flowing air that caused it.

Young immediately draws a musical corollary from his description of beats. Returning to the addition of two almost equal sounds, "the momentum of the joint sound is double that of the simple sound only at the middle of the beat, but not throughout its duration." Therefore, "the strength of sound in a concert will not be in exact proportion to the number of instruments composing it." Surprisingly, two violins playing in unison will still not consistently sound twice as loud as one alone, given that the players, however skilled, will inevitably deviate minutely in timing, volume, and pitch. Young reached this counterintuitive result from his thought experiment, rather than any actual observation, but he now realizes its possible significance as evidence of the wave theory, were it made observable. "Could any method be devised for ascertaining this by experiment, it would assist in the comparison of sound with light," evidently by demonstrating the palpable reality of beats in waves, whether of sound or light.[25] In this insight, Young expresses what now will be his quest: to find visible evidence of the "beating" of light waves that will be as clear as the evidence for the beating of sound.

Indeed, his whole plate of diagrams (figure 11.2) richly illustrates the way he juxtaposes light and sound. Where the diagrams on the right illustrate various possible sound-forms, those on the left show "the affections of light," its behavior in reflection, refraction, and passing "near an inflecting body," perhaps a string or knife's edge. The very layout of the plate invites us to contemplate sound and light together. After comparing them, Young returns to acoustic matters, particularly the problem of determining the frequency of vibrations, shape, and state of motion of what he calls a "chord," a stretched string. Here the

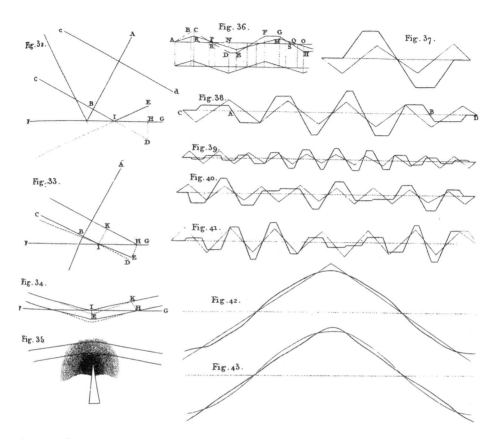

Figure 11.2
Plate 4 from Young's "Experiments and Inquiries Respecting Sound and Light" (1800). Figs. 32–35 show "affections of light," its reflection (32), refraction (33), total reflection (34), and light passing "near an inflecting body" (35); figs. 36–43 show the waveforms of various combinations of two sounds: an octave (37), major third (38), major tone (39), minor sixth (40), a fourth "tempered by about two commas" (41), a fourth further tempered by "subordinate vibrations of the same kind in the ratios of 3, 5, and 7" (42), and a vibration "corresponding with the motion of a cycloidal pendulum" (43).

visual appearance of a sounding body becomes of interest as he looks more and more closely at the vibrating string.

In fact, Young may have been among the first to use the piano, a rather recent arrival among musical instruments, as a *scientific* instrument. He uses "one of the lowest [wire-wrapped] strings of a square piano forte" to make an optical experiment: "Contract the light of a window, so that, when the eye is placed in a proper position, the image of the light may appear small, bright, and well defined, on each of the convolutions of the wire [due to its wrapping]. Let the chord be now made to vibrate, and the luminous point will delineate its path, like a burning coal whirled round, and will present to the eye a line of light, which, by the assistance of a microscope, may be very accurately observed."[26] Though his primary object was to gauge the shape of the vibrating string, the details of Young's experimental arrangement are, in fact, very close to what would turn out to be his crucial demonstration of light interference: a thin string illuminated by a small, well-defined light source. Young's own illustration of light passing "near an inflecting body" (in figure 11.2) gives evidence that he was aware of this parallelism, even though in this paper he does not take the next step, to allow the vibrating string to come to rest and then to see the vibrations of light surrounding it, as if that were silence made visible.

Young connects his studies of pipes with the problem of the human voice, "the object originally proposed to be illustrated by these researches." This recalls the physiological and medical aspects of his Göttingen dissertation, though here Young seems more interested in purely musical aspects of timbre and resonance. He connects the voice with his smoke pipes by noticing that, analogous to his rhythmic pipe-puffs, the human glottis can produce a slow vibration "making a distinct clicking sound" that can be made more continuous "but of an extremely grave pitch: it may, by a good ear, be distinguished two octaves below the lowest A of a common bass voice, consisting in that case of about 26 vibrations in a second" (♪ sound example 11.1). Young connects this glottal clicking with the methods used by ventriloquists to "throw" their voices and also (at still higher pitches) with falsetto singing. Though intriguing, his investigations are allusive and tentative, given the complexities of human vocal production. Though he refers to anatomy and physiology, he more often relies on "a good ear" that can (he tells us) hear four harmonics above the fundamental sung by "a loud bass voice."

The finale of this remarkable paper is even more purely musical and mathematical. Young, like so many before him, became fascinated with the question of temperament and here offers his own solution to the age-old problem, an astutely practical variant of continuing use in performances of late eighteenth-century music that seek authenticity (♪ sound examples 11.2, 11.3).[27] Young illustrates his own temperament in a diagram comparing various systems of tuning, showing the depth of his study not only of the question of musical temperament but of their varieties throughout history (figure 11.3), using spatial visualization to illustrate sonic issues. His comparative investigations closely

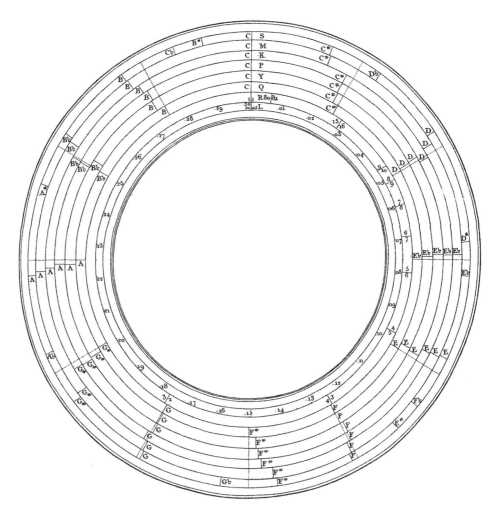

Figure 11.3
Young's comparison of different schemes of musical temperament (1800), including his own (the ring labeled *Y*); the entire circle spans an octave around *C*, shown at the top.

resemble, in scope and structure, his concurrent comparative work on languages, as if they were various possible "temperaments" of living speech.

Only four months later (April 1800), Young published "An Essay on Music," giving important evidence of his ongoing interest in music during the height of his optical researches. He begins by acknowledging "the agreeable effect of melodious sounds, not only on the human ear, but on the feelings and on the passions," yet Young considers music far more than "delicate titillation" or even than "giving expression to poetical and impassioned diction," which Coleridge and other Romantic thinkers emphasized. *Contra* Kant, Young argues that the study of music is not "amusement only" but reveals a science "that, in its whole extent, it is scarcely less intricate or more easily acquired than the most profound of the more regular occupations of the schools." Those who show "superior brilliancy" in music "seem almost to require the faculties of a superior order of beings." Young's essay shows considerable familiarity with the history and theory of music, as well as the importance he ascribed to it. He emphasizes the role of harmonics or overtones for the common triads and scales of contemporary musical practice. Finally, he discusses the terminology of musical tempo and gives a detailed table of the number of measures per minute sounded by composers such as Handel, Haydn, and Mozart. This table shows acquaintance with J. J. Quantz's attempt several decades earlier to standardize tempo using as a standard the common resting pulse rate of eighty beats per minute. As a physician, Young well knew the variability of human pulse and chose instead a more objective standard given by the number of measures per minute taken at various tempi. His table gives valuable evidence of performance practices around 1800.[28]

Seven months later (November 1800), Young presented his paper "On the Mechanism of the Eye" to the Royal Society.[29] Revisiting his maiden discovery about the accommodation of the eye, Young argued that he had been fundamentally right that changes in the shape of the lens were responsible for accommodation; only after his lifetime was the mechanism identified with the ciliary muscles surrounding the lens, rather than (as Young had initially thought) muscular fibers inside the lens itself. But in 1800, Young established conclusively that the lens alone was responsible for accommodation, not the cornea or the length of the eyeball, as had been suggested by others.

Young's argument is a tour de force of persistent experimentation and deduction that refutes the suspicion that he was an inspired dilettante who merely guessed discoveries without ever exhaustively demonstrating them. His experiments required not only ingenuity but courage, for he experimented on himself, following an old medical tradition and also Newton's more disturbing example. Newton had inserted a bodkin (a thin knife) behind his own eyeball to demonstrate that visual perception could be caused by ocular pressure without any incoming light.[30] Young, in his turn, performed no less invasive tests of his own eye and its functioning by fashioning a compass whose ends were keys that he pressed against his sclera, the whites of his own eye (figure 11.4).

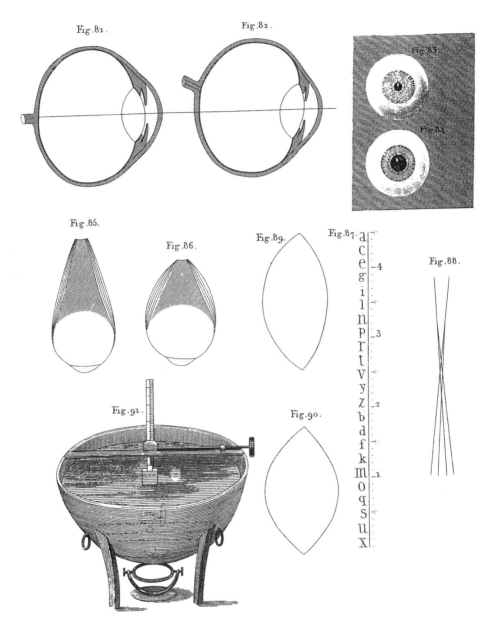

Figure 11.4
Young's illustrations of measurements made on his own eyes, from "On the Mechanism of the Eye" (1800).

Though these excruciating measurements and Young's ensuing physiological deductions are the bulk of his paper, he first lays his groundwork on another extended comparison of sound and sight, of ear and eye. He judges that the ear is "the only organ that can be strictly compared" with the eye, for the other senses operate through more immediate contact of their objects with the nerves. Thus, Young uses the ear as a comparative touchstone that illuminates the eye's different functioning. He calculates the quantitative difference between the ear's ability to discriminate the angular direction from which sounds are coming (only within about 5 degrees) and the eye's far sharper directional abilities (90,000 times finer). On the other hand, the eye's "field of perfect vision, for each position of the eye, is not very great," whereas "the sense of hearing is equally perfect in almost every direction." Using these comparisons between eye and ear as an initial point of reference, Young then goes on to devise what he calls a new optometer that will allow precise measurement of the eye's focal distances, as well as the other parameters needed to make his argument about accommodation fully detailed and complete.[31]

Thus, these three papers of Young's *annus mirabilis* 1800 all invoke sound, hearing, and music in fundamental ways that inform and shape his arguments about seeing and light. In August 1801, he published a letter reaffirming his account of sound and his new temperament against the criticisms of a Professor Robinson in Edinburgh. In November, his paper "On the Theory of Light and Colours" juxtaposed excerpts from Newton's writings with Young's own series of new propositions, presented in Euclidean-style hypotheses and demonstrations.[32] Young's rhetoric enlists Newton on the side of the wave theory of light, defusing Newton's objections to it by juxtaposing them with the many passages in Newton's own works where he recognized its merits.

As the essential background for his argument in favor of an ether carrying the vibrations of light, Young assumes the prior case of air as the medium for sound vibrations. "Every experiment, relative to sound, coincides with the observation already quoted from Newton, that all undulations are propagated through the air with equal velocity"; Young thought this a capital point in favor of the wave theory of light that Euler himself did not seem to understand when he maintained incorrectly that waves of higher frequency travel faster. Here and throughout, Young uses the wave theory of sound to establish the essential results he will apply to light; returning to his earlier arguments against Smith, he notes that "it is obvious, from the phenomena of elastic bodies and sound, that the undulations may cross each other without interruption" by "uniting their motions," though different frequencies of wave will not intermix. Likewise, he relies on the example of sound to establish that waves expand spherically through a homogeneous medium.[33]

Though Young claims not to "propose any opinions which are absolutely new," he offers an important suggestion that color vision relies on only "three principal colours, red, yellow, and blue," which he chooses because their "undulations are related in magnitude nearly as the numbers 8, 7, and 6," whose ratios he will shortly relate to music theory. Thus, green light, whose frequencies are about 6.5 in terms of these ratios, "will affect equally the particles in unison with yellow and blue, and produce the same effect as a light composed of those two species: and each sensitive filament of the nerve may consist of three portions, one for each principal colour."[34]

Young continues to follow closely what Newton had called "the analogy of nature," noting that, on the basis of his own argument, "any attempt, to produce a musical effect from colours, must be unsuccessful, or at least that nothing more than a very simple melody could be imitated by them" because the ratios of the primary colors limit the range of any such "color melody" to less than an octave because anything larger would go "wholly without [outside] the limits of sympathy of the retina, and would lose its effect; in the same manner as the harmony of a third or a fourth is destroyed, by depressing it to the lowest notes of the scale." That is, musical melodies would not translate directly to colors because musical intervals become indistinguishable when transposed to the extreme limits of audible frequencies. The analogy between the ear and the eye guides Young's hypothesizing even when he becomes aware of their important differences, which are no less significant to him than their similarities. "In hearing, there seems to be no permanent vibration of any part of the organ," implying its greater simplicity and unity, compared to the eye as a two-dimensional field of sensors that, at every point, cannot possibly have the range of vibrations available to the ear in its single canal. His three-color hypothesis emerges under the direct pressure of the pitch-distinguishing capabilities of the ear.[35]

Young goes on to offer additional evidence in favor of the wave theory of light, drawing especially on the arguments about the superposition of waves he had earlier made against Smith, culminating in "Proposition VIII. *When two Undulations, from different Origins, coincide either perfectly or very nearly in Direction, their joint effect is a Combination of the Motions belonging to each.*" Young notes that he had earlier "insisted at large on the application of this principle to harmonics; and it will appear to be of still more extensive utility in explaining the phenomena of colours." He applies it now to "Mr. Coventry's exquisite micrometers; such of them as consist of parallel lines drawn on glass, at the distance of one five hundredth of an inch," what we now call diffraction gratings.[36]

From Proposition VIII, Young derives a simple mathematical criterion for the light waves of a given monochromatic wavelength (coming from a point source of red light, say) to combine constructively and yield a bright red spot whenever the sine of the angle of that spot is an integral multiple of the ratio of the spacing between lines on the grating and the wavelength of light. Because the incident red light can reflect constructively off the grating at a whole series of angles, we will see not one but a series of red spots, each corresponding to a different integer in Young's formula. He notes that the particle theory of light would not produce any such periodic and recurrent spots, so that "it is impossible to deduce any explanation of it from any hypothesis hitherto advanced; and I believe it would be difficult to invent any other that would account for it. There is a striking analogy between this separation of colours, and the production of a musical note by successive echoes from equidistant iron palisades; which I have found to correspond pretty accurately with the known velocity of sound, and the distances of the surfaces." Once again, music gives the point of departure for his optical analogy. As he contemplates the lines of the grating, he analogizes them as "echoing" the light, as if audition and vision had merged.[37] Once again, a sonic, temporal phenomenon translates into a spatial, optical one.

Young's account of his sound experiment also suggests that he could have used it to connect the speed of sound with its wavelength and the spacing between the iron palisades. Though Young is quite aware of the significance of determining the wavelength of light experimentally, he does not do it here, reserving it for his reconsideration of Newton's rings, which (as noted above) Young had earlier instanced as the linchpin of his analogy with the recurrent frequencies of organ pipes. In "On the Theory of Light and Colours," Young obviously attaches special significance to determining the wavelength of light from Newton's own data, as if to show what was lying right in front of Newton all along, had he only realized it. Here, as elsewhere, Young both challenges and retroactively co-opts Newton, enlisting his posthumous support for the wave theory, though he had resisted it during his life. As he self-consciously stepped beyond Newton, Young always looked back to him, seeking his support even in the process of overthrowing his conclusions.

Newton had framed his spectral colors by assuming that they formed an octave; he did not seem to recognize that his own ring data contradicted such a 2:1 ratio.[38] But now Young corrects Newton's musical mistake, as had Euler before him: "The whole visible spectrum appears to be comprised within the ratio of three to five, which is that of a major sixth in music; and the undulations of red, yellow, and blue, to be related in magnitude as the numbers 8, 7, and 6; so that the interval from red to blue is a fourth."[39] Thus, Young specifically returns to the same musical analogy that Newton had used, though Newton had mistakenly substituted the octave for the major sixth. By getting right what Newton had mistaken, Young is able to retrieve the accurate wavelengths of the optical spectrum, which he goes on to state in musical terminology: "The absolute frequency [of light] expressed in numbers is too great to be distinctly conceived, but it may be better imagined by a comparison with sound. If a chord [vibrating string] sounding the tenor c, could be continually bisected 40 times, and should then vibrate, it would afford a yellow green light: this being denoted by c^{41}, the extreme red would be a^{40}, and the blue d^{41}."[40] Even the identity of these colors is "better imagined" by giving their musical note names, as if Young preferred to "hear" than to see them, though the "pitches" involved are enormously higher than any audible sound. The resultant synesthesia goes far beyond our normal senses: Young concludes that C is "yellow-green" and D is "blue," as if we were able to hear forty octaves above middle C. He also provides a table stating the "absolute length and frequency of each vibration" of different colors of light (figure 11.5), thereby reminding us of their sheer physical reality in space and time. Young does not seem to notice that orange and indigo do not really appear in the spectrum, or perhaps he simply bows to Newton's musically inspired definition of the spectral colors.[41]

Colours.	Length of an Undulation in parts of an Inch, in Air.	Number of Undulations in an Inch.	Number of Undulations in a Second.	
Extreme - -	.0000266	37640	463	millions of millions.
Red - -	.0000256	39180	482	
Intermediate - -	.0000246	40720	501	
Orange - - -	.0000240	41610	512	
Intermediate - - -	.0000235	42510	523	
Yellow - - - -	.0000227	44000	542	
Intermediate - - -	.0000219	45600	561	($=2^{48}$ nearly)
Green - - -	.0000211	47460	584	
Intermediate - - -	.0000203	49320	607	
Blue - - -	.0000196	51110	629	
Intermediate - -	.0000189	52910	652	
Indigo - - -	.0000185	54070	665	
Intermediate -	.0000181	55240	680	
Violet - - -	.0000174	57490	707	
Extreme - -	.0000167	59750	735	
Mean of all, or White	.0000225	44440	547	

Figure 11.5
Young's table showing the wavelengths and frequencies of different colors of light, as he calculated from Newton's experiments on thin plates. From "On the Theory of Light and Colours" (1801).

Young presented new evidence to the Royal Society in "An Account of Some Cases of the Production of Colours not Hitherto Described," delivered in July 1802.[42] In it, he further distills the content of his principle of superposition from its somewhat less quantitative form in his Proposition VIII to what he now calls "a simple and general law": "Wherever two portions of the same light arrive at the eye by different routes, either exactly or very nearly in the same direction, the light becomes most intense when the difference of the routes is any multiple of a certain length, and least intense in the intermediate state of the interfering portions; and this length is different for light of different colours."[43] With his general law in hand, Young returns to simple experiments mentioned by Newton and Grimaldi, from which he can now deduce the exact wavelengths they themselves did not calculate. Observing the "fine parallel lines of light which are seen upon the margin of an object held near the eye," Young notes that "they were sometimes accompanied by coloured fringes, much broader and more distinct." To make them more distinct, he observes a horse hair, then a wool fiber, then a single strand of silk, which gave the clearest, broadest pattern. Young made a rectangular hole in a card and bent its edges to support a hair parallel to the sides of the hole, a stabilizing mounting that allowed him to measure the deviations of the various colored fringes, which coincided with those he had measured in Newton's rings.[44]

Young takes these experiments a step further in his final paper before the Royal Society (November 1803), which begins by noting that "fringes of colour … produced by the interference of two portions of light" prove "the general law of the interference" and hence the wave theory in a "decisive" way.[45] His new experiment is even simpler: making a small hole in a window shade, on which a mirror directs the sun's light, he used his artificial sunbeam to illuminate "a slip of card, about one thirtieth of an inch in breadth, and observed its shadow, either on the wall, or on other cards held at different distances." Young now proves that the fringes were the *joint* effects of light passing on *both* sides of the card, not just one. He uses "a little screen" to block the light coming on one side of the card and then notes that "all the fringes which had before been observed in the shadow on the wall immediately disappeared, although the light inflected on the other side was allowed to retain its course." Therefore, the fringes could only be produced by the joint action of light "passing on each side of the slip of card, and inflected, or rather diffracted, into the shadow."[46] He goes on to show that his results are quantitatively consistent with his "general law" and that the distances between the dark lines in his fringed shadows agree accurately with analogous distances that he calculates from Newton's own observations of the shadow of a knife's-edge and of a hair.[47]

Young concludes that light "is possessed of opposite qualities, capable of neutralising or destroying each other, and of extinguishing the light, where they happen to be united," so that light plus light may yield darkness. As he emphasizes, this seemingly paradoxical conclusion is the essence of the wave theory, which gives it the power to explain the recurrences, fringes, and inner rainbows he had identified. His arguments also contradict Newton's hypothesis that (particulate) light speeds up in denser media. Thus, "the advocates for the projectile [particle] hypothesis of light must consider, which link in this chain of reasoning they may judge to be the most feeble; for, hitherto, I have advanced in this paper no general hypothesis whatever"; here, he clearly signals the failure of the particle view. Young's conclusion takes him full circle, back to the musical hypotheses with which he had begun: "But, since we know that sound diverges in concentric superficies [surfaces], and that musical sounds consist of opposite qualities, capable of neutralizing each other, and succeeding at certain equal intervals, which are different according to the difference of the note, we are fully authorized to conclude, that there must be some strong resemblance between the nature of sound and that of light."[48]

In 1801, in the midst of this series of papers, Young became professor of natural philosophy at the Royal Institution, founded the year before by the flamboyant Count Rumford as "a great metropolitan school of science" that would also undertake grand practical projects and to which we shall return in a succeeding chapter.[49] Though instruction in mathematics and natural philosophy had been the province of the ancient British universities, the Royal Institution addressed a broad educated public in London, whose fascination with science it fed and profited from. This audience also included women, still excluded from the universities. Within a few months, Rumford quarreled with the other directors and abandoned his fledgling institution; Young and Humphrey Davy carried it forward.

Though among the first in this eminent succession, Young was a less dramatically successful public figure than those who followed him; he thought his public presentations were too "compressed and laconic," not "very popular or very fluent."[50] On the other hand, *A Course of Lectures on Natural Philosophy and the Mechanical Arts* (1807) was the most comprehensive yet given in England, one of the first attempts of general synthesis in the aftermath of Newton. Addressing a broad audience, Young presented a general picture, emphasizing the leading concepts and omitting mathematical details. Reading him now, we can see how Young's synoptic project reflected his own work synthesizing sound and light through the wave theory. His papers showed the importance of music and sound as he discovered his new insights; his lectures showed how he continued to rely on sound and music not only in the context of the discovery of his ideas but also in the context of their public justification and popularization.[51]

After 1803 and the remarkable series of papers considered above, Young left the Royal Institution and active research in optics, discouraged by vitriolic attacks on his papers by Lord Brougham, a fanatical adherent of the particle theory of light. Young then wrote on medical subjects and increasingly worked on deciphering Egyptian hieroglyphics. Later, he was greatly encouraged by the recognition and praise of the younger French researchers in optics, especially Dominique Arago and Augustin-Jean Fresnel. The "Young–Fresnel theory," as it came to be called, prevailed by the 1820s, having converted all except for a few stubborn partisans of Newtonian orthodoxy (such as Brougham). In 1817, Young surveyed these confirmations in a magisterial article for the *Encyclopaedia Brittanica* entitled "Chromatics."[52]

Indeed, many new things had emerged after 1803, particularly the discovery of the polarization of light by Étienne-Louise Malus in 1807, which added a new level of puzzlement that the undulatory theory of the time could not illuminate. Gazing through an Iceland spar (calcite) crystal at the reflected sunlight from a neighboring glass window, Malus noticed that the two images of the window would alternately disappear and appear as he rotated the crystal. Somehow, the reflected light had some kind of directionality that the crystal could transmit only when correctly oriented. The crystal would split the incoming reflected light into two separate beams, each "polarized" differently, as Malus phrased it. If indeed light is a wave, how could it exist in the different states of orientation Malus had discovered?[53]

By 1815, Young doubted that his theory could account for this new phenomenon, as he wrote in his private correspondence at the time. But in a letter of 1817, he himself proposed a solution that both used and reversed the analogy with sound. Writing to Arago, he noted that "it is a principle in this [wave] theory, that all undulations are simply propagated through homogenous mediums in concentric spherical surfaces like the undulations of sound, consisting simply in the direct and retrograde motions of the particles in the direction of the radius [i.e., the direction of propagation of the wave], with the concomitant condensation and rarefactions."[54]

In modern terminology, Young was pointing out that sound is a *longitudinal* wave, causing fluctuations of density of the air along the wave's direction of propagation. In his 1807 lectures, he noted that "Dr. Chladni has discovered that solids, of all kinds, are capable of longitudinal vibrations," though "the vibrations which most bodies produce are, however, not longitudinal but lateral."[55] As we will consider in the next chapter, Ernst Chladni's vibrating plates showed Young visible evidence of both longitudinal and lateral (transverse) motion. In 1817, though Young clearly understood the force of the example of sound, he now realized that light waves might operate in an importantly different manner: "And yet it is possible to explain in this theory a transverse vibration, propagated also in the direction of the radius, and with equal velocity, the motions of the particles being in a certain constant direction with respect to that radius: and this is a *polarization*."[56]

That is, if the vibrations of the light wave occur in the transverse (perpendicular) plane to their direction of propagation, they can then be polarized in that plane. The two split beams transmitted by Iceland spar turned out to exemplify the two orthogonal directions in that plane: Malus's images appeared and disappeared as the crystal was rotated, first transmitting the polarized light, then not.[57] Thus, Young suggested, as did André-Marie Ampère, Arago, and Fresnel independently, light could be a transverse wave, compared to sound waves as longitudinal.[58] Though several of Young's biographers assert at this point that he and Arago had been "blinded" by the analogy with sound, Young's letter suggests the opposite, for he says that he was led to his new suggestion precisely by sound itself.[59] Note that he speaks, in both the case of transverse and of longitudinal waves, of "this theory" in the singular, indicating that the general characteristics of "undulatory theory" are shared by both, including the concepts of wavelength, frequency, velocity, and direction of propagation.

Returning to this issue in 1823, Young again represents himself as "strongly impressed with the analogy of the properties of sound," but he now notices that the possibility of transverse light waves lead to a "perfectly *appalling*" consequence: because they had always been formulated in terms of the vibrations of a solid, "it might be inferred that the lumeniferous ether, pervading all space, and almost all substances, is not only elastic, but absolutely solid!!!"[60] Though Young's biographers take this, too, as even stronger evidence of his being "blinded" by the analogy to sound, it has not been noticed that his objection indicates the very difficulties with the ether to which we shall return in chapter 13. Here too, Young credited this final contribution to optics to his reflections on "undulations of sound," though he might have said the same for many of his prior insights connecting music, sound, and light.

This concluding example confronts us with the full richness of Young's translation of sound vibrations into light waves. As with his youthful rendition of Shakespeare into classical Greek, he was alive both to the possibilities and the perils of such translation. In the present case, his "translation" yielded both the possibility of transverse light waves but also the attending paradox of the ether. Young was content to follow this translation from sound to light far enough to contemplate these new, "appalling" implications; characteristically, he left to Fresnel and Arago the detailed mathematical exploration of the new terrain.[61] Similarly, in his subsequent work on Egyptian hieroglyphics, Young discovered that the language was phonetic and correctly identified many characters on the Rosetta Stone, leaving to Jean-François Champollion the full decryption of the rest of the text and the attendant *réclame*.[62] Ironically, Young's French acclaim for his light theories was accompanied by British neglect; conversely, the British magnified and the French minimized his achievements in hieroglyphics, compared to Champollion. In the tumult of the Napoleonic era, Young experienced the frustrations of a cosmopolitan polymath traversing the British–French divide.

Experiencing the crucial moment of breakthrough in translation may have been more satisfying for Young than the subsequent labor to fill out the gaps and continue the work to the bitter end. Ultimately, he may have been most hampered by his aversion to the "too wide and too barren" mathematical language Fresnel used so powerfully. Ironically, though a polymath suspected of speaking too many tongues, Young may have had one too few, insofar as he eschewed the Continental mathematical language. Perhaps his disinclination may reflect his idiosyncratic education, steeped in Newton's intentionally archaizing, anti-Cartesian geometrical language, rather than the algebraic symbology associated with Leibniz. Rather than merely imitating Newton, though, this may have reflected Young's (and Newton's) deep respect for antiquity. Both were curious about *prisca sapientia*, the primal wisdom of the ancients, as became manifest in Young's work on hieroglyphics and in Newton's on ancient chronology. However one reads his own wide-ranging quest, Young himself thought that "it is probably best for mankind that the researches of some investigators should be conceived within a narrow compass, while others pass more rapidly through a more extensive sphere of research."[63] Though this fluent statement does not make explicit the difficulties and frustrations involved, Young was the exemplar of this second path, poised between languages in ways that paralleled his fundamental role in translating the wave theory between sound and light.[64]

12   Electric Sounds

Thomas Young's translation of sound into light provided a crucial example for parallel work connecting sound with electricity and magnetism that emerged in the decades just before and after him. The early connection that Georg Christoph Lichtenberg made between electricity and its visual trace led directly to Ernst Chladni's vibrating plates, which gave visual form to sound. Félix Savart continued the exploration of the electricity–sound connection, as did Hans Christian Ørsted and Johann Wilhelm Ritter in their own ways. In all these cases, sound represented a parallel venue for ideas and experimental approaches that contributed to the Biot–Savart law of magnetic action and to Ørsted's discovery of what he called electromagnetism. The complex interweaving of these studies of sound, light, electricity, and magnetism aptly reflects the traveling vibrations they all pursued.

Earliest in this network is Lichtenberg, a remarkable polymath, writer, and wit, and a friend of Goethe and Kant, whose aphorisms have resonated for generations in the German-speaking world. The first professor of experimental physics in a German university, whom we have already met as Young's teacher, Lichtenberg was active in geodesy, volcanology, meteorology, astronomy, and mathematics, to name only a few of his endeavors. These manifold topics, though, he understood as part of a search for *Ganzheit*, the wholeness and integrated unity of nature: "There is only *one* natural science before God; man makes isolated chapters out of it, and *must* make them, in accordance with his limitations. As long as the chapters do not fit together, an error lurks hidden somewhere, in the different chapters separately, or in all of them."[1] This quest for the unity of nature characterized German *Naturphilosophie* as a whole, though Lichtenberg stood apart from many of its more visionary adherents in his hard-headed empirical and experimental orientation. For him, "unity" was a watchword for tough-minded science, not a mystic slogan.

In that spirit, music and sound take their place among many other significant fragments of the great whole. Convinced that "everything is in everything," Lichtenberg sought connections between magnetism and electricity, magnetism and light, light and heat, heat and sound—for instance: "Has one ever produced heat through sound?"[2] "Does music make plants grow, or are there among the plants some that are musical?"[3] His most famous intervention in physics made invisible electricity take a visible form. In 1777, he completed construction of a large electrophorus (figure 12.1a), whose resinous "cake," two meters in diameter, when rubbed with fur generated static electricity, whose polarities he symbolized as + and −. By placing the insulated metal plate in contact with the charged "cake" and then touching the plate's upper surface to allow positive charge to escape ("charging by induction"), the bottom of the plate had sufficient negative charge to generate sparks up to 40 cm long (about 20,000 volts in modern units).

He noticed that the charged plate, when discharged, would cause the resin dust on his table to assume striking shapes, now called *Lichtenberg figures* (figure 12.1b,c), "innumerable stars, galaxies, and large suns ... finely formed branches, similar to those made by frozen steam on window panes."[4] Consistent with his quest for the unity of nature, he described the electrical patterns in the language of astronomy or botany, as if the electrified dust visibly disclosed the deep affinity between all these realms. Further, he could clarify the figures by scattering more dust on them, bringing out their fine details, which remained clear even after several days.

Lichtenberg felt he had discovered a means of investigating the "motion of electrical matter" similar to the long-known use of iron filings to make visible the action of magnets. He noted the pronounced difference between the figures generated by positive (figure 12.1b) and negative (figure 12.1c) charge, taking this as evidence that electricity is composed of two different fluids, not one (the subject of much controversy into the early nineteenth century). He also realized that pressing black paper on his figures allowed them *to print themselves* in far greater and more perfect detail than he could achieve in a drawing. Some historians thus credit Lichtenberg with discovering the essential process of xerography, now used in every copier or printer.[5]

Lichtenberg's figures found immediate resonance in the work of Ernst Chladni. The son of a law professor, he quit the law after his father's death to follow the "study of nature, which had always been my secondary and therefore dearest occupation. As an amateur of music, whose elements I had begun to learn a bit late at age nineteen, I noticed that the theory of sound was more neglected than many other branches of physics, which gave rise to my desire to remedy this lack and be useful to this part of physics through some discoveries." In 1785, Chladni began "very imperfect" experiments striking plates of glass or metal at different points and trying to understand their sounds, compared to the familiar vibrating string, which he studied from Bernoulli and Euler. Chladni then read of "a musical instrument made in Italy by abbé Mazzocchi, consisting of bells to which one applied one or two violin bows," which he then decided to try on his plates. He obtained sounds whose pitches were in the ratios of the squares of 2, 3, 4, 5, ... but felt he did not understand the underlying motions. Reading Lichtenberg's experiments made him "presume that the different vibratory motions of a sounding plate also ought to show different appearances if a little sand or a similar substance were spread on the surface." His first trials on a round plate yielded ten- and twelve-pointed stars (figure 12.2a),

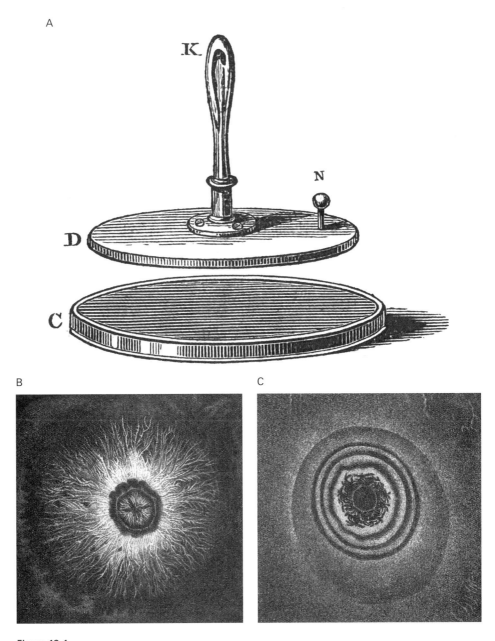

Figure 12.1
(a) An electrophorus, composed of a "cake" of resinous material (bottom) and a metal plate with an insulating handle (top). (b) Positively (+) and (c) negatively (−) charged Lichtenberg figures produced using the static electricity generated by induction from the electrophorus.

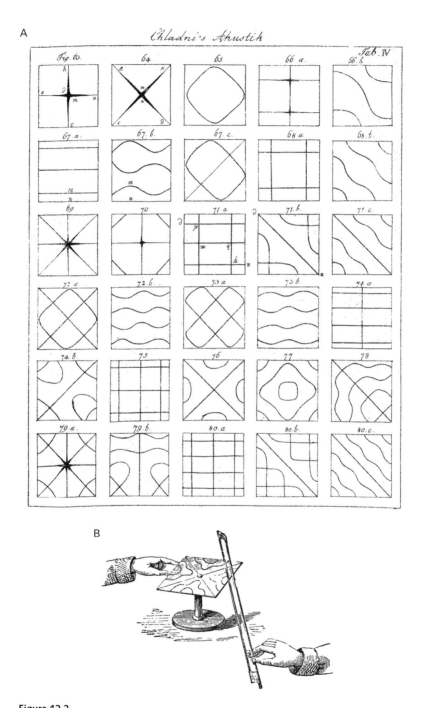

Figure 12.2
(a) Chladni's figures (*Acoustics*, 1830). (b) John Tyndall's illustration of the process of forming a Chladni figure (*Sound*, 1871).

whose attendant "very high sound" changed according the squares of the ratios he had found: "Imagine my astonishment in seeing this phenomenon that no one had ever seen before."[6]

These figures make visible the spatial patterns produced by sound waves, the correlation between them brought home by the coincidence of sight and sound, especially as the patterns suddenly change when the bow audibly excites a new standing wave in the plate. Chladni's discovery transcribed Lichtenberg's electrical figures into sound on the implicit presupposition that an analogous "vibration" lay behind the action of electric charge and behind the sounding plate. Chladni went on to exhibit his figures throughout Europe. He also invented two musical instruments (the euphonium and the clavicylinder), each an offspring of his vibrating plates.[7]

Chladni visited Paris in 1808 and demonstrated his figures to the Académie des Sciences, including Napoleon himself (figure 12.3), showing the enormous interest evoked by the wonderful spectacle of visible sound, audible sight. In the painting, Napoleon and his entourage gaze thoughtfully on the demonstration, showing his celebrated interest in exact science as he scrutinized a talisman of the new physics. Several French savants took up Chaldni's experimental program, particularly Félix Savart, who began as a physician but became "truly Chladni's professional successor."[8] Where Chladni had applied the violin bow to various shapes of plates, Savart used the same techniques to study the violin itself as a special kind of vibrating plate, exploring the relation between its structure and its sound. In this way, Savart clarified the complex functions of the violin's bridge and sound post, part of artisanal violin-making that previously lacked theoretical explanation. By locating the nodal lines, where sound waves would interfere and allow sand to settle quietly, Savart showed that the best placement of the sound post—in French the "âme," the violin's "soul"—avoided those nodes. After studying models based on Stradivari and Guarneri violins, Savart thought he could improve the instrument's basic design. To maximize the violin's symmetry, he built a trapezoidal violin with rectangular sound holes, thereby simplifying the violin's traditional curves and *f*-holes (figure 12.4). Savart's novel violin was studied carefully by an eminent committee formed jointly by the Académie des Sciences and the Académie des Beaux-Arts, including the composer Luigi Cherubini and also Savart's older colleague, the physicist Jean-Baptiste Biot, who wrote the report. Their unanimous opinion was that "the new violin could pass for an excellent violin," its tones even more suave than a standard instrument, though a bit more subdued when heard close by.[9]

The parallel strands in this intertwined story takes us to yet another country in which a young savant made a new connection with Chladni's work. In Denmark, Hans Christian Ørsted had drawn inspiration from Kant's attempt to deduce natural philosophy from basic forces of attraction and repulsion, which Ørsted and others interpreted in the spirit of contemporary *Naturphilosophie*. Charles-Augustin de Coulomb's work in the 1780s had offered persuasive evidence that electricity and magnetism involved two utterly

Figure 12.3
Chladni demonstrating his figures to Napoleon (1808). (Courtesy Deutsches Museum, Munich.)

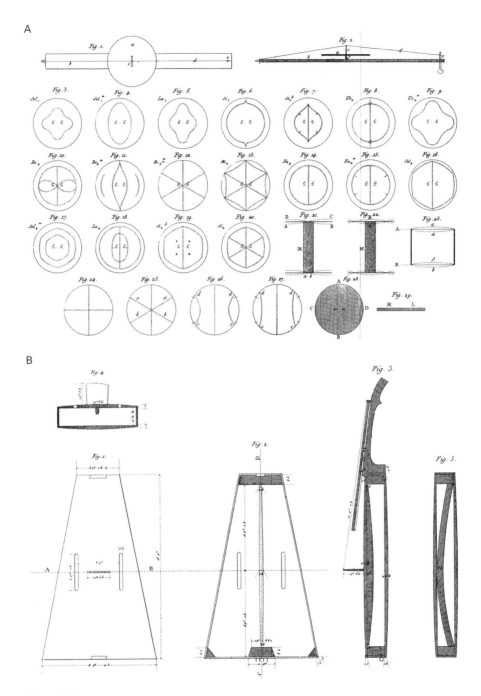

Figure 12.4
(a) Savart's illustration of the nodal pattern of a violin in relation to its structure; (b) his design for a trapezoidal violin (*Memoir on the Construction of String Instruments*, 1819).

different species of matter. Yet the unitive spirit of *Naturphilosophie* led Ørsted to hypothesize that they were, in fact, connected. In the course of his attempts to demonstrate this decisively, Ørsted began to work with Chladni's acoustic figures. In an 1804 letter to his friend and fellow *Naturphilosoph* Johann Wilhelm Ritter in Jena, Ørsted expresses his view that Chladni's figures not only "may offer important insights into the theory of sound" but could offer the hope of discovering "electric phenomena in the production of the acoustic figures." Thus, the impulse that came from Lichtenberg's figures recording electricity subsequently shaped Chladni's acoustic figures, then circled back to influence Ørsted's work on electricity. To implement his plan, Ørsted (like Lichtenberg before him) used fine lycopodium power, more responsive to electric charge than the sand or rosin dust Chladni had used. Ørsted observed that "a number of small waves or nodal points developed with each stroke of the violin bow" applied to this electrified Chladni setup, so that "each acoustic oscillation is composed of a number of smaller ones." Thus, "each tone in itself would be an organization of oscillations just as any music is an oscillation of tones," thereby unifying the structure of music with the physical structure of sound in general.[10]

The following year (1805), Ørsted continued his train of thought in a letter to Marc Auguste Pictet in Geneva, which, like his letter to Ritter, was published and thus shared more widely. Ørsted observed the fine detail of the motion of his tiny piles of lycopodium powder as they moved on the vibrating plate, giving a dynamic quality to the movements underlying the formation of the visible patterns and the concomitant sound. He then directed attention to the friction involved in these processes, which "produces not only heat but electricity." Using Coulomb's electrometer, Ørsted reports his tentative finding of electric charge "on the edges and corners" of his vibrating bodies, which he proposes to investigate further.

At this point, Ørsted involves the work of his friend Ritter, surely the most "Romantic" of the *Naturphilosophen*: pursuing their common search to unveil the unity of Nature (*Einheit der Natur*), Ritter took this quest to extremes that tested the limits of that unity and even of his own safety.[11] His boldness and radical imagination clearly fascinated his friends, who included Goethe, Alexander von Humboldt, and the poets Novalis (Friedrich von Hardenberg) and Clemens Brentano, as well as Ørsted. For instance, Ritter applied the principle of the unity of nature to argue that, corresponding to the "heat rays" (infrared radiation) just discovered by William Herschel (1800), by symmetry there should be cooling "chemical rays" at the opposite end of the spectrum (ultraviolet). In 1801 (the year he met Ørsted), Ritter showed that these rays darkened silver chloride. In 1800, using the new voltaic pile Ritter had independently discovered the electrical decomposition of water, which he again had deduced from the implications of *Naturphilosophie* for the interrelation between positive and negative electricity. At the time, he was twenty-four, a dropout from medical school, basically self-taught as a scientist. Not all his deductions about the unity of Nature were confirmed by others, such as his claim that the Earth had electric (as well as magnetic) poles or that magnets (as well as electric currents) could decompose water, not to speak of his investigations of occult practices such as metal witching and sword swinging.

In his letter on acoustic vibrations, Ørsted mentions that the "celebrated Ritter ... had long ago discovered that Volta's electric pile [battery] is capable of producing sound when the ear receives a shock from it."[12] Ørsted's surprising statement is far paler than the reality of Ritter's experiments, which applied electric currents (some quite large) to his own body, including his eyes, ears, "organs of evacuation," "organs of reproduction," and "other choice parts of the body." Though Ritter probably went further in self-experimentation than anyone else, in 1795 Alexander von Humboldt (who enlisted Ritter's collaboration) devised a galvanic circuit that connected some frogs via open wounds he had caused on his own back.[13] Their horrifying fascination aside, Ritter's detailed accounts describe his self-experimentation as brave explorations of a terrain of experience he dared not inflict on anyone else, yet considered important to reveal the dimensions and implications of unified nature, including human beings. Though one wonders about his exact relation to these transgressive experiments and their possibilities for superlative pain and pleasure, Ritter explicitly used these electrical stimulations to probe the exact relation between the different (yet presumably unified) modes of sensation.

For instance, in one series of experiments in 1803, Ritter stimulated various organs with the positive pole of a voltaic column, resulting "in the eye: increased influence of light, bluish color, diminution of objects, narrower than usual field of vision; in the ear: sound with a deeper tone than g; in the nose: suppression of scent as with *acide muriatique oxygéné*; on the tongue: acidic taste; and finally in all of these, as in every other part of the body, *expansion*." In contrast, application of the negative pole yielded "in the eye: diminished influence of light, reddish color, and wider than usual field of vision; in the ear: sound with a higher tone than ḡ [presumably an octave higher]; in the nose: impulse to sneeze; on the tongue: alkaline taste; and finally the general feeling [*Empfindung*] in all of these, as in every other part of the body: *contraction*."[14] The symmetry of these extremes in fact shows the larger rationale for Ritter's expectation of "chemical rays" lying beyond the violet. Beyond his general expectations from *Naturphilosophie* about the polarity of expansion and contraction throughout the sensorium, Ritter's findings reveal high and low pitches (as well as blue and red colors, acid and alkaline tastes) as expressions of *electrical* activity. As he notes, with some wonder, "the same thing that produces colors in the eye, produces tones in the ear,—as though colors were *mute tones*, and tones, in turn, *speaking colors*.—That may well seem to be only a manner of speaking, but it could be more than one may have allowed oneself to believe."[15]

Writing to Pictet, Ørsted interprets Ritter's findings to mean "that in each sound there are as many alternatives of positive and negative electricity as there are oscillations, but the union of the two electricities produces a shock. ... The perceptible effect of the union of all these imperceptible shocks is sound."[16] Because of their overriding commitment to *unity*, Ritter and Ørsted understood electricity to underlie the senses, no less than the senses provide evidence of electricity: Ritter felt he was *hearing* electricity, no less than Ørsted was *seeing* it move the powder on his vibrating plates. For these Romantics, hearing and music enabled the deepest and most inward knowledge. Ørsted wrote a dialogue about music. Ritter's *Fragments from the Posthumous Writings of a Young Physicist* (1810), which Walter Benjamin considered the "most significant confessional prose of German Romanticism," ends with a visionary interweaving of Chladni's and Ørsted's sound figures, electricity, light, and music. For Ritter, "there must absolutely be no human relation, no human history, which could not be expressed through music. ... All life is music, and all music as life itself—at least its *image*."[17] Ritter's provocative thoughts on music made deep impressions on Novalis, E. T. A. Hoffmann, and Robert Schumann.[18] Calling light "the bond that binds together all and every thing," Ritter nonetheless considers that "every tone is the *life* of the sounding body and in it, as long as it holds, as tone is extinguished with it. Every tone is a whole organism of oscillation and figure, shape, as is also every organic living thing. It expresses its existence [*Er spricht sein Daseyn aus*]."[19] He often connects music with the deepest sources of language, just as he treats electricity as a kind of "fire-writing [*Feuerschrift*]" that inscribes its primordial glyphs in Lichtenberg figures.[20] Ritter considers these expressive shapes to be the originals for written language itself, taken as visibly recording the emergent shocks of consciousness: "Music is also language, general language, the first of mankind. ... Music decomposes into languages," rather like the decomposition of water into oxygen and hydrogen Ritter was one of the first to achieve. "Thus every one of our spoken words is a secret song, for music from within continuously accompanies it."[21]

In closely harmonizing ways, Ritter and Ørsted sought the inner music of electricity. During 1802–1803, Ørsted visited Paris to represent their very different approach in the capital of mechanistic science whose champion was Pierre-Simon Laplace.[22] Ørsted kept returning to his electric version of Chladni's experiment, for instance in his essay "On the Harmony Between Electrical Figures and Organic Forms" (1805), an extended meditation on the striking differences he noted between the polarities of the electrical figures: the positive charge patterns' "striking resemblance to vegetation" (such as figure 12.1b) versus "the internal form of the plant" seen in the negatively charged patterns (figure 12.1c), with their womb- or egg-shaped contours. Ørsted connected these contrasting forms with chemical phenomena and with repulsive and attractive forces in general, ending with the behavior of light. Though he notes that "the first fundamental law of light is that its effect spreads along straight lines, or that it is in the form of the first dimension," when light strikes a partly opaque body "another force must act against it, as its opposite." Thus, a straight ray of light passing through a prism give rise to a new effect, the spectrum, and "the direction of this effect is precisely perpendicular to that first straight line." Ørsted further speculates a third dimension, a "penetration or chemical process" like the effect Ritter discovered from "chemical rays" acting beyond the violet end of the spectrum.[23]

Prefacing his discussion of the multidimensionality of these phenomena, Ørsted mentions a general philosophical argument from Schelling correlating the three dimensions of space with "the construction of matter by the attractive and repulsive forces." Schelling's general observation, however, only sets the stage for Ørsted's own specific argument, phrased against the background of his sound-electricity figures and ending with a specific identification of magnetism as "longitudinal," electricity as "latitudinal," and "depth" as chemical. Thus, in 1805 Ørsted was already thinking about the relation between electricity and magnetism in terms of perpendicular directions.

The relation between this train of thought and sound phenomena comes forward clearly in his "Experiments on Acoustic Figures" (1810). Ørsted notes that, when observed more closely, what had been taken as simple straight lines in Chladni figures are actually hyperboloidal curves. He connects this with the varying transmission of sound waves through the plate, as excited by a violin bow, speculating that "there is nothing to prevent" all the other conic sections from forming on very large plates. Ørsted contrasts this with the various patterns seen and sounds (more or less dull) heard when one taps different points along the plate. Thus, he treats the action of the bow as compounding many small taps into a more continuous excitation. To test this, Ørsted first records the "crude" image after one stroke of the bow, then again after the pattern has been completed after further strokes (figure 12.5). This emphasizes the *dynamic* quality of the sound figures, not merely their *static* appearance once they have been formed. Ørsted was struck by the living, organic quality of this dynamism, which he takes not simply as transient phenomena superseded by the completed figure but as clues to the "life" recorded by the figures' changing forms.

Consistent with his writings on sound since 1804, Ørsted in this 1810 essay treats electricity and sound as essentially connected: the motions of the particles on a plate indicate its electrical configuration. He shares Ritter's idea that convex curvature of the plate would lead to positive charge, concave to negative.[24] At this point, Ørsted's thought becomes sharper, akin to the very pattern he was watching clarify on the plate. The motion of the powder particles elucidates the inner activity of electricity/sound, moving in undulations around the plate, in which an initial impulse (say from some point on the edge) spreads out sideways as it travels, generating further motion *perpendicular* to its original direction. Here Ørsted has gone past the generalities of Schelling to indicate a nascent argument about how electricity, like the sound it generates and that in turn generates it, may cause effects that transduce between one and two dimensions. Ørsted observes various degrees of harmoniousness generated by different "dimensionalities" of tapping: a "dull thud" at one point, a "clatter" when a whole side of the plate is struck, a "proper tone" when the whole surface is excited (as by a violin bow), at which point Chladni's striking patterns emerge. From this, Ørsted connects music with these visible manifestations: "the most perfect and internally harmonious motion of bodies is also the one which, through the ear, produces the deepest impression on our internal sense of beauty."[25] Ørsted also echoes

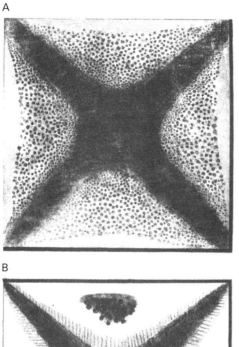

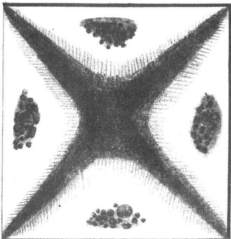

Figure 12.5
Ørsted's figures of the pattern seen after one stroke of the violin bow (above) and then "completed" after further strokes (below).

Ritter's speculations, "the greatest ever said about tones" and their felt musical effects, "how sorrow and joy each has its own, the former in minor, the latter in major." Ørsted considers light a higher frequency of vibration than sound: "According to this conception, one sense would become an octave of the other on the grand scale of sensations, and all would be subject to the same laws. Thus all sensations spring from the same original force, which in light works *in puncto* but in galvanism spreads in space, where, however, it runs through all forms of vibration so that it becomes perceptible to every sense."[26] Marveling at this all-encompassing unity, Ørsted hears in musical harmony, as in Chladni's sound figures, "the mark of an invisible Reason" that is far more than "mechanical sensory stimulation."

Ørsted's major discovery emerged within this nexus of electricity and sound. In 1815, he had demonstrated "that heat and light consist of the conflict of the electricities," implicitly including sound, as his other comments confirm.[27] In 1820, he gives specific form to "electromagnetism" (as he christens this new unity) by exhibiting the essentially *perpendicular* effect of currents on magnets through the intermediacy of the "conflict of electricities," which he defines as "the effect which takes place in this conductor and in the surrounding space."[28] Ørsted's description of the circular lines of "conflict" surrounding the conductor directly recall the transverse outlines he had provided for the dynamic motions of sound and electricity (figure 12.5) as well as his detailed descriptions of the motions induced by those forces. He does not discuss the contrast with longitudinal motions noted in the previous chapter, though he was probably aware of Chladni's use of this term to describe the propagation of sound in his figures. Ørsted seems to take as self-evident that the symmetry of the source (the linear conducting wire) should be reflected in the form of the "conflict" that surrounds it, whose transverse (rather than radial) symmetry is far from obvious. The only precedent we can find in Ørsted's earlier work is the transversality of the sound/electric effects on his vibrating plates, though his 1820 papers announcing the new electromagnetic effect do not explicitly direct attention to its acoustic prehistory.

Ørsted, however, always mentions the unity of forces that guided all his endeavors, among which sound figures in so many cases. He certainly takes his 1820 discovery to confirm his guiding presupposition. By comparing his new work to the response of magnets during storms, Ørsted implicitly includes it in the interrelated framework of thunder and lightning, as if such weather provided a macroscopic precedent for his discovery, dramatically combining sound and light. His work on sound becomes the all-pervasive background and matrix in which he explores the unification of electricity, magnetism, and light.

Ørsted's publications spread news of his discovery, which first found its experimental confirmation during one of his lecture demonstrations; his first direct observations were thus preceded and prepared by a long period in which the example of sound and the quest for unification of the forces of nature focused his attention. In Paris about at the same time, Savart turned from his investigations of violins to studies of magnetic action, paralleling Ørsted's path from sound to electromagnetism. After finishing his work on the committee judging Savart's trapezoidal violin, his colleague Biot joined him in the detailed study of electromagnetic action that became their most famous work. Ørsted's orientation toward the holistic generalities exalted by *Naturphilosophie* shaped his qualitative description of the circular lines of force surrounding a conducting wire. In contrast, Biot and Savart's immersion in the decidedly mathematical orientation of French physics is manifest in the quantitative law for the magnetic field strength, known by both their names (though in its mathematical detail very much also the work of André-Marie Ampère): the Biot–Savart law.[29] The larger French research tradition informed both their work on sound and on electromagnetism, but for them, as for Ørsted, sound came first.

13 Hearing the Field

In the wake of Ørsted's discovery, the entwined stories of Charles Wheatstone and Michael Faraday likewise interwove sound and electromagnetism. Starting out as an apprentice bookbinder, Faraday altogether lacked mathematical education; he said he could not understand a single equation. From his earliest work as a laboratory assistant to Humphrey Davy, Faraday thought in terms of experiments, in the felt reality of observation and manipulation, his visual turn of mind manifested in the constant sketches he put in his diaries, essential adjuncts to his hands-on experiences. His cultural awareness was far more sonic; he never mentions paintings or the visual arts but at several points makes clear his strong love of music.

Writing in 1813 to his close friend Benjamin Abbott, the young Faraday begins by quoting Shakespeare's famous encomium of music: "'He that hath not music in his heart &c' confound the music say I.—it turns my thoughts quite round or halfway round from the letter" that Faraday is trying to write. His joke implies that Shakespeare's line scarcely does justice to Faraday's infatuation with the music he hears in the night. "You must know Sir that there is a grand party dinner at Jacques hotell which immediately faces the back of the [Royal] institution and the music is so excellent that I cannot for the life of me keep from running at every new piece they play to the window to hear them—I shall do no good at this letter tonight and so will get to bed and 'listen listen to the voice of' bassoons violins clarinets trumpets serpents and all the other accessories of good music—I cant stop good night."[1] By this time, Faraday had just become Davy's "chemical assistant" at the Royal Institution, whose fame Faraday later augmented through his popular lectures, alongside his ceaseless stream of experiments and publications. In 1813 he was only beginning along that path, using his letters to improve his writing, studying elocution several times a week to rectify his pronunciation and overcome the signs of his lower-class background, though phonetic problems dogged him all his life (he could not pronounce the letter "r" and called his brother "Wobert").[2] His struggle with his own phonemes paralleled his attention to sonic questions.

Beginning his scientific life with Davy, chemistry occupied the first phase of Faraday's activity, though he increasingly devoted his attention to electricity and magnetism, as did many others. After the announcement of Ørsted's discovery in 1820, André-Marie Ampère wrote his son that "since I have heard of the beautiful discovery of M. Oersted ... I have thought of it constantly."³ Ampère was surprised because Coulomb's work had seemed to demonstrate that any influence of electricity on magnetism was impossible because they were two essentially dissimilar fluids, and hence were incapable of interacting. Further, circular lines of force had no precedent in Newtonian physics, which was usually understood to exclude transverse forces. Ampère gave precise mathematical form to Ørsted's discovery, bringing the quantitative power of French science to this child of *Naturphilosophie*. Reconsidering Coulomb's arguments, Ampère theorized that magnetism was caused by electric currents and described the mutual forces between two parallel current-carrying wires in what came to be called "Ampère's law."

In 1821, at the request of his friend Richard Phillips, Faraday published an extensive historical survey of the new field of electromagnetism, which showed that he considered history an essential part of what he called "philosophy," not fond of the new coinages "physicist" and "scientist." Indeed, Faraday's historical reconstruction helped him correct his initial misunderstanding of Ørsted's discovery as simple attraction or repulsion, rather than a circular force acting transversely.[4] As was the case with Descartes, Faraday's correspondence shows him moving between many fields, addressing questions of chemistry, sound, and electromagnetism. For instance, his letters to Charles Gaspard de la Rive began in 1818 with their shared interest in "singing tubes," heated vials of hydrogen and other gases that produce roars and even musical tones, about which Faraday had written a paper.[5] In September 1821, Faraday's subsequent letter to de la Rive combined descriptions of investigations of steel alloys with discussion of Ampère's theories, alongside Faraday's experimental demonstration that electromagnetism could rotate a wire: the first electric motor (figure 13.1).[6]

Faraday's correspondence with Ampère showed their mutual respect as well as Faraday's hesitation to accept the literal physicality of Ampère's microscopic theory of electromagnetism, in which each molecule was a tiny current loop. For Faraday in 1821, "we have no proof of the materiality of electricity, or of the existence of any current through the wire."[7] As an alternative to "the passage of matter" being the cause of the electromagnetic effects, as Ampère had argued, Faraday now suggests "the induction of a particular state of its [the conducting wire's] parts," the first mention of the intermolecular "state" that would figure so significantly in his later thoughts on what he eventually called the "electro-tonic state."[8]

Having learned that electric currents give rise to magnetic effects, Faraday and many others sought what seemed the symmetric and complementary effect: if indeed electromagnetism were a unified, symmetric whole, magnetism ought likewise to cause electric effects. Yet despite many attempts in the following decade, no one could find any such effect. Its eventual discovery as electromagnetic induction was preceded by Faraday's return to questions of sound, which clearly helped him find this elusive phenomenon,

Figure 13.1
Faraday's palm-sized demonstration of electromagnetic rotation (1821). An electric current passes through the apparatus via liquid mercury (at the bottom of the tube), setting the rod into rotation.

though his progress was slowed by the press of many other duties and experimental programs. In 1825, he tried the fundamental experiment suggested by Ampère's theory: the magnetic forces produced from current-carrying wires should in turn generate electric effects in a nearby wire loop, but Faraday registered no effect, which increased his skepticism about Ampère's theory.[9] Returning to Faraday's quest for an alternative "state" of a current-carrying wire, beyond Ampère's material currents, L. Pearce Williams noted that "it clearly had to be something more complicated than a mere arrangement of particles, for it was difficult to see how a static arrangement could cause a dynamic rotation."[10] Averse to the imponderable "fluids" favored by Ampère and others, Faraday turned to the wave theory of light advanced by Young and taken up by Fresnel, who in 1827–1829 published a nonmathematical account in English, which Faraday studied carefully. The example of light waves offered him another possible way of thinking about the "state" surrounding the wires as able to transmit force via waves without involving the transfer of matter.[11]

Yet sound was the most obvious pattern for such wave transmission, for Young as for those who came after him; in the 1820s, the wave theory of sound was far better established than the still controversial wave theory of light. Accordingly, Faraday's turn in 1828–30 to investigations of sound form a plausible staging-ground for his reconsideration of the "state" that might unlock the electromagnetic problem he could not solve in 1825. In the years immediately preceding his breakthrough with electromagnetic induction, Faraday concentrated on examples of sound transmission that offered suggestive and helpful avenues he then pursued in electromagnetism.

This new phase began in 1828 with Faraday's involvement in public lectures on sound at the Royal Institution that were curious exercises in ventriloquism, in which Faraday did not speak on his own behalf but as the voice of Wheatstone, whose shyness inhibited him from speaking publicly. A decade younger than Faraday, Wheatstone came from a family of musical instrument makers and dealers and had no formal scientific education. At age fifteen, apprenticed to his uncle the year before, Wheatstone composed two songs that were published. He spent most of his earnings on books, such as a work on Volta through which he learned of the recent electrical discoveries. Though involved in the music business after 1823, Wheatstone did not care much for its commercial side and spent most of his effort on musical inventions; in 1824, he published a "Harmonic Diagram" to help the public understand key signatures (figure 13.2).

He later noted that "as an admirer of music … I remarked that the theory of sound was more neglected than most of the branches of natural philosophy, which gave rise in me to the desire of supplying this defect."[12] In 1823, he published his first scientific paper, "New Experiments on Sound," which begins by distinguishing longitudinal from transverse modes of vibrations, using Chladni's technique on a glass plate covered by a layer of various fluids (water, oil, mercury) to induce "crispations," slight undulations that Wheatstone interprets as the "vibrating corpuscles" or "phonic molecular vibrations" of

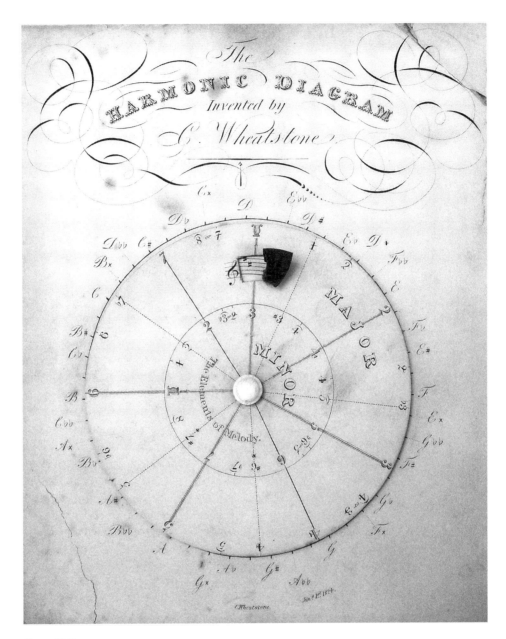

Figure 13.2
Wheatstone's "Harmonic Diagram" (1824), whose rotating wheel can be pointed to a pitch (here D), so that when the flap is raised, the correct key signature for D major appears.

the plate.[13] Wheatstone also excited such crispations by blowing a flute close to the moistened glass plate, forming "a circle round the end of the tube, and afterwards appearing to radiate in right lines; on the harmonics [overtones] of the tube being sounded, the crispations were correspondingly diminished in magnitude." He thus applied his own instrument to extend Chladni's static experiments and, in the process, evoked dynamic, transient wave phenomena that will be of increasing importance in what follows, as will Wheatstone's insistence that "the molecular vibrations pervade the entire substance of the phonic [vibrating body]," using "molecular" in the then-current sense of minute portions of substance. He also showed his experiments to Ørsted, who told Wheatstone of his own very similar earlier experiments, discussed earlier.[14] Here, as at many other points in this story, the protagonists reencounter and reinforce each other.

Wheatstone also shows his continuing interest in the problems of transmission of sound. In this first paper, he described his experiments transmitting sound along a long rod, noting his astonishment that "all the varieties of tune, quality, and audibility, and all the combinations of harmony, are thus transmitted unimpaired, and again rendered audible by communication with an appropriate receiver." Wheatstone exploited this device in his Enchanted Lyre or Acoucryptophone ("hearing a hidden sound"), first shown in London in 1821 (figure 13.3), in which a rod through the ceiling connects a piano in a room above to a lyre whose sympathetic resonance mysteriously transmits the hidden instrument. Its eerie sounds moved Wheatstone to note that "so perfect was the illusion in this instance from the intense vibratory state of the reciprocating instrument [the lyre], and from the interception of the sounds of the distant exciting one [the piano], that it was universally imagined to be one of the highest efforts of ingenuity in musical mechanism."[15]

In 1827, Wheatstone presented his new kaleidophone (literally "hearing beautiful forms"), which made visible the wave motion of a vibrating rod traced out over time by a luminous point at its end, a silvered glass bead (figure 13.4).[16] Though Wheatstone deprecatingly called this a "philosophical toy," the sonic counterpart to a kaleidoscope, he traced it back to Young's experiments making visible the vibrations of a piano wire and noted that "this instrument possesses higher claims to attention; for it exemplifies an interesting series of natural phenomena, and renders obvious to the common observer what has hitherto been confined to the calculations of the mathematician," namely the wave theory of light.[17]

Surely these words, and the striking instrument as well (still to be seen in science museums), must have resonated with Faraday's concern to penetrate the mathematical fog and reach the concrete details of the underlying "state" of wave motion. Indeed, in what follows Faraday himself acted as the resonator responding to and retransmitting Wheatstone's original impulse, as if they together constituted an Enchanted Lyre. In February, 1828 Faraday delivered Wheatstone's new communication "On the Resonances, or Reciprocated Vibrations of Columns of Air," thus allowing his friend to "speak" through Faraday's own resonantly vibrating column of air, transmuting the agony of Wheatstone's shyness into felicitous ventriloquism.[18]

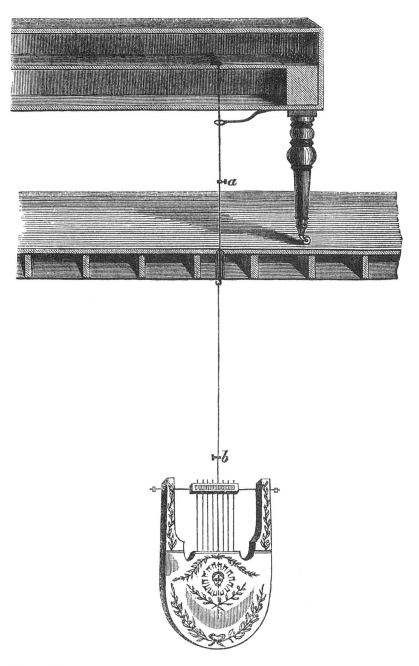

Figure 13.3
Wheatstone's Enchanted Lyre (1821); the rod transmits the sound of the piano to the lyre in the room below.

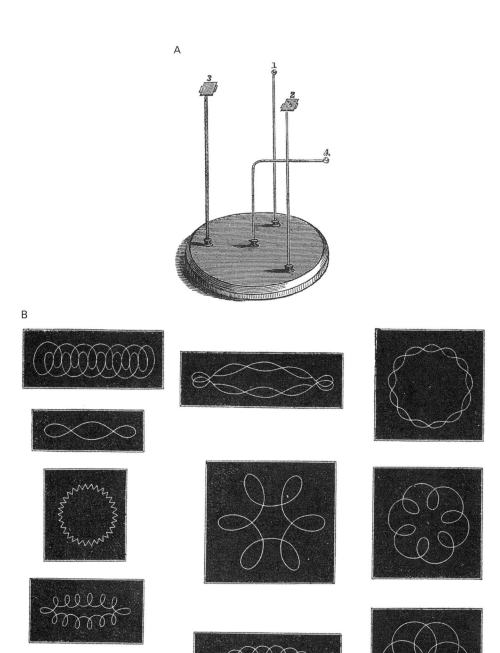

Figure 13.4
(a) Wheatstone's kaleidophone (1827), whose several rods exhibit various modes of vibration by means of a luminous point at the end of each; (b) patterns produced by the different modes of the kaleidophone.

The very first words spoken through Faraday brought forward elements that became particularly important in his subsequent work. Wheatstone began the text of their joint address by noting that an elastic body "may be made to assume a vibratory state" either "immediately, by any momentary impulse," which alters the natural position of its particles, allowing them afterward to return to their former state; secondarily, another sounding body may cause it to "reciprocate" via resonance. His historical overview goes back to Bacon's "experiments of sympathy," to Galileo making a pendulum move by "the least breath of the mouth" repeated at the resonant frequency, to Newton's comparison between a shining body's component colors and "the several pipes of an organ inspired all at once." Wheatstone notes that these sympathetic phenomena exhibit the classic Pythagorean ratios, which Faraday demonstrated by a Javanese musical instrument, a *génder*, whose tuned bamboo columns resonate vibrating plates (figure 13.5a). This exotic instrument had the aura of far-flung British colonial exploits; Faraday borrowed it from Lady Sophia Raffles, wife of the colonial governor in service of the East India Company.[19] But Wheatstone's purpose in using it went beyond mere exoticism and colonial display; he emphasized that no European instrument had yet used resonant columns of air to augment the intensity of sounds as did the *génder*, along with other Asiatic and African musical instruments. Thus, non-European musical traditions that had previously been treated condescendingly here taught techniques and designs from which European science and music might profit.

Wheatstone used the *génder* to show how resonances may occur at different frequencies than the original sound. To demonstrate this, he sounded a tuning fork next to a tube with a sliding piston, noting the resonance of the air column *"when the number of its own vibrations are any multiple of those of the original sounding body,"* but not the nonexistent undertones Euler had hypothesized.[20] Wheatstone uses these multiple resonances to explain the sound production of the humble guimbarde (sometimes called "jew's harp," though lacking any substantive connection to the Jewish people), a folk instrument of Asiatic origin (figure 13.5b). Here, the vibrating body is the steel tongue in the middle of a metal frame, which is held against parted front teeth, so that one's mouth becomes the resonating cavity, alterable in shape through changing the positions of tongue and lips. By considering the various resonances, Wheatstone explained how this simple instrument, seemingly restricted to the one fundamental frequency of its steel tongue, can produce a scale.

One imagines Faraday quite challenged to demonstrate these effects himself, serving as a human resonator for the guimbarde's metal tongue. Surely Faraday could not have managed the virtuosity Wheatstone ascribed to a certain Mr. Eülenstein, who used sixteen guimbardes "to modulate through every key, and to produce effects truly original and of extreme beauty."[21] Instead, Wheatstone had Faraday sound the instrument over resonators with movable pistons, thus achieving the scale formerly reserved to the virtuosic mouth. Such a setup could also duplicate Mr. Eülenstein's feat of sounding a major triad using three instruments simultaneously over a suitably chosen resonating tube. On a subsequent Wheatstone–Faraday lecture, a Mr. Mannin rounded out the evening by performing "some

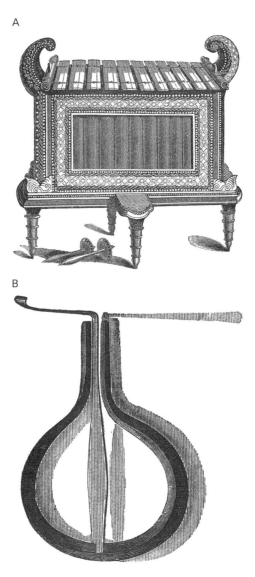

Figure 13.5
Wheatstone's illustrations of (a) "the Génder. Musical Instrument of Java"; (b) "the guimbarde or Jew's harp," from his 1828 paper.

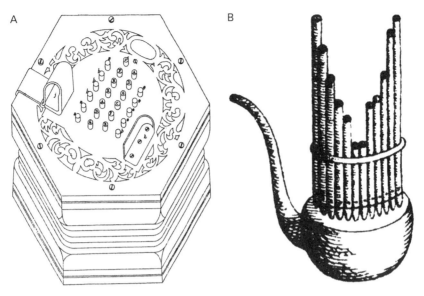

Figure 13.6
(a) Wheatstone's patent diagrams for his concertina. (b) Chladni's 1821 diagram of a Chinese *shēng*.

airs which he whistled first as solos, and then as duets," showing that the mouth itself could be divided into two differently resonating chambers.[22] Though Williams treats these as amusing, even silly, spectacles, they had a serious intent; most of the sessions of the Royal Institution included "serious" science alongside a potpourri of curiosities.[23] We might better understand them as a kind of performative *Wunderkammer*, a collection of remarkable and strange musical feats, inspired by the desire not merely to wonder but to understand, according to the emerging lights of Young, Davy, and Faraday. In the present instance, polyphonic whistling gave evidence of the multiplicity of modes of resonance.

Over the following two years, Wheatstone remained active as an inventor of musical instruments, including the concertina (1829), still in use today (figure 13.6a). This innovation was directly inspired by demonstrations of the Chinese *shēng* (figure 13.6b), a mouth-blown free reed organ brought to the West by Joseph Amiot in 1777, which Chladni described in 1821 and which led to many new Western instruments such as the harmonica and harmonium.[24]

Extending his work on the transmission of sound, in July 1830 Wheatstone announced via Faraday a method to determine the velocity of electricity in wires, which Wheatstone published in 1834.[25] Wheatstone's apparatus involved an ingenious revolving mirror that could render observable times otherwise too small to measure (figure 13.7). To test the stability of the speed of rotation of the mirror itself, Wheatstone relied on a musical device: he used the rotating arm to power a small siren, the stability of whose pitch accurately

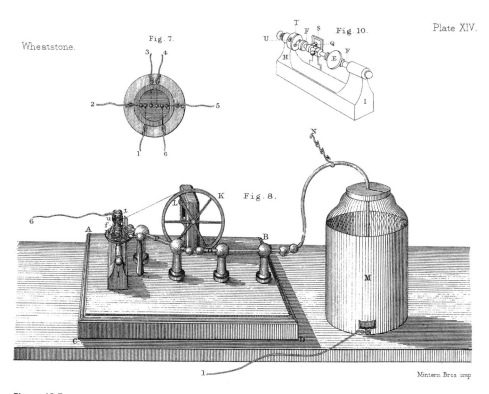

Figure 13.7
Wheatstone's rotating mirror apparatus to measure the speed of electricity in wires (1834). To test the stability of the rotation, he held a piece of paper touching the rotating arm (Q) to produce a pitch.

tested the constancy of rotation. Hippolyte Fizeau and Léon Foucault subsequently used this device to measure the speed of light in water and air, important experiments leading up to the theory of relativity.

In 1831, Wheatstone and Faraday returned to the transmission of sound. To illustrate the much greater speed of sound through solid media than through air, Wheatstone called for linked rods reaching forty feet into the cupola above the auditorium. A tuning fork applied to the upper end transmitted its vibrations to a sounding board at the bottom end with striking clarity. Faraday also brought Wheatstone's Enchanted Lyre (see figure 13.3) to the Royal Institution, its eerie resonance demonstrating the difference between longitudinal and transverse transmission by changing the angle of the lyre with respect to the rod connecting it to the hidden piano. Wheatstone described similar experiments with violins, flutes, and other instruments, finally transmitting the sound of an entire orchestra through a sounding board linked to a rod, resulting in a faint sound at the distant reciprocating soundboard: "but on placing the ear close to it, a diminutive band is heard, in which all the instruments preserve their distinctive qualities. ... Compared with an ordinary band, heard at a distance through the air, the effect is as a landscape seen in miniature beauty through a concave lens."[26] This miniature but faithful sonic transmission clearly was a step on Wheatstone's quest for telephony and telegraphy.

A decade later, he achieved long-distance telegraphy via electromagnetism, not solid rods, which required the step Faraday himself was about to take in discovering electromagnetic induction. Their joint presentations had steeped Faraday in what would prove to be two crucial elements: the relation between longitudinal and transverse vibrations (already important to Ørsted's discovery) and the essentially transient nature of the effects they had demonstrated. As Williams perceptively notes, "Although Faraday or Wheatstone did not remark the fact at the time, they were dealing with what might be called acoustical induction. An arrangement of particles on one plate could be effected by another plate thrown into a vibratory state."[27] Extending this insight (and giving more weight to Wheatstone's contribution), I would like to emphasize the specific parts played by the two crucial elements just listed.

Though Williams considers Wheatstone's work merely "suitable for the amusement of the audience at the Royal Institution," serious substance ran through that work, particularly the two persistent elements that moved Faraday to pursue sound and acoustical figures for the six months immediately preceding his discovery of electromagnetic induction in 1831.[28] His diary makes clear the scope of his acoustical work and helps us understand its relation to what follows, as does a paper he published in the midst of this work, "On a Peculiar Class of Acoustical Figures; and on Certain Forms Assumed by Groups of Particles upon Vibrating Elastic Surfaces."[29]

Faraday began his 1831 experiments by further investigating Chladni figures, using many kinds of powder and liquid, as Wheatstone had begun to do. Faraday wanted to know the fine details of exactly how and why these substances form their patterns on the plate, not just the resulting shape. He tested and found wanting a 1827 paper by Savart giving a simple account of the particles finding quiet resting places on the plate. To go more deeply, Faraday set up all kinds of baffles and partitions, designed to show exactly where the particles move, and when (figure 13.8). Looking at the patterns formed on some of his plates, Faraday surely compared them with the noticeably similar iron filing patterns he had seen produced by magnets (figure 13.9).

Above all, Faraday investigated the three-dimensional formation of these sound patterns, which previously had been treated as purely two-dimensional. He studied the air currents above the plate and how they affected the particles or liquids. Using a pump to lower the air pressure above the plate allowed still further control over the conditions in the medium forming the patterns. Faraday even did upside-down experiments, suspending liquids *underneath* the vibrating plate. Where Savart based his model only on the motion of the plate, Faraday concluded that "the nature of the medium in which those currents were formed ought to have great influence over the phenomena."[30] His observations provide a

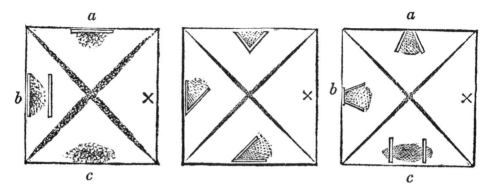

Figure 13.8
Illustrations from Faraday's 1831 paper "On a Peculiar Class of Acoustical Figures," showing the build-up of heaps of particles blocked by various paper-card obstacles; × shows where the violin bow excites the plate.

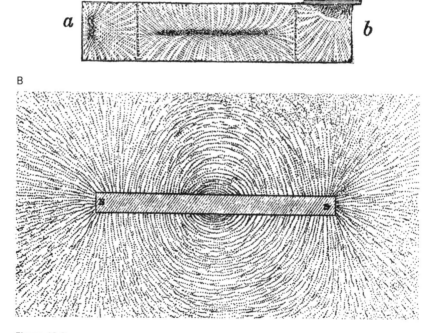

Figure 13.9
(a) Faraday's 1831 illustration of the pattern formed by lycopodium powder on a vibrating glass plate, supported by bridges or strings at the two vertical lines shown. (b) Faraday's illustration of the lines of force of a bar magnet, as made visible by iron filings.

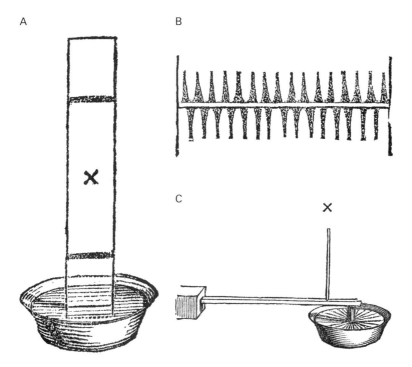

Figure 13.10
Faraday's 1831 illustrations of (a) tapping a plate at × to excite (b) the comb-shaped ripples on the surface of the water; (c) another device to excite strong radial ripples from a transverse tap at × on a rod attached to a cork extending into a bowl of water.

detailed picture of those currents in relation to the density and composition of the medium, extending several centimeters above the plate.

This thickly substantial view of the medium supported Faraday's consideration of the interplay within it of longitudinal and transverse forces, "considered as a pendulum vibrating to and fro under a given impulse," thereby giving rise to heaps of particles or crispations of the liquid surface. Perhaps remembering his apprentice days in the work-yard, Faraday instances the comb-shaped surface of "the water in a pail placed in a barrow, and that on the head of an upright cask in a brewer's van passing over stones," which he duplicated by tapping his plate (figure 13.10a,b). He was particularly pleased with "a very simple arrangement [that] exhibits these ripples beautifully" (figure 13.10c); tapping transversely created radial ripples, "the results of that vibrating motion in directions perpendicular to the force applied."[31] This significant transduction between *perpendicular* directions of motion—singularly important in Ørsted's discovery—Faraday attributes to the properties of the medium.

Faraday dated his draft paper on acoustical figures March 21, 1831, and sent it to Wheatstone, who responded in a detailed letter two days later, showing that they remained in contact during this new phase in which Faraday was conducting and writing his own version of these phenomena. Wheatstone's reply emphasizes the correctness of Faraday's experiments and inferences, *contra* Savart. In passing, Wheatstone mentions that he had shown his own experiments to Young, thus closing the circle with the older generation of wave theorists. Wheatstone also reminds Faraday of one of the techniques they had used in an 1830 joint lecture, which Wheatstone uses to give another disproof of Savart. Wheatstone ends by noting the "twofold importance" of Faraday's experiments both for "the investigation of the residual phenomena of elastic surfaces" and "further valuable information from the application of similar considerations to other phenomena with which they are intimately connected."[32] Given their shared history, these "other phenomena" may well have included the velocity of electricity.

Though many of the effects collected in his 1831 paper concern steady-state phenomena, Faraday often notes the transient quality of their onset or disappearance. For instance, he notices how "a strong steady wind" excites "stationary undulations" forming uniform ridges on the surface of shallow water. Such ridges can also have a transitory quality and are also seen "on the pavements, roads, and roofs when sudden gusts of wind occur with rain." Faraday deduces that these are not ordinary deep-water waves but physically different in causation and form, "due to the water acquiring an oscillatory condition … , probably influenced in some way by the elastic nature of the air itself and analogous to the vibration of the strings of the Aeolian harp, or even to the vibration of the columns of air in the organ-pipe and other instruments with embouchures." Faraday also thinks gases and vapors can show analogous effects, "their elasticity supplying that condition necessary for vibration which in liquids is found in an abrupt termination of the mass by an unconfined surface."[33] This spatial abruptness echoes the temporal suddenness he remarked in the gusts of wind causing ridges on a wet surface.

The collocation of temporal and spatial effects is evident in Faraday's descriptions of the striking changes in the geometry of the patterns, whose beauty he remarks at many points. Using a rectangular plate, he notices a characteristic time sequence beginning with circularly symmetric patterns centered on the source of excitation; increasing the force of vibration leads eventually to a surprising shift to a quadrangular pattern, first diagonal, and finally square (figure 13.11). Faraday notes that the reflected image of these patterns is not stationary, but rather "moves so as to re-enter upon its course, forming an endless figure, like those produced by Dr. Young's piano-forte wires or Wheatstone's kaleidophone, varying with the position of light and the observer, but constant for any particular position and velocity of vibration."[34] Evidently, Wheatstone's little instrument (and its filiation with Young) had remained in Faraday's mind.

Faraday's diary records his attention to the interplay between circular and rectangular symmetries in his experiments. On June 17, he drew the circular crispations on a round

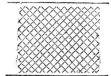 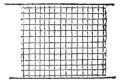

Figure 13.11
Faraday's 1831 illustration of the typical sequence of patterns shown on a rectangular plate excited near its center as the force of vibration increases (read left to right).

piece of glass barely covered in water (figure 13.12a). On July 2, he rang a glass by moving a moistened finger around its lip (à la Bacon and Mersenne) and drew its crispations (figure 13.12b) as "little ridges *apparently* permanently standing out along the surface of the water perpendicular to the glass," in contrast to "places where the crispations were weakest," marked ×, which "were breaking into confused heap like crispations. By diminish[ing] the force of vibration the former almost entirely disappeared and the latter became simple linear heaps perpendicular to the glass. *Very good.*"[35]

In his singing glass, Faraday approvingly noticed the interaction between vibrating *linear* patterns, only "*apparently* permanent," in its *circular* environment, thereby linking the medium's connection of circular and linear modes with the transience of these effects. On July 18, he wrote his diary entry at the beach at Hastings (where he was on vacation with his wife), noting the "peculiar series of ridges produced by steady strong wind on water on sandy shore." Going indoors, he "vibrated round plate on lath with water and sand so as to obtain circles and then square arrangement; the numbers of intervals between the circles and between the heaps were the same for the same plate, water, vibration, etc. etc."[36] Faraday dated the appendix to his paper on sound figures July 30, 1831, showing that his work and thinking continued past his final diary entry on this subject.

The very next entry in Faraday's diary is dated August 29, 1831 (four weeks later): "Expts. on the production of Electricity from Magnetism, etc. etc." During the intervening weeks, he had made an iron ring (about 2 cm in thickness and 91 cm in diameter) with two sets of wire windings, labeled *A* and *B* in his sketch (figure 13.12c), separated and insulated from each other so that no direct conduction of current could occur between them. The *B* coil he connected to a galvanometer to measure the current; the *A* coil could be connected to a battery. The moment of discovery was at hand, decisive yet subtle: "When the contact was made, there was a sudden and very slight effect at the galvanometer, and there was also a similar slight effect when the contact was broken."[37] But when the current through *A* was steady, there was no effect in *B*; the effect was "evident but transient," for "it continued for an instant only, and partook more of the nature of the electrical wave passed through from the shock of the common Leiden jar than of the current from a voltaic battery."[38] Earlier researchers had overlooked this effect because of its transience:

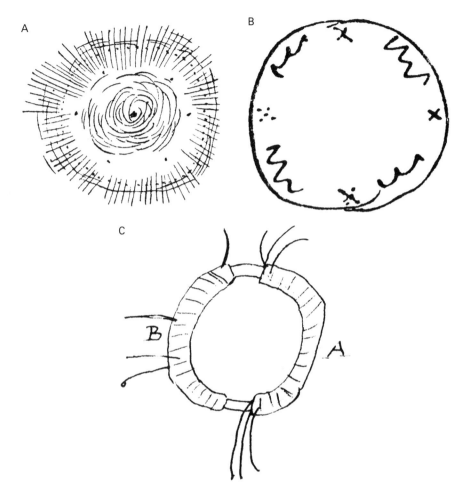

Figure 13.12
Drawings from Faraday's diary: (a) Modes of vibration excited on a circular plate barely covered by water (June 17). (b) The surface of a water glass rung with a moistened finger, showing crispations at the squiggly lines, and quiet water at × (July 2). (c) The iron ring Faraday designed for the first experiments seeking electromagnetic induction (August 29).

it vanished in the steady state and only appeared the instant contact was made or broken, and thus could easily be dismissed as a meaningless glitch.

Yet Faraday recognized it on the *very first day* he began his experiments; his mind was evidently thoroughly prepared to recognize the effect, however transient. Though his diary makes no explicit connection with his preceding experiments on sound, several clues indicate their significance for his parallel experience in electromagnetism. His published account compares the effect to an "electrical wave"; he had just spent six months studying in great detail the behavior of sound waves in various media. His careful consideration of the state of those media and their effects on the resulting patterns seems to inform the "new electrical state or condition of matter" he designates as "the *electro-tonic* state" that is "altogether the effect of the induction effect, and ceases as soon as the inductive force is removed. ... This peculiar state appears to be a state of tension, and may be considered as *equivalent* to a current of electricity, at least equal to that produced either when the condition is induced or destroyed."[39] Faraday's electro-tonic state clearly is the "state" he had sought in answer to Ampère; he would go on to use it as an important element in his further work on electromagnetism. Even his drawings of his vibrational and electro-dynamic setups show similar geometries (figure 13.12).

Six months after his discovery, on March 12, 1832, Faraday deposited in the safe of the Royal Society a statement that would "take possession as it were of a certain date, and a long right, if they are confirmed by experiments, to claim credit for the views of that date," even though he had already published the full experimental details (including his comments on the electro-tonic state) in November, 1831. The crux of this sealed statement is first that "magnetic action is progress, and requires time," as does "electric induction (of tension) [electrostatic induction]." Faraday continues:

I am inclined to compare the diffusion of magnetic forces from a magnetic pole, to the vibrations upon the surface of disturbed water, or those of air in the phenomena of sound; i.e. I am inclined to think the vibratory theory will apply to these phenomena, as it does to sound and most probably to light.[40]

Here the strong relation between sound, electromagnetism, and now also light becomes patent: Faraday summarizes his vision of physics, which he correctly anticipated would take decades of his life to complete; his entries on electromagnetic induction became the first paragraphs of his *Experimental Researches in Electricity* that eventually stretched to 3,362 paragraphs by 1855.

For his part, Wheatstone's work on the velocities of transmission of sound and electricity, combined with electromagnetic induction, led directly to his discovery of telegraphy (1837) with William Cooke.[41] Wheatstone reentered Faraday's story memorably on April 10, 1846. Scheduled to give a solo lecture on his electromagnetic chronoscope, at the very last moment Wheatstone panicked and fled, as legend has it; it is said he spotted a notorious heckler in the audience, though his own shyness may have been overwhelmed by the prospect of speaking without Faraday's sympathetic ventriloquism. To fill the evening after Wheatstone's flight, Faraday described his own speculations about the nature of electric and magnetic phenomena. His impromptu lecture was published as "Thoughts on Ray Vibrations," "one of the most singular speculations that ever emanated from a scientific mind," as John Tyndall put it.[42] As the story has come down, Wheatstone's part has tended to be portrayed as pitiable or even risible, his skittishness merely the embarrassing occasion for Faraday's profound discourse. In fact, Faraday began his own presentation by giving an account of Wheatstone's chronoscope, to which he said his further remarks were "incidental."

Wheatstone's chronoscope was a descendant of his work in the years immediately following Faraday's discovery of electromagnetic induction. In 1840, he published a "Description of an Electro-Magnetic Clock," which accurately linked a master clock with several slaves, "enabling a single clock to indicate exactly the same time in as many different places, distant from each other, as may be required."[43] To do this, Wheatstone used similar techniques to those he employed for telegraphy, incorporating the master clock into the telegraphic circuit using "Faraday's magneto-electric currents." Though Wheatstone instances the use of these linked clocks in an astronomical observatory to synchronize observations, he also notes that a modified form could be used "to act at great distances." Wheatstone's electromagnetic clock was thus the first step in the large process of synchronizing time throughout the world, so important for commercial and military purposes.[44]

In 1844, Wheatstone extended this invention to allow automated measurements of barometric pressure and temperature every half hour, controlled by his electromagnetic clock and recorded automatically on paper.[45] His setup "did not need any attention during an entire week, during which it recorded 1,008 observations," one of the first completely mechanized experiments. Though the emergent processes of industry had often relied on prior scientific advances, now Wheatstone applied the automated techniques of industry to science. In 1845, Wheatstone described his electromagnetic chronoscope, an extension of his 1840 clock allowing measurements of very small times, such as the duration of a bullet's passage through a gun.[46] Wheatstone's stopwatch was the direct precursor of the Hipp chronoscope, which became the standard instrument for precise measurements of duration.

Faraday was long aware of Wheatstone's work in all these directions, especially his results for the velocity of electricity, which seemed (perplexingly) larger than the velocity of light. Yet both were *finite* speeds: both light and electricity took *time* to propagate, as did sound.[47] The analogy with sound seemed to imply that electromagnetism and light should also propagate through some kind of medium, the ether. Young had begun to discern the problems with this imponderable substance, which Faraday now proposed to reject.[48] Speaking in 1846 in lieu of the missing Wheatstone, Faraday suggested that matter is nothing "but forces and the lines in which they are exerted" and that light radiation is "a high species of vibration in the lines of force which are known to connect particles and also masses of matter together." Faraday wanted "to dismiss the aether, but not the vibration." His new concept of lines of force suited the "lateral" (transverse) quality of light (especially its polarization), rather than the "direct" (longitudinal) transmission of sound through air or water. Relying on the finite velocities of all these phenomena, Faraday preserved the basic analogy with sound while removing the necessity of a material substratum, whose "appalling" consequences so troubled Young. Faraday also speculated that gravitation likewise propagates over time and involves curved lines of force without needing underlying matter.[49] No wonder Einstein kept Faraday's portrait above his work table. Wheatstone and Faraday's studies of sound constantly accompanied and inspired their work on electromagnetism and light, even as they stepped beyond the material framework to grasp a new world of fields.

14 Helmholtz and the Sirens

In the decades after 1850, Hermann von Helmholtz undertook extensive investigations into the nature of vision and hearing that rested on his deep interest in music and visual art. His unfolding conception of the "manifolds" or "spaces" of sensory experience radically reconfigured and extended Newton's connection between the musical scale and visual perception via Young's theory of color vision. In the process, Helmholtz's studies of hearing and seeing led him to compare them as differently structured geometric manifolds.

Helmholtz's life trajectory, spanning activity and mastery in many fields, was legendary in his own time. Though deeply interested in physics from early youth, family circumstances dictated his initial career as an army surgeon (1843–1848). Even while performing his onerous duties, he completed his seminal essay "On the Conservation of Energy" (1847), which was of great importance in establishing the fundamental status of that principle.[1] In his ensuing activities as professor of physiology at Königsberg (1849–1855), Helmholtz undertook an extensive study of many aspects of nerve action, which began with innovative experimental studies. He succeeded in measuring the velocity of propagation of nerve impulses (1850), a feat others had doubted was even possible, given the great celerity of those impulses.[2] To accomplish this, Helmholtz had to invent a myograph that allowed a frog's muscle to record itself (figure 14.1). This led, later that year, to his general study of methods of measuring the extremely small time intervals involved in this new arena of experimental physiology, for which *time* itself became both an experimental desideratum and an avenue to the attendant theoretical and philosophical questions to which he and many others had been alerted by the work of Immanuel Kant.[3] Thus, Helmholtz designed his tachistoscope (figure 14.2) to obviate the extraneous effects of eye movement by illuminating the eye with an extremely short burst of light, giving a nearly instantaneous image of the eye's position.

The *annus mirabilis* 1850 also included Helmholtz's most famed optical invention, the opthalmoscope, still in use today to examine the retina and the fundus of the eye.[4] But besides this well-known medical instrument, he also introduced many others, including the opthalmotrope, a mechanical model to demonstrate eye movements (figure 14.3). Such

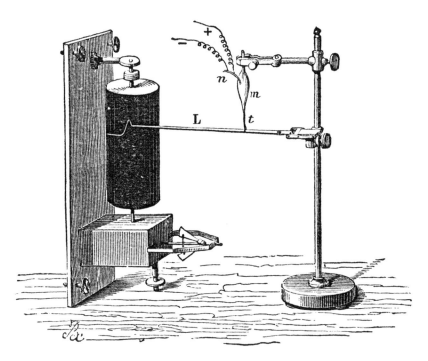

Figure 14.1
Helmholtz's myograph, used to measure the time required for nerve conduction in the thigh muscle of a frog (1873).

devices helped him develop a "sign theory" that associated each such movement and its muscular state with the attendant visual perceptions, no longer considered as realities in themselves but as symbols of underlying physiological states and their external correlates.[5]

Here, as throughout his career, Helmholtz used his experimental findings to ground his theoretical work.[6] Thus, his work on the mechanisms of vision led to his paper "On the Theory of Complex Colors" (1852), in which he revived the three-color hypothesis of Young and gave it new and fuller support from his own investigations.[7] This substantial outpouring of specialized researches on many aspects of human vision finally led to his massive *Handbuch der physiologische Optik* (*Handbook of Physiological Optics*, whose first edition appeared in three parts during 1856–1866), a *summa* whose synthetic breadth and systematic rigor put the whole field of physiological optics on a new plane of activity by applying physical principles to anatomical structures.

In part, Helmholtz accomplished this by including a historical dimension in his work, both to establish its sources and to make explicit its fundamental presuppositions. In the midst of his experimental studies, he was constantly looking to the larger theoretical

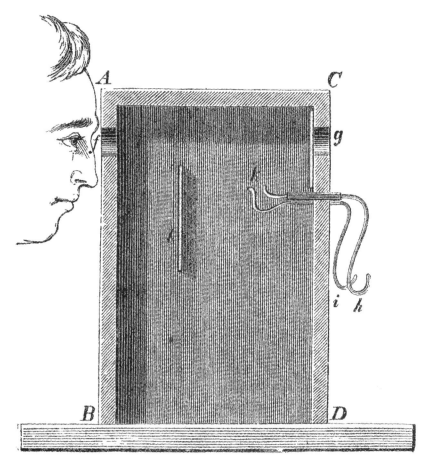

Figure 14.2
Helmholtz's tachistoscope, used to avoid involuntary movement of the eye by very brief illumination of test images (1866).

questions he hoped to resolve, which historical awareness helped him formulate more pointedly. Thus, his awareness of Young's three-color hypothesis helped him formulate the relation between human physiology and the purely physical theory of color presented by Newton. Helmholtz also provided various geometrical representations of color perception (figure 14.4), for which he used the terms "curve" (*Curven*), "color circle" (*Farbenkreis*), "color cone" (*Farbenkegel*), or "color pyramid" (*Farbenpyramide*) in his 1866 *Handbuch*.[8] In this edition, he does *not* use the terms "manifold" or "space" (*Raum*), to which we will return.

In such diagrammatic representations, Helmholtz was endeavoring to define three independent parameters of perceived color, which we now call hue, value, and saturation and

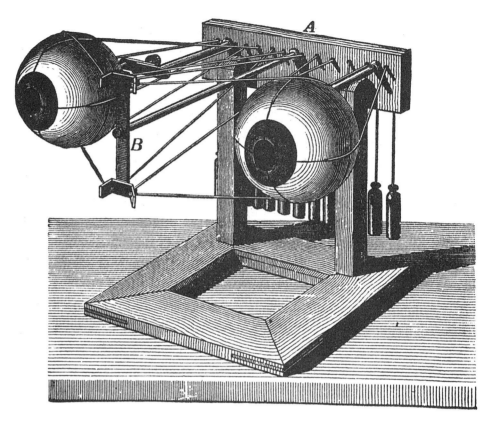

Figure 14.3
Helmholtz's opthalmotrope, a model used to study basic mechanisms of eye movements (built in 1857 by Christian Theodor Ruete).

which his work was extremely important in clarifying so as to address pervasive confusion about the exact meaning of these terms and the nuances between them. In brief, the linear sequence of the Newtonian spectrum, arranged from red to violet, is perceived by the human eye in a decidedly nonlinear way. Helmholtz's diagram (figure 14.4b) shows that, to mix colored lights to form white, a different amount of yellow must be mixed with indigo, as compared with the relative amounts of orange and cyan-blue needed to produce white. In this diagram, these differences show up in the asymmetric shape of the overall curve, whose skew toward the red-orange side reflects the higher sensitivity of human daytime vision to those colors, as compared with the blue-violet side.

In the course of this work, Helmholtz also devoted attention to the possibility of describing the perceived distances between colors "on the principle of the musical scale, because this seemed to be the best method for physiological reasons. Thus, colors whose

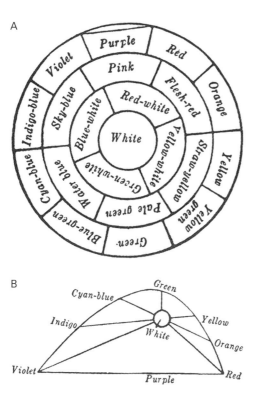

Figure 14.4
(a) Helmholtz's representations (from his *Handbook of Physiological Optics*, 1866) of Newtonian color theory using a "color circle" in which more saturated colors are near the circumference; this leaves out differences in luminosity. (b) Helmholtz contrasts this with a markedly asymmetric curve showing the relation between colors of equal luminosity.

wave-lengths are in the same ratio as the interval of a semi-tone between two musical notes are always at equal distances apart in the drawing [figure 14.5a]."[9] Helmholtz approached this parallelism in terms of Newton's imposition of the musical scale on the chromatic spectrum (figure 14.5b).

The different sensations of color in the eye depend on the frequency of the waves of light in the same way as sensations of pitch in the ear depend on the frequency of the waves of sound; and so, many attempts have been made to divide the intervals of color in the spectrum on the same basis as that of the division of the musical scale, that is, into whole tones and semitones. Newton tried it first. However, at that time the undulatory theory was still undeveloped and not accepted; and not being aware of the connection between the width of the separate colors in the prismatic spectrum and the nature of the refracting substance, he divided the visible spectrum of a glass prism, that is, approximately the part comprised between the lines *B* and *H* [in figure 14.5a], directly into seven intervals, of widths proportional to the intervals in the musical scale … ; and so he distinguished seven corresponding principal colors: *red*, *orange*, *yellow*, *green*, *blue*, *indigo*, and *violet*.[10]

Helmholtz's spectral diagram (figure 14.5a) shows only about nine semitones (hence slightly more than a major sixth) between red (*B*) and violet (*H*), rather than the twelve needed to span an octave between them, the overall interval Newton had assumed. In his diagram, Helmholtz's entire spectrum (*A–R*) spans sixteen semitones, almost an octave and a fourth, because his experimental work had shown that the ultraviolet wavelengths (*L–R*) "are not invisible, although they certainly do affect the eye comparatively much less than the rays of the luminous middle part of the spectrum between the lines *B* and *H*. When these latter rays are completely excluded by suitable apparatus, the ultra-violet rays are visible without difficulty, clear to the end of the solar spectrum."[11] Thus, his "scale of colors analogous to the notes of the piano," with yellow as middle C, extends from the "end of Red" as the F♯ below middle C to the highest visible ultraviolet frequency as the B above it.[12]

These investigations showed him that "this comparison between music and color must be abandoned," because "the spectrum is broken off arbitrarily at both ends," and hence its divisions into colors are "more or less capricious and largely the result of a mere love of calling things by names."[13] Most of all, the eye's sensitivity varies greatly: "At both ends of the spectrum the colors do not change noticeably for several half-tone intervals, whereas in the middle of the spectrum the numerous transition colors of yellow into green are all comprised in the width of a single half-tone. This implies that in the middle of the spectrum the eye is much keener to distinguish vibration-frequencies than towards the ends of the spectrum; and that the magnitudes of the color intervals are not at all like the gradations of musical pitch in being dependent on vibration-frequencies."[14] As remarkable as visual perception is, Helmholtz's critique brought forward important respects in which it falls short of the ear's capabilities to discriminate between audible frequencies.

With this in mind, starting in 1852 and overlapping with his ongoing visual researches, Helmholtz began a no less sustained and exhaustive series of investigations into the physiology of hearing. This was close to his own personal inclinations, for he had played the piano since childhood, growing up in a musical household in a music-loving country and era.[15] When he went off to university (taking his piano with him), his father warned him not to allow "his taste for the solid inspiration of German and classical music be vitiated by the sparkle and dash of the new Italian extravagances."[16] Of course, Helmholtz, as a true *Kulturträger*, a bearer of cultural tradition, was also well acquainted with the masterworks of visual arts and later wrote a popular lecture "On the Relation of Optics to Painting" (1876).[17]

Helmholtz's investigations into music, sound, and hearing began during his Königsberg period and grew after he became the professor of anatomy and physiology at Bonn (1855–1858), where he wrote "On Combination Tones" (1856), and then professor of physiology at Heidelberg (1858–1871), where he wrote "On Musical Temperament" (1860) and "On the Arabic-Persian Scale" (1862).[18] These few samples show something of the breadth of his investigations, for his interest in music led him to explore beyond the confines of European practice in order to study the cultural determination of hearing. The whole project eventuated in his masterwork, *Die Lehre von den Tonempfindungen als physiologische Grundlage für die Theorie der Musik* (*On the Sensations of Tone as a Physiological Basis for the Theory of Music*, first published in 1863), whose title proclaims *music* as the true object of his study; in contrast, his *Handbook of Physiological Optics* makes no mention of painting or the visual arts.[19] His central term *Empfindung* includes the meaning "sensation" but is also the standard term for "expression" in the artistic sense.

As with his studies of vision, Helmholtz developed or improved many instruments to undertake experimental examination of the issues that emerged, such as the glass resonators he used to isolate overtones and render them more audible (figure 14.6). The resonator acted to amplify a sonic phenomenon to make it more amenable to careful scrutiny. In other cases, Helmholtz devised means of translating and recording sonic events in a visual form, including their time dependence (figure 14.7). In this way, a tuning fork can be made to inscribe its sinusoidal vibrational pattern along a moving strip of paper, producing a visible trace that diagrammatically graphs space against time.[20]

So far, Helmholtz's sonic investigations had stayed with the study of vibrating bodies, but he realized (following the earlier example of Young) that sound was not restricted to them, however lucid was the classic mathematical analysis of their motion dating back to Euler, whose basic connection between the complexity of excitation and dissonance Helmholtz acknowledged and confirmed.[21] Where Young had reduced sound to pure puffs of air, without any vibrating body as their source, Helmholtz used the nascent technology of sirens to "mechanize" this process. He began with such instruments as the Seebeck siren, which used a rotating disc to interrupt an air stream to produce its wails (figure 14.8).[22]

Though he did not invent this instrument, Helmholtz explored and exploited its implications far beyond earlier investigators, particularly because he understood the *theoretical* implications of its construction and operation:

The sensation of a musical tone is due to a rapid periodic motion of the sonorous body; the sensation of a noise to non-periodic motions. ... [The siren] is constructed in such a manner as to determine the pitch number of the tone produced, by a direct observation. ... It is clear that when the pierced disc of one of these sirens is made to revolve with a uniform velocity, and the air escapes through the holes in puffs, the motion of the air thus produced must be *periodic* in the sense already explained. The holes stand at equal intervals of space, and hence on rotation follow each other at equal intervals of time. Through every hole there is poured, as it were, a drop of air into the external atmospheric ocean, exciting waves in it, which succeed each other at uniform intervals of time, just as was the case when regularly falling drops impinged upon a surface of water.[23]

Helmholtz, like Young before him, understood that music and noise formed a continuum, distinguished by the periodicity of the sound, or the lack thereof. The siren renders this periodicity manifest because we *see* it in the pierced disc whose rotation modulates the air stream: "equal intervals of space" between holes directly generate "drops" of air over

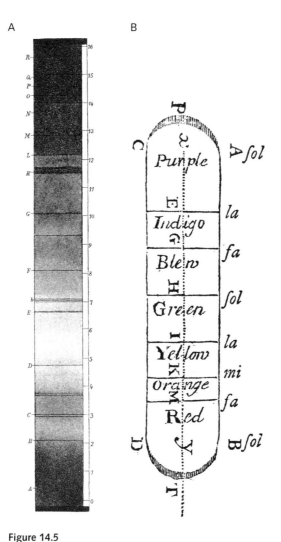

Figure 14.5
(a) Helmholtz's plate from his *Handbook of Physiological Optics* (Part II, 1860) showing the solar spectrum with the more prominent Frauenhofer lines indicated in capital letters (to the left of the corresponding dark lines) and a numerical scale (to the right) showing the correspondence between musical intervals of a semitone (labeled by successive numbers) and the spectral colors. The Frauenhofer line C roughly corresponds to red; E, green; F, "cyan-blue"; H–L, violet. (b) Isaac Newton, "An Hypothesis Explaining the Properties of Light…" (1675–76), showing his comparison between the musical scale and the spectral colors; he introduced indigo and orange in order to fill out the analogy between a complete spectrum and the seven diatonic notes in an octave.

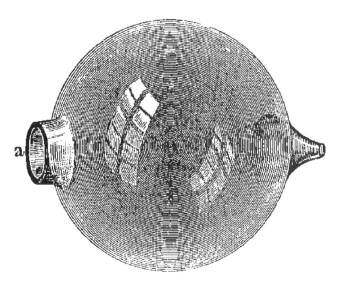

Figure 14.6
Resonator to isolate an overtone, from Helmholtz, *Tonempfindungen* (1863).

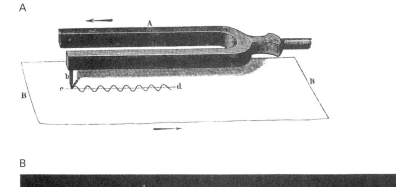

Figure 14.7
Helmholtz's illustration of the visual trace made by the motion of a tuning fork, from *Tonempfindungen* (1863).

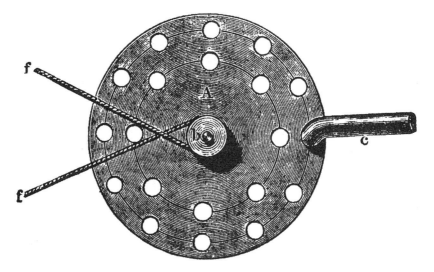

Figure 14.8
Seebeck siren, from Helmholtz, *Tonempfindungen* (1863); *c* shows the source of the air stream that is periodically interrupted by the holes in disc *A*, which the cord *f* rotates.

"equal intervals of time," audible as pure tones. Thus, Helmholtz uses the siren to map visible hole spacings into audible pitches (figure 14.9), bridging space and time through the spinning disc and the concept of *frequency*, both as the siren's rotational frequency and the sound frequencies its disc thereby generates.

Helmholtz also advanced the technology of the siren so that it could sound two pitches simultaneously, making possible comparisons in perception (figure 14.10). Such a double siren could produce "combination tones," sounding the difference or sum of two pitches more powerfully than any other instrument. Helmholtz himself discovered the faint sum tones, which he could produce only with a siren or special harmonium; the stronger difference (or "Tartini") tones had long been known. Helmholtz argued that "the greater part of the force of the combinational tone is generated in the ear itself," which combines the pure superposition of the incoming pitches, heard as two distinct tones, with their difference or sum, as predicted by nonlinear differential equations derived from Newtonian mechanics.[24] Helmholtz's use of mathematics shows its essential role in his argument here and in acoustics in general, as he conceives it. Helmholtz ascribed the failure of superposition and the resultant combination tones to "the unsymmetrical form of the [ear] drumskin itself," and, more importantly, to "the loose formation of the joint between the hammer and anvil" ossicles of the middle ear. "In this case, the ossicles may *click*," which he hears as a "mechanical tingling in the ear" when "two clear and powerful soprano voices executed passages in Thirds, in which case the combinational tone comes out very distinctly."[25] Here, his musical experience impinged strongly on the formation of his mathematical acoustics.

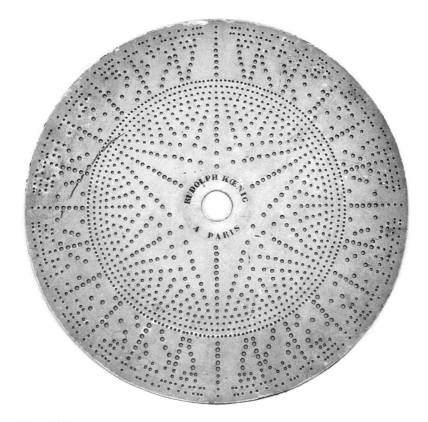

Figure 14.9
Disc for an Oppelt siren, made by Rudolph Koenig (ca. 1865) (Collection of Historical Scientific Instruments, Harvard University).

Using the double siren, Helmholtz could produce other varieties of "intermittent" or "beat tones," whose sum or difference lies below the frequencies of audible pitches, and hence are not hearable as a combination tone but felt viscerally as "a jar or rattle." Such subsonic phenomena probe the differences between hearing and seeing:

A jarring intermittent tone is for the nerves of hearing what a flickering light is to the nerves of sight, and scratching to the nerves of touch. A much more intense and unpleasant excitement of the organs is thus produced than would be occasioned by a continuous uniform tone. …

When the separate luminous irritations follow one another very quickly, the impression produced by each one lasts unweakened in the nerves till the next supervenes, and thus the pauses can no longer be distinguished in sensation. In the eye, the number of separate irritations cannot exceed 24 in a single second without being completely fused into a single sensation. In this respect the eye is far surpassed by the ear, which can distinguish as many as 132 intermissions in a second, and probably even that is not the extreme limit. …

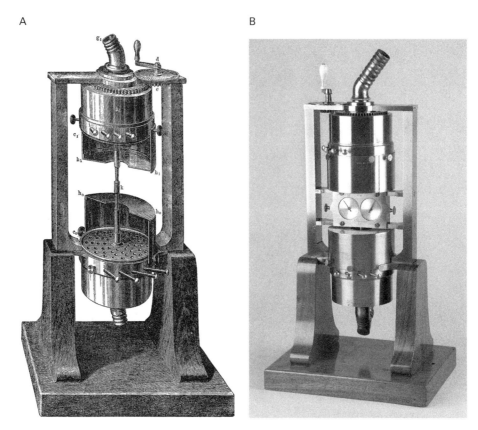

Figure 14.10
(a) Helmholtz's double siren, from *Tonempfindungen* (1863). (b) A double siren built by Sauerwald, ca. 1870 (Collection of Historical Scientific Instruments, Harvard University).

> The ear is greatly superior in this respect to any other nervous apparatus. It is eminently the organ for small intervals of time, and has been long used as such by astronomers.[26]

This striking comparison shows how far he took comparisons between hearing and seeing to illuminate their shared domains of space and time.

In an essay entitled "The Recent Progress of the Theory of Vision" (1868), Helmholtz drew attention to another fundamental contrast: vision blends several incoming colors into one perceived hue, whereas hearing always leaves several notes distinctly *separate*: "The eye cannot tell the difference if we substitute orange for red and yellow; but if we hear the notes C and E sounded at the same time, we cannot put D instead of them, without entirely changing the impression upon the ear. … The practiced musician is able to catch the separate notes of the various instruments among the complicated harmonies of an entire orchestra, but the optician cannot directly ascertain the composition of light by means of the eye; he must make use of the prism to decompose the light for him."[27] Unaided hearing, then, can perceive the precise underlying mathematical ratios within a certain harmony in ways that sight cannot perform without auxiliary instruments. Thus, his essay "On the Physiological Causes of Harmony in Music" (1857) apostrophized "Mathematics and music! The most glaring possible opposites of human thought! And yet connected, mutually sustained! It is as if they would demonstrate the hidden consensus of all the actions of our mind, which in the revelations of genius makes us forefeel unconscious utterances of a mysteriously active intelligence."[28]

Because of the ear's direct access to these mathematical underpinnings, Helmholtz did not rely completely on such mechanical devices as the siren, as useful as they are for isolating and illustrating the periodicities that underlie pitch. He constantly turned back to music itself as his touchstone of sonic experience, to which all his other experiments and speculations refer. As noted above, a sizable part of *Tonempfindungen* is devoted to a rather technical exposition of musical theory, including the sophisticated harmonies of augmented sixth chords that were important in the contemporary music of Wagner and Brahms. Among the deductions Helmholtz made from music theory, quite apart from acoustics, is a kind of principle of relativity, phrased in terms of recognizing a particular kind of *invariance*:

> We recognize the resemblance between the faces of two near relations, without being at all able to say in what the resemblance consists. ...
>
> When a father and daughter are strikingly alike in some well-marked feature, as the nose or forehead, we observe it at once, and think no more about it. But if the resemblance is so enigmatically concealed that we cannot detect it, we are fascinated and cannot help continuing to compare their countenances. And if a painter drew two such heads having, say, a somewhat different expression of character combined with a predominant and striking, though indefinable, resemblance, we should undoubtedly value it as one of the principal beauties of his painting. ...
>
> Now the case is similar for musical intervals. The resemblance of an Octave to its root is so great and striking that the dullest ear perceives it; the Octave seems to be almost a pure repetition of the root, as it, in fact, merely repeats a part of the compound tone of the root, without adding anything new.[29]

This passage comes in the final pages of the work, in its section entitled "Aesthetic Relations," as it stood in the first two editions of the book (1863, 1865). In the next chapter, we will return to his later (1870) additions that amplify this image; here already Helmholtz recognizes a special quality of spatial "resemblance" or "recurrence" in related shapes and musical intervals that are aesthetically fascinating even (or especially) when "enigmatically concealed." This quest echoes Helmholtz's favorite citation from Friedrich Schiller's poem "Der Spatziergang": the wise man "seeks a stable pole amid the flight of phenomena" (*sucht den ruhenden Pol in der Erscheinungen Flucht*).[30] Following this advice, Helmholtz sought the stability of invariance in the welter of visual and musical forms.

15 Riemann and the Sound of Space

Already in 1862, in the midst of his detailed investigations of vision and hearing, Helmholtz became interested in the more general question of the problem of space itself.[1] At first, he was unaware of the seminal work done decades before by Carl Friedrich Gauss and Bernhard Riemann. Beginning with practical problems in geodesy that originated partly in his work surveying the duchy of Brunswick, in 1827 Gauss had formulated a mathematical criterion that calculated the degree of curvature of a two-dimensional surface (its *intrinsic* or *Gaussian curvature*) only from surveying data collected within that surface.[2] Gauss proved the "remarkable theorem" (*theorema egregium*) that this curvature is invariant no matter what coordinate system is chosen in the surface.

In his 1854 lecture "On the Hypotheses That Lie at the Foundations of Geometry," Riemann generalized these ideas to what he called a "manifold" having an arbitrary number of dimensions, not just the two dimensions Gauss had considered.[3] Riemann drew the term "manifold" from Kant, who had already used it in his first published work, "Thoughts on the True Estimation of Living Forces" (1747), continuing through his celebrated discussion of space and time in his *Critique of Pure Reason*.[4] Riemann's lecture ends by indicating that his argument leads from geometry and its hypotheses "into the domain of another science, the realm of physics."[5]

Riemann based his argument on a comparison between manifolds, which he defines as comprising "multiply extended quantities," such as the coordinates of ordinary space generalized to arbitrary dimensions or the parameters describing the mixture of colors: "The general concept of multiply extended quantities, which include spatial quantities, remains completely unexplored. ... Opportunities for creating concepts whose instances form a continuous manifold occur so seldom in everyday life that color and the position of sensible objects are perhaps the only simple concepts whose instances form a multiply extended manifold."[6] Though he does not make explicit his sources, Riemann was probably referring to Helmholtz's early 1852 paper on color vision, as well as to Young's seminal work.[7] Riemann's wording also raises the question of whether or not the manifold of color perception is Euclidean in its geometry, though he does not make this explicit. His general concept of manifold included the non-Euclidean possibilities that had been revealed decades before by the work of Gauss, Nicolai Ivanovich Lobachevsky, and János Bólyai.

Indeed, the whole point of Riemann's lecture was to show that the concept of intrinsic curvature can be carried forward into manifolds no longer restricted to three dimensions. To do so, Riemann needed to express the "line element," the length of a line in the multidimensional manifold. The easiest way to do so is to generalize the Pythagorean theorem, which relates the square of the length of any line to the sum of the squares of its components along the orthogonal coordinate axes. In three-dimensional space, there are three such components, and likewise there are n components for n-dimensional space. At this point, however, Riemann paused to wonder whether there were any other possible expressions for the distance between points on a line. For instance, what about using fourth powers instead of squares? "Investigation of this more general class would actually require no essentially different principles, but it would be rather time-consuming and throw proportionally little new light on the study of space, especially since the results cannot be expressed geometrically," so Riemann restricted himself to the Pythagorean distance relation.[8] His reasoning seems to have been that because the geometry in an infinitesimal neighborhood around any point eventually approaches a flat tangent plane at that point, the distance function should *locally* always obey the Pythagorean form.

The implications of this visionary lecture excited and startled its 1854 audience, including Gauss himself, who had chosen this very topic from Riemann's list of proposals. Between then and his death from tuberculosis at the age of forty, Riemann worked intensively on several projects. He had made important strides in understanding electromagnetism and in 1858 was the first to formulate a partial differential equation expressing the propagation of the electric potential with the velocity of light, thus providing an electrodynamic wave equation.[9] By comparison, Maxwell derived such a wave equation only in 1868, *after* having set forth the field equations that today bear his name and having duly acknowledging Riemann's priority.[10] Yet Riemann was able to reach his wave equation without having completed what, for Maxwell, was necessary groundwork.

It is tempting to speculate that Riemann might have been able to complete an independent deduction of the full electromagnetic field theory, had he lived longer. As it was, his wave equation explicitly linked the *time* and *space* behavior of the electric potential. His 1854 lecture had positioned him to consider higher-dimensional manifolds; his electromagnetic wave equation offered him a link between the "dimensions" of space and time. If so, one could imagine him entertaining a four-dimensional space-time manifold long before Einstein and Minkowski.

Though this did not, in fact, happen, this thought experiment in counterfactual history may illuminate what Riemann did do. During the period 1854 to 1861, in which he could, imaginably, have discovered the full electromagnetic field equations, he produced the mathematical work on distribution of prime numbers and the zeta function (see box 9.2), which became known as the Riemann Hypothesis (1859), arguably his most famous initiative and the premier unsolved mathematical problem up to the present day.[11] This, by itself, surely helps to explain why he might not have placed electromagnetism higher on his list of priorities, though his surviving drafts and papers show his continuing interest in physics, not to mention his other important mathematical projects. The speculative writings and drafts in his posthumous papers show that his attention in natural philosophy was directed toward the possible unification of gravitation and electricity.[12] Given the general framework of his 1854 lecture, Riemann's project seems to have envisaged using his many-dimensional curved manifolds as the framework for a grand unified theory of all physical forces.

These private theoretical drafts give the context for what remained, at his death, his major uncompleted paper entitled "The Mechanics of the Ear."[13] For both Riemann and Helmholtz, the problem of hearing was a significant part of their larger enterprises, an intermediate zone in which waves, geometry, and sensation met. Riemann's choice to study the ear (rather than the eye) is also noteworthy; surely questions of hearing must have seemed very important to him if he set them next to or even ahead of his other ambitious projects in electrodynamics, gravitation, and number theory. By comparison with Helmholtz, little evidence survives that would give biographical insight into Riemann's choice. The son of a pastor and himself deeply religious, Riemann considered "daily self-examination before the face of God" to be "the main point in religion." Alongside this austere, contemplative persona, Riemann evidenced considerable love of art. According to his friend Richard Dedekind, Riemann's long stays in Italy after 1862, seeking to recover his health, "were a true luminous point in his life … looking at the glory of this enchanting land, of nature and art, made him endlessly happy." The newly married Riemann took "great interest" in the "art treasures and antiquities" of Italy, also greatly admired by other *Kulturträger*, such as Helmholtz.[14] Like most of them, Riemann probably felt deeply the power of music.

At any rate, Riemann's deep interest in understanding the ear shines through his essay. Riemann praises Helmholtz's ingenious experimental work on hearing, while criticizing its findings and basic methodology. In Riemann's view, Helmholtz *synthesizes* the anatomical structures of the ear into the functioning of the whole organ, but only at the cost of making questionable assumptions about the goals of those structures. Instead, Riemann advocates an alternative process of *analysis* that begins with the observed behavior of the whole organ and then constructs a mathematical model that would explain those functions in necessary, not merely sufficient, terms. By emphasizing the central functions of the organ as a whole, Riemann strives to avoid Helmholtz's suppositions about the purposive interrelation of its anatomical subunits. Riemann uses anatomical knowledge for clues to guide his model-building, not as a definitive level of explanation.

The post-Kantian language of analysis and synthesis, the contrast between necessity and sufficiency, marks Riemann's approach as essentially mathematical and hypothetical in spirit. "We do not—as Newton proposes—completely reject the use of analogy (the 'poetry of hypothesis'), but rather afterwards emphasize the conditions that *must* be met to account for what the organ accomplishes, and discard any notions that are not essential to the explanation, but that have arisen solely through the use of analogy."[15] In contrast to Newton's famous avoidance of "feigning" hypotheses, Riemann's remarkable expression "the 'poetry of hypothesis' [*Dichten von Hypothesen*]" rhetorically emphasizes the creative freedom of imagination, its suggestive power in the formation of analytic representations of phenomena, whether aural or geometric, in the form of hypotheses that are not restricted by anatomic presuppositions.

With this in mind, we can read Riemann's "Mechanism of the Ear" as a nascent essay "On the Hypotheses That Lie at the Foundations of Hearing," comparable to his earlier work on the hypotheses he considered fundamental to geometry. Enough remains of Riemann's draft to show some general features of his proposed analysis. Against Helmholtz's assertion that the ossicles click, Riemann notes that "the apparatus within the tympanic cavity (in its unspoiled condition) is a mechanical apparatus whose sensitivity is infinitely superior to everything we know about the sensitivity of mechanical apparatuses. In fact, it is by no means improbable that it faithfully transmits sonic motions that are so small that they cannot be observed with a microscope." For instance, "the call of the Portsmouth sentry is clearly audible at night at a distance of 4 to 5 English miles," so that "the ear does pick up sounds whose mechanical force is millions of times weaker than that of sounds of ordinary intensity."[16] This, he feels, negates Helmholtz's claim about the noisiness of the ossicles, which Riemann judges a supposition introduced primarily to support Helmholtz's theory of combination tones.

Instead, Riemann's approach is much closer to what now is called systems theory: he treats the ear as a "black box" whose overall functioning can be mathematically modeled based on its essential parameters, especially its high sensitivity and fidelity.[17] His modeling involves pointed comparisons with vision: "I find nothing whatsoever [in hearing] analogous to the eye's response to the degree of illumination of the visual field, and have no idea what a continuously variable reflex activity of *M. tensor tympani* is supposed to contribute to the exact comprehension of a piece of music."[18] Here Riemann refers to the tensor tympani muscle that attaches to the hammer bone of the middle ear and can dampen the vibrations of the tympanic membrane. Though Helmholtz had not explicitly extended his sign theory to hearing, Riemann seems to take him to imply that the varying states of the tympanic muscle are "local signs" of the associated sounds, as the movements of the eye muscles are signs of what it sees.[19] If so, the variable activity of the tensor tympani muscle would correlate with auditory response to varying musical sounds. In contrast, Riemann argues that a *constant* tension of this auditory muscle should accompany the activity of "the alert ear—the ear deliberately prepared for precise perception," whose acuity depends on the tympanic muscle to maintain steady contact between the ossicles and the inner ear.[20]

Riemann's analytic program required that "we must now derive from the empirically known functions performed by the organ, the conditions which must be met in this transmission ... [by] seeking a mathematical expression for the nature of the pressure fluctuation upon which timbre depends."[21] Though in his 1854 lecture, Riemann held that "color and the position of sensible objects are perhaps the only simple concepts whose instances form a multiply extended manifold," by 1866 he seems poised to treat hearing as a further example of such a manifold. Riemann does not provide any mathematical details of his approach to hearing, but, based on his work on geometrical hypotheses and his work on shock waves in fluids, we may infer that he intended to use some kind of multidimensional manifold, analogous to those he proposed to represent the geometric effect of physical sources.[22] Where Helmholtz took evidence from hearing and seeing into his geometric investigations, Riemann traversed an opposite course, applying geometric insights to model the functioning of the ear.

In its unfinished form, Riemann's "Mechanism of the Ear" was published posthumously in a medical journal in 1867. Helmholtz responded in two papers, both entitled "On the Mechanism of the Ossicles of the Ear" (1867, 1869), whose titles once again reflects the fundamental contrast between the two men: Helmholtz's "facts" (or ossicles) versus Riemann's "poetry of hypothesis" (which treated the ear as a high-sensitivity sound transducer, regardless of its anatomical details).[23] Though publicly Helmholtz wrote respectfully of the "great mathematician's" foray into his own domain, privately he expressed irritation at Riemann the "amateur."[24] In his printed response, Helmholtz did not engage Riemann's philosophical contrast between analytic and synthetic, but argued that the ossicles can act "practically, as absolutely solid bodies" that thereby can transmit sound with the high sensitivity Riemann had emphasized. To show that his anatomical model could meet Riemann's critique, Helmholtz gave a detailed account of the fine structure of the ossicles and their subtle interconnections, as well as of the tensor tympani muscle (figure 15.1). Rhetorically, Helmholtz swept away Riemann's theorizing under a deluge of anatomical observations, implicitly arguing that only in such terms can any physiology of the ear be responsibly phrased. For the time being, Riemann, the defunct "amateur," was quietly buried under a mountain of Helmholtz's "professional" anatomy.[25]

This controversy about hearing led Helmholtz to devote much attention to Riemann's work, though he received Riemann's 1854 lecture only in May 1868, the year after it finally appeared in print.[26] Yet even *before* he had read it, Helmholtz had already inferred "that Riemann came to exactly the same conclusion as myself," as he wrote Ernst Schering on April 21, 1868:

My starting-point is the question: What must be the nature of a magnitude of several dimensions in order that solid bodies (i.e. bodies with unaltered relative measurements) shall everywhere be able

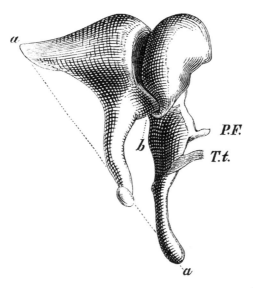

Figure 15.1
A diagram showing Helmholtz's response to Riemann regarding the precise functioning of the hammer and anvil, from "Mechanism of the Ossicles of the Ear" (1873). *T.t.* is the tendon of the tensor tympani, *P.F.* the processus Folianus (part of the hammer), *b* the cog of the anvil, and *aa* shows the straight line transmitting the action of the hammer to the tympanum (ear drum).

to move in it as continuously, monodromously, and freely, as do bodies in actual space? Answer, expressed according to our analytical geometry: "Let x, y, z, t be the rectangular co-ordinates of a space of four dimensions, then for every point of our tri-dimensional space it follows that $x^2 + y^2 + z^2 + t^2 = R^2$, where R is an undetermined constant, which is infinite in Euclidean space."[27]

This extraordinary statement has received little notice, though (to my knowledge) it may be the first explicit use of four dimensions to address the problem of space, aside from a few speculative remarks by Jean le Rond d'Alembert (in 1754) and Joseph Louis Lagrange (in 1797).[28] As we shall see, this formulation remained in Helmholtz's mind.

Helmholtz's pursuit of invariance, whether resemblances in the visual field or recurrences in music, led directly to his paper "On the Factual Foundations of Geometry" (1868), which begins with an explicit connection to his work on the physiology of vision:

Investigations into how localization in the visual field comes to pass have led the author also to reflect on the origins of spatial intuition in general. This leads first of all to a question whose answer definitely belongs to the sphere of exact science, namely, which propositions of geometry express truths of factual significance and which, on the contrary, are only definitions or consequences of definitions and their particular manner of expression? …

One could follow this direction and find out which analytical characteristics of space and spatial magnitudes must be presupposed in order to ground the propositions of analytic geometry completely from the beginning.[29]

For Helmholtz, questions about "the origins of spatial intuition in general" emerge from studies of the visual system and lead directly to considerations about the nature of geometry. As a result, he breaches the customary barrier between the propositions of geometry and physical reality, previously considered separate from one another.

Though during this period he had been mainly working on experimental physiology, Helmholtz reveals that he had gone remarkably far in his own self-directed reconsideration of the mathematical and philosophical problems concerning the nature of space: "The author had already begun such an investigation and had completed it in the main when Riemann's habilitation lecture 'On the Hypotheses That Lie at the Foundations of Geometry' was made public, in which an identical investigation is carried out, having only a slightly different formulation of the question. On this occasion, we learned that Gauss had also worked on the same subject matter, of which his famous essay on the curvature of surfaces is the only published part of that investigation."[30] Riemann's argument assumed a generalized quadratic line element but did not prove its necessity. Helmholtz asked whether there is some fundamental reason that would necessarily mandate this assumption, rather than other, more general possibilities.

Helmholtz's 1868 paper summarized his response to this problem.[31] Though he shared with Riemann the fundamental idea that geometry ultimately rested on physics rather than on transcendental ideas, Helmholtz replaced Riemann's "hypotheses" with "facts." Steeped in Goethe, like his educated contemporaries, Helmholtz knew by heart Faust's emendation of the Gospel of St. John's opening line from "In the beginning was the Word" to "In the beginning was the Deed" ("*Im Anfang war die That*"); like Faust, Helmholtz moved from the Word (or Riemann's "hypotheses") to the Deed, understood as the Fact.[32]

Helmholtz argued that fundamental physical facts necessitate the quadratic form of the line element. Specifically, he assumes "(1) *continuity and dimensions*" (each point in space is determined by n continuous, independent variables), "(2) *the existence of moving and rigid bodies*," "(3) *free mobility*" ("each point can pass over into any other along a continuous path"), and finally "(4) *the invariance of the form of rigid bodies under rotation*." From these premises, he deduced that "if we desire to find the degree of rigidity and mobility of natural bodies attributable to our space in a space of otherwise unknown properties, the square of the line element ds would have to be a homogenous second-degree function of infinitely small increments of the arbitrarily chosen coordinates u, v, w. This proposition ... [is] the most general form of the Pythagorean Theorem. The proof of this proposition vindicates the assumption of Riemann's investigations into space."[33] In his original draft, Helmholtz thought this meant that quadratic form had to correspond to *Euclidean* geometry, but Eugenio Beltrami and Sophus Lie soon objected to this erroneous overspecialization of a more generalized result that Helmholtz should have found: in fact, the quadratic form was, in general, non-Euclidean, as Helmholtz acknowledged in a note appended to his 1868 paper.[34] In the subsequent literature, this issue became known as the Helmholtz–Lie *Raumproblem*, the so-called problem of space; not a merely technical matter or a fine point of mathematical rigor, this has deep implications for Einstein's geometric account of gravitation because it dictates the fundamental form of the metric, the geometrical field created by the bodies immersed in it, which in turn move along its shortest (geodesic) paths.[35]

Helmholtz's oversight probably implies his initial lack of knowledge about non-Euclidean geometry, confirming that he may not have been aware of the non-Euclidean import of his color diagrams (such as figure 14.4) when first he published them in 1866, before his enlightenment by Beltrami and Lie.[36] In his 1868 paper, he emphasized that

> the independence of the congruence of rigid point-systems from place, location, and the system's relative rotation is the fact on which geometry is grounded.
>
> This becomes even clearer when we compare space with other multiply extended manifolds, for example the system of colors. In this case, as long as we have no other method of measurement than through the law of color mixing, there exists, unlike in space, no relation of magnitudes between any two points that can be compared with that between two other points. Instead, there exists a relation between groups of any three points that also must lie in a straight line (that is, in groups of any three colors, among which any one is mixable into the other two).
>
> We find another difference in the field of vision of a single eye, where no rotations are possible so long as we confine ourselves to natural eye movements.[37]

Under the influence of Riemann's conception of manifold, Helmholtz now reinterprets his earlier diagrams of "the system of colors" as a "threefold-extended manifold" akin to three-dimensional space (*Raum*).[38] Though we have become used to the notion that non-spatial magnitudes can be described as if they constituted a "space," the broadening of the concept of space should be credited to Riemann's manifolds.[39] Following on Helmholtz's pioneering experimental studies of vision, Riemann adduced space and color as comparable manifolds, terminology Helmholtz then used to categorize the "system of colors" more deeply.

Overlapping his work on the "problem of space," Helmholtz returned to musical concerns as he prepared a third edition of his *Tonempfindungen* (1870). His new additions clarified the significance of music for his thinking about geometry in the process of developing his nascent ideas of resemblance and invariance. In the 1870 version, he expanded the concluding passage of the work concerning visual resemblance and musical recurrence, adding that this is "by no means a merely external indifferent regularity," compared to the way "rhythm introduced some such external arrangement into the words of poetry." Instead, he showed "that the equality of two intervals lying in different sections of the scale would be recognized by immediate sensation. ... This produces a definiteness and certainty in the measurement of intervals for our sensations, such as might be looked for in vain in the system of colors, otherwise so similar, or in the estimation of intensity in our various sensual perceptions."[40]

The invariance of musical intervals or melodies, when transposed, has no precedent in the "space" of color; we can transpose a Beethoven sonata up a half step and still recognize the work as in some sense still the same, yet we cannot likewise "transpose" all the colors of a Rembrandt (say by shifting all reds to orange, orange to yellow, and so forth): though the basic line and surface contours of the painting remain unchanged, its color harmony cannot be "transposed" and remain recognizably identical. Helmholtz extended the special quality of spatial "resemblance" that can be seen in related shapes (the similar profiles of father and daughter) and to the characteristic melodic contour of a certain piece of music, but *not* to colors.

Helmholtz goes on to emphasize the consequences of this resemblance or invariance in music:

Upon this reposes also the characteristic resemblance between the relations of the musical scale and of space, a resemblance which appears to me of vital importance for the peculiar effects of music. It is an essential character of space that at every position within it like bodies can be placed, and like motions can occur. Everything that is possible to happen in one part of space is equally possible in every other part of space and is perceived by us in precisely the same way. This is the case also with the musical scale. Every melodic phrase, every chord, which can be executed at any pitch, can be also executed at any other pitch in such a way that we immediately perceive the characteristic marks of their similarity. On the other hand, also, different voices, executing the same or different melodic phrases, can move at the same time within the compass of the scale, like two bodies in space, and, provided they are consonant in the accented parts of bars, without creating any musical disturbance. Such a close analogy consequently exists in all essential relations between the musical scale and space, that even alteration of pitch has a readily recognized and unmistakable resemblance to motion in space, and is often metaphorically termed the ascending or descending *motion* or *progression* of a part. Hence, again, it becomes possible for motion in music to imitate the peculiar characteristic of motive forces in space, that is, to form an image of the various impulses and forces which lie at the root of motion. And on this, as I believe, essentially depends the power of music to picture emotion.[41]

Because music relies on the recognition of analogy, resemblance, and invariance, Helmholtz deduces that it therefore can "imitate the peculiar characteristic of motive forces in space"; though not itself spatial or extended, music can *move* in precise analogy to spatial motion, from which Helmholtz boldly identifies the emotive force of music: its virtual motion is felt as emotion precisely because of the isomorphism between musical and physical space.

Over the next few years, Helmholtz extended the implications of his 1868 arguments about the *Raumproblem*. He presented popular lectures and essays, addressed to a wider audience, concerning larger philosophical issues emergent from his own work.[42] Immediately after completing the additions we have just considered to the 1870 edition of his *Tonempfindungen*, Helmholtz delivered "On the Origin and Meaning of Geometrical Axioms," which discussed "the philosophical bearing of recent inquiries concerning geometrical axioms and the possibility of working out analytically other systems of geometry with other axioms than Euclid's."[43]

For the first time in his writings on the problem of space, Helmholtz described in detail non-Euclidean geometries and the pseudosphere of Beltrami, whose criticisms had first moved Helmholtz to address this issue directly. Helmholtz also brought forward a striking device for comparing and contrasting these different geometries: "Think of the image of the world in a convex mirror." In the mirror-world, the theorems of Euclidean geometry would instantly be translated into non-Euclidean image-theorems, at least as seen from our side of the mirror. "In short I do not see how men in the mirror are to discover that their bodies are not rigid solids and their experiences good examples of the correctness of Euclid's axioms. But if they could look out upon our world as we can look into theirs, without overstepping the boundary, they must declare it to be a picture in a spherical mirror, and would speak of us just as we speak of them; … neither, so far as I can see, would be able to convince the other that he had the true, the other the distorted relations." As further evidence, Helmholtz also adduces the eye's ability to accommodate seeing through "convex spectacles," of the sort he had experimented with in the course of his visual studies: "After going about a little the illusion would vanish. … We have every reason to suppose that what happens in a few hours to anyone beginning to wear spectacles would soon enough be experienced in pseudospherical space. In short, pseudospherical space would not seem to us very strange, comparatively speaking," once we had gotten used to it, just as our eyes would quickly get used to those "distorting" spectacles.[44] Helmholtz's penetrating insight into the relative consistency of these seemingly antithetical geometries, Euclidean and non-Euclidean, was directly indebted to his studies of visual physiology.

Looking back at his recent work, he remarks that "while Riemann entered upon this new field from the side of the most general and fundamental questions of analytical geometry, I myself arrived at similar conclusions, partly from seeking to represent in space the system of colors, involving the comparison of one threefold extended manifold with another, and partly from inquiries on the origin of our ocular measure for distances in the field of vision."[45] As in his 1868 paper, Helmholtz locates his own "facts" as confirming Riemann's "hypotheses."

In his 1870 exposition, besides adducing the three-dimensional manifolds of "the space in which we live" and "the system of colors," Helmholtz adds that "time also is a manifold of one dimension."[46] Here, for the first time, time enters the discussion as a manifold, albeit one-dimensional.[47] Nor did Riemann include time explicitly in his geometrical (hence implicitly spatial) manifolds. Immediately after his mention of time, Helmholtz goes on to include the manifold of musical tones, whose time-dependence he had studied so closely:

In the same way we may consider the system of simple tones as a manifold of two dimensions, if we distinguish only pitch and intensity and leave out of account differences of timbre. This generalization of the idea is well-suited to bring out the distinction between space of three dimensions and other manifolds. We can, as we know from daily experience, compare the vertical distance of two points with the horizontal distance of two others, because we can apply a measure first to the one pair and then to the other. But we cannot compare the difference between two tones of equal intensity and different pitch. Riemann showed by considerations of this kind that the essential foundation of any system of geometry is the expression that it gives for the distance between two points lying in any direction from one another.[48]

Helmholtz's concept of manifold includes music and sound in the same arena as space, time, color, and vision. Though *simple* tones may be described as a manifold of two dimensions, Helmholtz had investigated the parameters of timbre that distinguish complex musical sonorities from simple tones. At this point, the question of dimensionality seems open: going beyond the two dimensions of simple tones, how many dimensions are needed to describe the full character of musical "space"? And what then of the dimensional relations between space and time?[49] Though he does not go further with these questions, Helmholtz leaves the *Raumproblem* as the shared heritage of the manifolds of music, vision, space, and time.

The conclusion of Helmholtz's 1876 revised version of his essay "On the Origin and Meaning of Geometrical Axioms" clarified his current understanding that

(1) The axioms of geometry, taken by themselves out of all connection with mechanical propositions, represent no relations of real things. ... They constitute a form into which any empirical content whatever will fit and which therefore does not in any way limit or determine beforehand the nature of the content. This is true, however, not only of Euclid's axioms, but also of the axioms of spherical and pseudo-spherical geometry.

(2) As soon as certain principles of mechanics are conjoined with the axioms of geometry we obtain a system of propositions which has real import, and which can be verified or overturned by empirical observations.[50]

Helmholtz stepped decisively beyond Kant by including Euclidean and non-Euclidean geometries on the same footing, each "a form into which any empirical content whatever will fit."[51] Hence, the axioms of geometry must meet "certain principles of mechanics" in ways that finally rest on empirical observations. Helmholtz's view of this empirical confrontation was informed both by optics (and visual physiology) and mechanics (and its connection to acoustics and music).

In this revised, 1876 version, Helmholtz also added a mathematical appendix on "the elements of the geometry of spherical space," the same four-dimensional manifold he had mentioned to Schering in his 1868 letter, cited above, described by the expression $x^2 + y^2 + z^2 + t^2 = R^2$. Though there is no hint that t is not a fourth spatial coordinate, its common identification as time pervaded contemporary mathematical physics; Helmholtz also allowed t to become an imaginary quantity, further increasing the similarity with the pseudo-Euclidean space-time later used by Einstein and Hermann Minkowski.[52]

Such beguiling speculations aside, it would go too far to conclude that Helmholtz had (even unknowingly) written down an expression from relativistic physics, fifty years in advance. His appendix, however, does illustrate his ability to invoke a four-dimensional manifold to describe mathematically our visual experience looking "through a pair of convex spectacles" that have been specially ground to give a negative focal length. Consonant with his empirical method, Helmholtz showed that we could thereby imagine a four-dimensional pseudospherical "space," contra Kant. Long before Edwin Abbott's *Flatland* (1892), in this essay Helmholtz was probably the first writer to describe "reasoning beings of only two dimensions" who "live and move on the surface of some solid body" in order to help us imagine the felt reality of higher dimensions.[53]

The influence of Riemann and Helmholtz remained crucial in subsequent developments of the problem of space. As noted above, Lie embedded his correction of Helmholtz's erroneous generalizations in the emergent structure of his theory of continuous groups.[54] Aside from William Kingdon Clifford's solitary (and visionary) response, Riemann's work lay dormant among his immediate successors. The philosophical implications of Helmholtz's work were important to Felix Klein in connection with his Erlangen program to characterize spaces by their characteristic groups of transformations and respective invariants.[55] As Klein remarked in 1893, "Our ideas of space come to us through the senses of vision and motion, the 'optical properties' of space forming one source, while the 'mechanical properties' form another; the former corresponds in a general way to the projective properties, the latter to those discussed by Helmholtz."[56]

Henri Poincaré emphasized Riemann's work and also responded strongly to Helmholtz's arguments, in connection with his own view that convention and convenience underlie the choice of a geometry for space.[57] Poincaré also carried forward Helmholtz's thought experiment of viewing the Euclidean world through convex mirrors or distorting spectacles, which Poincaré phrased in terms of a "dictionary" that would translate the terms of Euclidean geometry into non-Euclidean terms, one for one, so as to make clear that each geometry was no less consistent than the other.[58] Thus, within purely Euclidean geometry, a model could be made using Euclidean figures that behaved in every respect like Lobachevskian geometry, once the fundamental elements (lines, angles, etc.) had been suitably redefined, corresponding to the action of the distorting mirrors or lenses; conversely, Lobachevskian geometry could be made to behave as if it were Euclidean by a similar set of redefinitions. Poincaré's argument and Klein's further activities in providing other such models were crucial steps in understanding the relationships between the different geometries as not only equally *possible* but equally *consistent*. This demonstrated equality of status in turn opened the possibility of addressing the empirical observations that (as Helmholtz suggested) might then ground the choice between geometries.

In the midst of a letter to Mileva Marić written in August 1899, the twenty-four-year-old Albert Einstein paused to tell his girlfriend that "I admire ever more the original, free thinker Helm[holtz]."[59] The protean activities of Helmholtz resonated sympathetically with Einstein; both were deeply interested in fundamental principles of science, such as the law of conservation of energy that Helmholtz advanced so powerfully and which Einstein inscribed in relativistic dynamics. Both were devoted to music; both were concerned with light, Helmholtz with its physiology, Einstein with its speed and interactions with matter. Both were engaged by the "problem of space," the general question about the possible geometries of space and experience. Around 1903, Einstein and his friends in the "Olympia Academy" read Helmholtz as well as Riemann.[60]

Einstein's general theory of relativity gave a precise form to Riemann's connection between the empirical world (understood as composed of stress-energy) and the geometrical (the invariant curvature of space-time).[61] Rather than ignoring the history of these concepts (as he sometimes is represented to have done), in fact Einstein was deeply conscious of them and drew not only general inspiration but specific guidance from what went before. As he wrote Robert Thornton in 1944, "A knowledge of the historical and philosophical background gives that kind of independence of prejudices of his generation from which most scientists are suffering. This independence of philosophical insight is—in my opinion—the mark of distinction between a mere artisan or specialist and the true seeker after truth."[62] Einstein's own essays contain a wealth of historical reflection and awareness, such as his observation that "only the genius of Riemann, solitary and uncomprehended, by the middle of the last century already broke through to a new conception of space, in which space was deprived of its rigidity and in which its power to take part in physical events was recognized as possible." Indeed, Riemann had worked out the curvature tensor (now named after him) that was all-important for Einstein's general theory.[63] Einstein's tribute pays what he recognizes as a major debt.

Einstein's words in praise of Riemann are far better known than his 1917 encomium of Helmholtz's Goethe essays—"Dear reader! Summarizing would be profanation. Read for yourself!"—or his 1925 *hommage*: "[that] all propositions of geometry gain the character of assertions about real bodies ... was especially clearly advocated by Helmholtz, and we can add that without him the formulation of relativity theory would have been practically impossible."[64] Einstein considered Helmholtz's connection of geometric hypotheses with empirical facts absolutely crucial for the general theory of relativity, whose field equations epitomize that connection. To reach that point, Helmholtz connected his work in music and vision, hearing and seeing, whose comparison lay at the grounds of his synthetic understanding. His dialogue with Riemann reflected and underscored the significance of their shared concern with hearing in the context of the problem of space and the physical foundations of geometry.

16 Tuning the Atoms

By the end of the nineteenth century, the Pythagorean quest might have seemed played out. With Newtonian and Maxwellian physics securely in place, the metaphorical language of harmony, not to mention the details of music theory, might appear to be only a historical vestige, a transitional scaffolding that by then could be left behind. But facing the puzzles and paradoxes of the nature of matter first raised by the study of spectra, physicists again resorted to the precise kinds of numerological-musical theorizing that had many precedents in the Pythagorean episodes discussed above. At such moments of trial and disorientation, it was as if the scaffolding reemerged from the buildings into which it had seemed absorbed, ready once again to help us scale what appeared to be unsurmountable obstacles. The story that unfolded was epitomized in "the bible of spectroscopists," Arnold Sommerfeld's *Atomic Structure and Spectral Lines* (1919), which begins by asserting that "what we are nowadays hearing of the language of spectra is a true 'music of the spheres' within the atom, chords of integral relationships, an order and harmony that becomes ever more perfect in spite of the manifold variety."[1] How, then, did the music of the spheres come to the atoms?

Because these developments were the work of many hands, assembling several individual portraits may elucidate the different musical facets of the emergence of quantum theory. After Helmholtz and Lord Rayleigh had brought the theory of sound and music to such a high degree of development, experimentally and mathematically, it stood ready, in its new formal generality, to address kinds of problems no one had anticipated might by informed by harmonic considerations. Given the universal success of continuum mechanics in the theory of light and sound, the discovery of discrete spectral lines characteristic of each chemical element was a shock. Such dark, discrete lines were observed in the sun's spectrum, first by William Wollaston (in 1802), who attributed them to the boundaries between Newtonian colors, and then by Joseph von Fraunhofer (in 1814), who began cataloging them using his newly invented spectrograph (figure 16.1).

The problem was how such discrete lines could be consistent with the continuous wash of spectral colors Newton observed in sunlight. When Anders Jonas Ångström measured

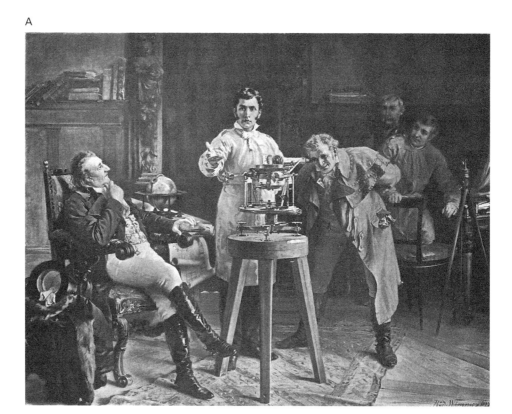

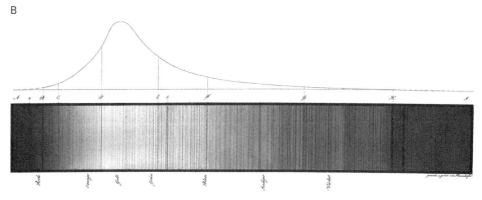

Figure 16.1
(a) Richard Wimmer, *Frauenhofer Demonstrating the Spectrograph*. (b) Frauenhofer's illustration of the dark lines in the solar spectrum.

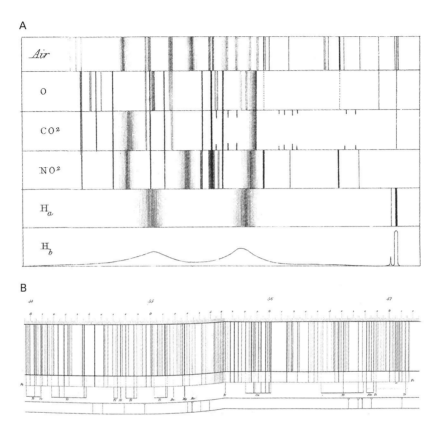

Figure 16.2
(a) Ångström's 1855 comparison of the spectrum of an electric spark passing through air with spectra of various elements and compounds. (b) Ångström's 1868 atlas of spectral lines from the sun (above), identifying them with various elements (below), in the range 5,400–5,700 Å.

the spectra induced by electrical sparks in tubes containing various elemental gases, beginning with hydrogen in 1853, he observed discrete *bright* lines (figure 16.2). To understand the relation between the absorption and diffusion of these dark and bright lines, Ångström relied on Euler's theory of light and color, itself (as we have seen) constructed in analogy to the wave theory of sound. Ångström began with "the fundamental principle of Euler" that "the color of a body is produced by the resonance of the oscillations, which can be assumed by the particles themselves," so that "the same body, when heated so as to become luminous, must emit the precise rays which, at its ordinary temperature, is absorbed."[2] Using this principle of resonance, Ångström was able to correlate the dark lines in the solar spectrum, considered as formed by absorption, with bright lines he observed in the emission spectra of individual elements. Thus, the solar dark lines are evidence for the presence of the corresponding elements in the sun (as emerged in subsequent contributions by Gustav Kirchhoff, Robert Bunsen, and others).[3] But that still avoids the question why elemental spectra are discrete, rather than continuous.

To address this, the mysterious thicket of spectral lines needed to be cataloged; Frauenhofer himself had noted no fewer than 570 dark lines in the solar spectrum, whose distribution seemed to follow some complex but definite pattern, as did the individual elemental spectra Ångström produced. Nor was the complexity of the pattern the product of randomness or experimental uncertainty, for Ångström's results were accurate to within one part in ten thousand, showing the precision that his optical methods could achieve even at the beginning of spectroscopic analysis. In the case of hydrogen, Ångström measured the wavelengths of the first four spectral lines (all in the visible range), which he called H_α, H_β, H_γ, H_δ.[4] Efforts to find some mathematical order in the lines began in the late 1860s.

In 1871, the physicist G. Johnstone Stoney argued that "the lines in the spectra of gases are to be referred to periodic motions within the individual molecules, and not to the irregular journeys of the molecules amongst one another"; as was common at the time, he used the word "molecule" where we would say "atom," treating it as vibrating in response to the incoming waves of light. To describe this, Stoney relied on Helmholtz's work on sound: "A *pendulous* vibration, according to the meaning which has been given to that phrase by Helmholtz, is such a vibration as is executed by the simple cycloidal pendulum."[5] Indeed, Helmholtz began his *Tonempfindungen* with those pendulous or simple vibrations "since they cannot be analyzed into a compound of different tones," and hence form the basis on which musical tones are built.[6] Stoney was not original in describing atoms in terms of such simple modes of vibration (which Maxwell and others had explored in the preceding decade), but he was the first to do so in the context of spectra, for which "one periodic motion in the molecules of the incandescent gas may be the source of a whole series of lines in the spectrum of the gas," using the exact mathematical form of the overtone series.[7] Stoney also used the musical analogy of atomic vibrations to explain why many in the "overtone series" of spectral lines seem to be absent, "analogous to the familiar case of the suppression of some of the harmonics in music," such as the clarinet, whose characteristic timbre is caused by the suppression of even-numbered overtones.[8] Stoney then takes the further step of applying this analogy to hydrogen and its visible spectral wavelengths H_α, H_β, H_δ, which he describes as "the 32nd, 27th, and 20th harmonics of a fundamental vibration," whose wavelength he calculates as 131,277.14 Å (where 1 Å = 10^{-10} m).[9] Put another way, Stoney expressed the ratio of the wavelengths of these lines as

$$H_\alpha : H_\beta : H_\delta = \frac{1}{20} : \frac{1}{27} : \frac{1}{32}$$

in terms of that implicit fundamental wavelength, which he calculates corresponds to a fundamental time of 4.4×10^{-14} seconds, considered as "very nearly the periodic time of

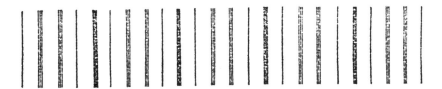

Figure 16.3
Stoney and Reynolds's spectrum of chlorochromic anhydride (CrO_2Cl_2), showing its evenly spaced lines in the green-yellow (1871).

one of the motions within the molecules of hydrogen." He notes that many other possible "overtones" of this fundamental vibration should be found in the spectrum, though they had not observed heretofore, for "if the law of this undulation were the same as that of the motion of a point near the end of a violin-string, and of a periodic time sufficiently long (as, for example, two million-millionths of a second), this undulation, when analyzed by the prism, would give a spectrum covered with lines ruled at intervals about the same as that between the two [sodium] D lines."[10] By considering the atom as essentially analogous to a violin string, Stoney could "hear" its spectra as comprising the full complement of its "overtones," thus also predicting the positions of spectral lines still unobserved.

Stoney continued to pursue the implications of his violin-string model, but he unfortunately left aside the case of hydrogen, seemingly forgetting to include the other visible line, H_γ; instead he and a collaborator went on to try their luck with a much more complex molecule (chlorochromic anhydride, CrO_2Cl_2), which gave a much more simple, regular spectrum than hydrogen (figure 16.3).[11] Though they were successful in applying the violin-string model in this case, Stoney left suspended the full relation between the musical model and the observed hydrogen lines; in 1880 he noted that the missing H_γ could be understood as a thirty-fifth harmonic of the same fundamental tone as the other hydrogen lines, but did not go further to address other possible overtones and their respective spectral lines.[12]

Specifying the structure of the hydrogen "overtones" was achieved by Johann Balmer, a Swiss mathematician who taught at a school for girls in Basel and who was sixty years old in 1885 when he published his account of the hydrogen spectral lines, his first paper on physics of any kind. Though he was a geometer, a friend directed him to this problem, for which Balmer's point of departure was a presentation Helmholtz gave in 1880 at the Royal Academy of new measurements of the hydrogen spectral lines, including several newly observed lines in the violet and ultraviolet, which also were prominent in the spectra of white stars. Balmer emphasizes the importance of "hydrogen, the atomic weight of which is by far the smallest of all substances known to date and characterizes it as the simplest chemical element, that substance through the refraction of analyzed light in solar spectrum gives us perceptible knowledge of the powerful motions and forces that the surface areas of our central bodies excite, seems more qualified than any other body to open new vistas in the investigation about the nature and properties of matter. In particular, the wavelengths of the first four hydrogen lines excite and arrest attention."[13] Where Stoney had preferred to study a simpler *spectrum*, Balmer chose the simplicity and fundamentality of the *element*, despite its more complex spectrum.

Seemingly aware of Stoney's conclusions about hydrogen, Balmer notes that "the relationships of these wavelengths allow themselves to be expressed with surprising accuracy through small numbers. Thus, the wavelength of the red hydrogen line is to the violet as 8 to 5; that of the red to the blue-green as 27 to 20 and that of the blue-green to the violet as 32 to 27."[14] Though in many texts Balmer is depicted as having guessed his results by mere trial and error, he explains that they came from analogies with sound:

This circumstance must necessarily recall analogous relations in acoustics, and one believes the vibrations of the same spectral lines of a substance ought to be understood almost as overtones of the same characteristic fundamental tone [*Grundton*]. Yet all attempts to find such a fundamental tone for hydrogen, for example, have not turned out satisfactorily. One would have come with such a calculation to such large numbers that would not have yielded thereby a clearer insight. For instance, one gets for the first, second, and fourth hydrogen lines the same fundamental tone of which the second line represents the twenty-seventh multiple. With every newly included line the sought-for fundamental tone will be represented with quite important augmented wavelengths. Nevertheless, the idea suggested itself that there should be a simple formula with whose help the wavelengths of the four indicated hydrogen lines could be represented.[15]

From the hypothesis that hydrogen wavelengths represent overtones and using Ångström's values for its four visible wavelengths, Balmer deduced that the fundamental wavelength (corresponding to the *Grundton*) is $3,645 \times 10^{-7}$ mm = 3,645 Å, "the *fundamental number* [*Grundzahl*] of hydrogen," from which he then deduces the observed wavelengths expressed as $3,645\text{Å}\left(\dfrac{m^2}{m^2 - n^2}\right)$, where m, n are integers (he assumes $m > n = 2$). Though Balmer does not detail the "various grounds on which it is likely" that the wavelengths should take this form, we can infer from what he has already said that he derived it from his general idea of overtones applied to the ratios between Ångström's wavelength values.

Based on this assumption, Balmer sought a common fundamental number as their *Grundton* but then must have realized that none of the observed lines gave this number directly, nor could he infer the wavelengths from the usual simple formulas for overtones in terms of the Pythagorean ratios for octave, fifth, and so forth. Even so, his musical assumption directed him to find a common number that, multiplied by simple fractions, would give the observed Ångström wavelengths, which his following description confirms: "the wavelengths of the first four hydrogen lines are given if the fundamental number h = 3,645 [Å] is multiplied sequentially by the coefficients 9/5, 4/3, 25/21, and 9/8. Clearly these four coefficients form no lawful sequence, but as soon as one multiplies the first and the last by 4 the lawfulness comes forward and the coefficients maintain as numerator the squares of the numbers 3, 4, 5, 6, and as denominator one of these less one of the four smaller numbers [i.e., the squares of 1, 2, 3, 4]."[16] Once having set out to find the fundamental number, and using the Pythagorean presumption that it would be multiplied by simple fractions to give the observed hydrogen lines, Balmer was then led to his coefficients, thence to their numerators (which he realized could all be expressed as squares of integers), making the inference that their denominators were the differences of squares a fairly small further step.

Nor would Balmer have simply expected that his coefficients all be simple Pythagorean ratios, though two of them (4/3 and 9/8) indeed were; his reading of Helmholtz would have assured him that more complex bodies than simple vibrating strings have more complex overtone structures, yet are still governed by the same mathematical principles. For instance, Helmholtz reviewed the harmonic series of a vibrating circular plate, already famous from Chladni's experiments and familiar to mathematicians via the special functions devised by Friedrich Bessel describing their modes of vibration.[17] In this case, Helmholtz noted that "there is no commensurable ratio between the prime tone and the other tones." Balmer could also have found precedent for his use of squared integers in Chladni's empirical law, which expressed the frequency of vibrating bodies as roughly proportionate to the squares of a series of integers.[18] Balmer was probably pleased and surprised that the vibrations of hydrogen seem comparable in complexity with those of a circular plate.

After obtaining these results, Balmer learned from his friend Eduard Hagenbach about the new spectroscopic discoveries of Edward Huggins and Hermann Vogel, which he mentioned at the beginning of his paper. Evidently, in hot white stars hydrogen reach states capable of producing lines not heretofore observed in earthly experiments. Where Stoney had merely noted the four Ångström lines as "overtones," without considering other possibilities, Balmer's formula led him to consider the potential infinitude of spectral lines as the integers m and n take on ever larger values. Already in his first publication, Balmer explored this possibility and found good agreement with the newly observed lines (figure 16.4). Although he did not go further still to use his formula to predict as-yet unobserved lines, Balmer ends by proposing that his general approach could be extended to other elements, which he discussed in more detail in his fourth and final publication (1897).[19]

Though Balmer's initial publication was in a Swiss journal (published in his hometown, Basel), when he republished these same results later that year in a well-known German journal, *Annalen der Physik*, he omitted all the explanatory material that shows the connection of his work with the theory of overtones.[20] Instead, his formula appears to come from nowhere, as if plucked from the ether by pure mathematical speculation; this expurgated version of his discovery conforms to what was increasingly a preference to erase the context of discovery from the published account. In the case of acoustics, music tended to recede further into the background as time went on. Whereas in Helmholtz's *Tonempfindungen* music figured very prominently, a decade and a half later Lord Rayleigh's

Tabelle der Wellenlänge für die Wasserstofflinien in $\frac{mm}{10^7}$.

Frauenhofers Bezeichnung:	$H\alpha=\frac{9}{5}h$ C	$H\beta=\frac{4}{3}h$ F	$H\gamma=\frac{25}{21}h$ vor G	$H\delta=\frac{9}{8}h$ h	$H\epsilon=\frac{49}{45}h$ nahe vor H_1	$H\zeta=\frac{16}{15}h$	$H\eta=\frac{81}{77}h$	$H\vartheta=\frac{25}{24}h$	$H\iota=\frac{121}{117}h$	Mittelwerthe der Grundzahl h.
Beobachter:					ultraviolett					
Van d. Willigen *)	6565,6	4863,94	4342,80	4103,8	(H_1=3971,3)					h=3647,821
Ångström	6562,10	4860,74	4340,10	4101,2	(H_1=3968,1)					h=3645,589
Mendenhall	6561,62	4860,16								h=3645,932
Mascart	6560,7	4859,8			(H_1=3967,72)					h=3644,842
Ditscheiner	6559,5	4859,74	4338,60	4100,0	(H_1=3966,98)					h=3644,460
Huggins		für die ultravioletten H linien weisser			Sterne:	3887,5	3834	3796	3767,5	h=3644,528
Vogel					3969	3887	3834	3795	3769	h=3644,379
Formel: $H=\frac{m^2}{m^2-2^2}h$	m=3	m=4	m=5	m=6	m=7	m=8	m=9	m=10	m=11	
h=3645,6	6562,08	4860,8	4340	4101,3	3969,65	3888,64	3834,98	3797,5	3770,2	
h=3645	6561	4860	4339,283	4100,625	3969	3888	3834,35	3796,875	3769,615	

Figure 16.4
Balmer's 1885 table showing his calculations, compared to contemporary observations.

Theory of Sound (1877) refers only occasionally to musical matters.[21] Rayleigh makes clear that his real subject is "the theory of Vibrations in general"; he stresses "the establishment of general theorems by means of Lagrange's method," a sophisticated mathematical technique, rather than the far more empirical, musically oriented approach of Helmholtz.[22] Rayleigh, and seemingly Balmer also, seemed to think that the sheer physicality of music was best sublimated into disembodied theory, as if the more mathematical representation were to be accounted more true, or at least most general.

It would be too facile to conclude that they and others were simply ashamed of music as atavistic or unscientific; arguably, they treated music as an empirical level on which the modern edifice of acoustics was built. But that completed building necessarily would cover and hide the foundations on which it rested. Thus, in Balmer's later publications on spectra, the musical analogy with which he began faded into the background as he became involved in more and more detailed calculations of spectra for elements besides hydrogen; the others who followed his lead into this field do not mention acoustics or overtones at all, as if they had been merely suggestive scaffolding no longer relevant once they had given birth to the mathematical theory, whose general terms carried no reference to its sonic beginnings.[23]

This process might be considered an example of what Edmund Husserl called "sedimentation or traditionalization," the modern scientist's constitutional proclivity to assume and subsume prior foundational work so as to incorporate "the constant presuppositions of his [own] constructions, concepts, propositions." He goes on: "Are science and its method not like a machine, reliable in accomplishing obviously very useful things, a machine everyone can learn to operate correctly without in the least understanding the inner possibility and necessity of this sort of accomplishment?"[24] Insisting that he does not thereby denigrate the achievement of modern science, Husserl nevertheless stresses, from the point of view of his phenomenological analysis, "that the true meaning of these theories—the meaning which is genuine in terms of their origins—remained and had to remain hidden from the physicists, including the great and the greatest." His geological metaphor implies that the later layers of scientific work necessarily bury and thereby render invisible the earlier "strata" on which they themselves were built. Yet though he thinks that, as they build ever further, scientific workers must remain increasingly unaware of the buried strata on which their edifice rests, Husserl notes that later readers "can make it self-evident again, can reactivate the self-evidence" by gradually desedimenting the impacted layers of assumptions, bringing to light their "intentional history" and their latent implications.[25]

In these terms, we could be said to be unearthing the sedimented role of music as a historically relevant stratum, though one that was, by the late nineteenth century, fairly well buried and increasingly irrelevant to contemporary scientific practice. To some extent, this is an accurate description of the process just discussed in the cases of Balmer, Rayleigh, and their successors. Yet in what follows, Husserl's metaphor and its implications will break down because, rather than simply lying buried and ever-more-forgotten, the musical theme will break out again more or less overtly and not through self-conscious attempts to "desediment" the layers. To return to the geological metaphor, strata may not always remain sedimented beneath the surface but can in fact emerge into view as outcroppings that not only indicate the buried past but also become manifest in the visible landscape, the very rocks we see around us.[26]

17 Planck's Cosmic Harmonium

Though Balmer and Rayleigh did not comment on the disappearance of music as part of the foundations of their work, other scientists indicated awareness and even intentionality about the process that Husserl summarized by saying that "sedimentation is always somehow forgetfulness."[1] In part, this reflected a widespread decision of scientists to reach past the "all-too-human," including music and sensation in general. Thus, in 1909 Max Planck argued that

> the characteristic feature of the actual development of the system of theoretical physics is an ever extending emancipation from the anthropomorphic elements, which has for its object the most complete separation possible of the system of physics and the individual personality of the physicist. One may call this the objectiveness of the system of physics. … Certainly, I might add, each great physical idea means a further advance toward the emancipation from anthropomorphic ideas. This was true in the passage from the Ptolemaic to the Copernican cosmical system, just as it is true at the present time for the apparently impending passage from the so-called classical mechanics of mass points to the general dynamics originating in the principle of relativity.[2]

Planck had nothing against the sensible grounds of physics—and, as we shall see, he was deeply concerned with music—but he felt that physics should aspire to a degree of generality that rises far above the sense data that originally evoked it or even the experimental data that can serve to test it. Given his call to remove anthropomorphic elements from physics, one might have inferred that he would have praised the greater generality of mathematical physics, which referred to music less and less.[3]

Planck's own case, however, shows the curious ways in which music could still figure in the intellectual life of someone so deeply committed to transcending the anthropomorphic and merely personal. An exemplary *Kulturträger*, highly cultivated and especially devoted to music, Planck was a pianist of considerable skill. As a student, he composed songs and even an entire operetta that was performed in the musical evenings that were fixtures of professorial life in those days; he conducted choruses and orchestras, played the organ at church services, and studied harmony and counterpoint.[4] He wondered whether he should pursue a career in music rather than physics. In 1877, he spent a student year in Berlin, where he studied with Helmholtz, reading thermodynamics and eventually becoming a close friend, participating in Helmholtz's musical evenings.[5] That year, Helmholtz published the fourth edition of his *Tonempfindungen*, including his latest thoughts on sound and space; Planck surely studied this work closely, both as a student and admirer of Helmholtz and out of his own deep-seated philosophical interests.

In later life, after Planck had returned to Berlin in 1889 as a professor, his own home music-making included collaborations with such outstanding musicians as the preeminent violinist Joseph Joachim and the no less remarkable Albert Einstein. Every other week Planck conducted an informal chorus that included his children, neighbors, and friends. Ironically, the very sensitivity that made him feel music so deeply also made it hard for him to endure anything less than absolute perfection in intonation. His pitch sense was especially acute as a child; he remembered not being able to play on a piano tuned to lower than normal pitch because of the strong tonal disorientation he felt between the nominal pitches and the actual sounds.[6] According to the recollections of his friends, Planck's sense of pitch was so acute that he could scarcely enjoy even a professional concert, but, in the view of John Heilbron, "like his politics and his thermodynamics, his ear gradually lost its absolutism and allowed him greater satisfaction."[7] What follows will examine further these interrelations between music, physics, and Planck's search for the absolute.

Beginning with his doctoral dissertation (1879), Planck was concerned with the status of the first two laws of thermodynamics as absolute laws, exemplars of the nonanthropomorphism he was later to state as a guiding principle. Until 1914, far longer than most of his peers, Planck maintained that the second law of thermodynamics was an absolute law of nature, not merely valid with high probability, as Ludwig Boltzmann had argued. Planck was long skeptical of the physical reality of atoms, compared to what he regarded as the absolute certainty of the laws of mechanics and thermodynamics. He was largely responsible for showing the practical consequences of the concept of entropy in physical chemistry, which occupied him from 1887 until 1893. In the process, Planck gradually became convinced that physicists had to rely on the atomic hypothesis to make progress, lacking any other fruitful fundamental theory, but in 1893 he still preferred not to make use of the atomic hypothesis if at all possible. Because atoms remained for him hypothetical, at that point he felt that mechanics and mechanical thermodynamics were "the deepest form of coherence."

At this turning point in his scientific biography, Planck entered a curious musical bypath, which Erwin Hiebert and Alexandra Hui's pioneering work has illuminated.[8] Planck recounts in his scientific autobiography that "by a sheer whim of fate, no sooner had I reported to my post in Berlin [in 1893] than I was temporarily assigned a task in a field quite remote from my self-chosen special branch of physics. Just at that time, the Institute for Theoretical Physics happened to receive a large harmonium, of pure untempered tuning, a product of the genius of Carl Eitz, a public school teacher in Eisleben, built by the Schiedmayer piano factory of Stuttgart for the Ministry. I was given the task of using this musical instrument for a study of the untempered, 'natural' scale."[9]

Though this represented a radical departure from any research work he had done before, evidently Planck received this task because of his well-known musical interests and acute sense of pitch, despite his lack of any prior experimental work in acoustics. Still, he was well aware of Helmholtz's reliance on the harmonium, which Helmholtz praised "on account of its uniformly sustained tone, the piercing character of its quality of tone, and its tolerably distinct combinational tones, [which] is particularly sensitive to inaccuracies of intonation. And as its vibrators also admit of a delicate and durable tuning, it appeared to me peculiarly suitable for experiments on a more perfect system of tones."[10] The Eitz instrument Planck used was ordered by Helmholtz from the same factory that had earlier provided him a two-manual harmonium so precisely tuned that he could experimentally compare equal-tempered with "perfect" just intonation, which he found decisively superior (see box 4.1).

Using equal temperament, Helmholtz had noted that "when moderately slow passages in thirds at rather a high pitch are played, [the resulting combination tones] form a horrible bass to them, which is all the more disagreeable for coming tolerably near to the correct bass, and hence sounding as if they were played on some other instrument that was horribly out of tune. They are heard most distinctly on the harmonium and violin. Here every professional and even every amateur musician observes them immediately, when his attention is properly directed."[11] Helmholtz became a strong advocate of "perfect" (meaning just intonation) rather than equal temperament, which was then becoming more and more standard. At the time, he was perceived as crotchety and even rather eccentric; despite his assertions about professional musicians sharing his views, when Helmholtz met Johannes Brahms and tried to persuade him, Brahms remained distinctly unmoved and even dismissive, remarking that "in musical things, he is an enormous dilettante," though Brahms's close friend, the physician and musician Theodor Billroth, was a great admirer of Helmholtz.[12] Nevertheless, Helmholtz's advocacy of older temperaments had its day much later as part of the movement to restore the "authentic" performance practices of the past.[13]

Planck thus was stepping into a lively, if somewhat outré, controversy. That a full professor of physics in Berlin would be seconded to this musical investigation for several years shows the continuing importance of Helmholtzian acoustics, with its strong connection to musical issues of theory and practice that impinged also on the young science of musicology. Even though tasked thus by the authorities, Planck said he took to his project with "keen interest," as if the whole episode were an idyll in which he could function simultaneously as performing musician and as physicist.[14] His work during that time shows his growing fascination with the possibilities that emerged. He wrote a brief 1893 paper describing the Eitz harmonium, which divided the octave into 104 steps (using 52 keys), thus allowing the performance of many kinds of temperament. Though the nonstandard keyboard required seems dauntingly complicated (figure 17.1), for Planck playing it was easy "with a little practice," as he modestly put it.

Thus far, Planck seemed caught up in Helmholtz's campaign for just intonation as more "perfect" or "natural," a loaded term that further underlined its special claim to legitimacy. Planck, like his mentor Helmholtz, thought that highly skilled musicians with good ears would naturally prefer "perfect" intervals to the vulgar compromises involved in the standard equal temperament commonly in use. Here both men extrapolated from their own sonic experience to what they thought would hold for other acute hearers. Planck decided, though, to put this hypothesis to experimental test, perhaps moved by such stories as Brahms's gruff refusal to worship at the altar of "natural" temperament. His ensuing investigation was the only piece of experimental work he ever did, and hence was especially remarkable at a time when theoretical physics as such had just begun to exist as a separate discipline from physics as a basically experimental science. Indeed, Planck himself was one of the first of the new breed of theoretical physicists and the first to occupy a special chair under that rubric at Berlin.

Planck's experiment did not involve (as Helmholtz's did) mechanical equipment of various sorts, only different scores and choruses, so that they were decidedly *musical* experiments. His 1893 paper "On Natural Tuning in Modern Vocal Music" appeared in a musicological journal, putting Planck alongside Helmholtz and Mach as having worked in that research world as well as in that of standard physics.[15] Planck's paper is an extraordinary document that shows the great depth of his musical knowledge and his experience as a choral conductor; he has at his fingertips a wealth of specialized knowledge about musical practice and temperaments that would be worthy of a professional. He begins by remarking on the universal acceptance of "tempered tuning," as he calls the equal-tempered practice of his time. In itself, this is a significant piece of evidence that by 1893 some form of equal temperament was expected (a conclusion that has been questioned by some of its most passionate critics).[16] On the other hand, Planck finds evidence that performing musicians deviate from equal-tempered tuning: perceptive violinists note that a double-stop (a two-note chord) sounds better if it is not exactly equally tempered, but "softened [flattened] a little"; likewise, directors of a cappella choruses note the tendency for the third in a major triad to sound better when very slightly flat, with respect to a perfect equal-tempered third. These instinctive practical adjustments seem to show that musicians revert to the natural, unequal temperament (now called "just intonation") that historically preceded equal temperament. Planck notes that "some theorists even go so far as to deny any justification to [equal] tempered tuning, because it is distant from natural conditions and somewhat lies to the ears, so to speak," compared to pure natural tuning.[17]

To this long-vexed controversy Planck brought the experimental sensibility of a physicist, theorist though he be. He used the Eitz harmonium to produce reliably the various equal or naturally tempered intervals and thereby test for himself this dispute between

Figure 17.1
A 1911 Eitz harmonium and its keyboard (photo courtesy of Deutsches Museum).

Figure 17.2
Planck's test passage from Heinrich Schütz's motet "So fahr ich hin zu Jesu Christ" (SWV 379, *Geistliche Chorwerke*, 1648). Text: "Thus I fall asleep and rest soundly" (♪ sound example 17.1).

tunings. In the more normal setting of a choral rehearsal, he used a piano to give an accurate starting pitch and then allowed a well-trained chorus to sing without any accompaniment, in order to observe toward which temperament it naturally tended, the natural or the tempered tuning. His first experiments used a passage from a sacred motet by the great composer Heinrich Schütz (figure 17.2, ♪ sound example 17.1), a sophisticated choice that reflects both Planck's knowledge of the repertoire of older masterworks as well as his ability to choose a telling example.

Schütz's collected works had appeared only a few years before, part of a new wave of interest in long-forgotten works that involved such serious musicians as Brahms and Philipp Spitta, and clearly also Planck. The passage he chooses is an exquisite example of Schütz's word-painting, setting long, lulling notes to the word *ruhe*, signifying the repose of the believer who is awaiting resurrection, using the minor subdominant at that point to achieve an effect at once moving and unearthly.[18] Planck attended many rehearsals of this work by the chorus of the Royal Musical Hochschule in Berlin, a highly trained ensemble. He noticed that, if the piano accompaniment were too soft to be heard, the chorus would sink in pitch enough that the conductor would then tap his baton, break off, and repeat the passage with the piano. Planck realized that the sound was not only "especially good" when supported by the tempered piano but that, in fact, though flat in pitch when unaccompanied, the chorus still moved toward a *tempered* triad, not the natural one, as Helmholtz would have expected.

Figure 17.3
Planck's first test composition, devised to check whether an unaccompanied chorus would gravitate toward natural or tempered tuning; using natural tuning, the tonic C should fall about five syntonic commas (about a half step) (♪ sound examples 17.2, 17.3).

To test this surprising result, Planck wrote a short composition (though he modestly does not use this term for what he calls his "series of chord progressions") specifically devised to test whether, when singing without accompaniment, a chorus would gravitate toward natural or tempered tuning (figure 17.3; ♪ sound examples 17.2, 17.3).

Planck constructed his first example so that, if the chorus allows every triad to adjust itself to natural tuning, then by the end, the final C would be five syntonic commas lower than the initial C, a bit more than a half step flat.[19] Planck assembled a chorus of "friendly musical ladies and gentlemen" who sang this passage to him many times, beginning with a normal-tempered piano chord at the beginning. He found, indeed, that the pitch sank about a half step, as he expected, showing that natural tuning is a significant effect on choral singing. But then he arranged a second "counter-experiment," as he called it, an inversion of the melodic motions of the first example that would, by symmetry, evoke a *rise* of five syntonic commas (about a half step) above the starting pitch (figure 17.4; ♪ sound examples 17.4, 17.5), were the chorus constantly to adjust according to natural tuning.

Using this inverted test-composition, Planck's results seemed contradictory; the pitch rise did not materialize as would have been anticipated had the singers always reverted to natural tuning. He reasoned that singers tend, when prolonging a note, to go flat, which would have canceled out the expected rise in pitch here, though it would have assisted the fall in pitch in the previous example (figure 17.3). From his observations, Planck concluded

Figure 17.4
Planck's second test composition, whose tonic should rise about five syntonic commas (about a half step), using natural tuning (♪ sound examples 17.4, 17.5).

that in certain specific passages such as his first example and especially in the older repertory (such as his Schütz example; figure 17.2), natural tuning is important to the musical effect intended by the composer and also comes naturally to the singers. In general, though, "modern vocal music relies almost completely on tempered tuning … through habituation to hearing tempered intervals."[20]

This was arguably the first really surprising result Planck had obtained in his research career so far, for his result contradicted that of Helmholtz, the great authority in the field, as well as Planck's own expectations. The whole episode was significant enough that he included it in his scientific autobiography: "These studies brought me the discovery, unsuspected to a certain degree, that the tempered scale was positively more pleasing to the human ear, under all circumstances, than the 'natural,' untempered scale. Even in a harmonic major triad, the natural third sounds feeble and inexpressive in comparison with the tempered third. Indubitably, this fact can be ascribed ultimately to a habituation through years and generations. For before Johann Sebastian Bach, the tempered scale had not been at all universally known."[21]

In his autobiography, Planck immediately turns from this description back to his involvement in thermodynamic issues, especially those that emerged the following year (1894) concerning black-body radiation, leading to his famous postulation of the quantum of action in 1899–1900. Though outwardly Planck's musical investigations seem unrelated to the culmination of his work in thermodynamics, he treats both as part of his ongoing stream of scientific activity. The connections and contrasts that emerge shed a new light onto Planck's development, helping us understand how this immensely cautious man found himself advancing the most controversial and consequential innovation of modern physics.

Until his musical work of 1893–94, Planck had been an unalloyed conservative, a devotee of absolute laws who (against the powerful arguments of Boltzmann) resolutely maintained the absolute validity of the second law of thermodynamics, upholding entropy alongside energy as mainstays of deterministic (as opposed to statistical) mechanics. His work on the battle between tempered and natural tuning was the first place in which his sense of the absolute was truly challenged. Initially prepared to believe in the absoluteness of natural tuning, at least for cultivated musicians, his own musical experiments led him to the opposite conclusion: the prevalent conventions of tempered tuning, however recent in the long view of music history, outweighed the claim of the "natural." His results were nuanced: though natural tuning does, in fact, have some sway over music, he realized that musicians gravitate toward tempered tuning out of habit, and (more surprising still) he himself found those "unnatural" tempered intervals more expressive. This moment of self-realization was mirrored in his experiences with others. Not only did he observe his singers sink into tempered triads, he realized that he himself was not immune, but had been formed by the conventions of tempering he too had grown up with.

For someone as devoted to the absolute as Planck, these musical results were disquietingly relativistic. Though he does not use that precise term, in fact he tested the absolutistic claims of the advocates of natural tuning and concluded that, far more than he or they ever expected, the conventional "frame of reference" of tempered tuning conditions our expectations and our felt experience. To put the matter provocatively, Planck's musical trials were a kind of Michelson–Morley experiment with respect to absolute versus relative tuning; he, like they, derived a null result, namely that musicians do not perceive absolute natural tuning any more than optical experimenters can observe absolute motion with respect to the ether.[22] On the other hand, though the claims of natural tuning had now been rebuked by experiment, he does not altogether disavow its power in musical experience, even though he showed that natural tuning was overshadowed by the greater practical force of habitual tempered tuning.

Planck's musical realization resonated through his work as he sought a unified conception of the forces of nature, in accord with more general Wilhelmian views about the unity of knowledge.[23] As Hui put it, his conception of sound sensation fit into this widening space between the world of physics and the sense world based on human mensuration. For one, this increasing distance reinforced Planck's framework by revealing the deceptiveness of the senses—a single listener could hear the same interval differently depending on her use of accommodation, and a vocalist could only maintain a pure interval with careful and sustained concentration.... If passive hearing could shift significantly in just a few generations due to material and aesthetic shifts in the music world with the spread of equal temperament (that is, very much as a result of human activity), and yet a listener could, with accommodation, toggle between tuning systems in their sensory perception of sound, certainly the sense world and the real world were not one and the same.[24]

In the end, Planck was not disillusioned but instead raised his quest for the absolute to another venue: that of the highest aims of musical art. This emerged as he confronted practical problems of performance: "How is one to proceed in such cases as, for example, the above-cited composition of Schütz? Should one, in order not to give up the absolute pitch level, let the choir sing the third in a triad not in natural tuning, as musical hearing suggests, but in tempered tuning? Or should one, yielding to everything, renounce performing a constant tonic?" His conclusion is a judicious triumph of musical sensibility:

> Above all, in such a question the composer must be consulted; he alone, through the composition given us, ought to speak the deciding word. If, though, as in the foregoing case, this court is no longer accessible, then other considerations enter in and here it cannot be sufficiently stressed that the last, highest decision once and for all ought to rest on the consideration of the artistic effect. For art finds its justification in itself and no theoretical system of music, be it ever so logically founded and consistently realized, is capable once and for all of meeting all the demands of the human spirit as well as of ever-changing art. In this connection, the natural system has absolutely no priority over the tempered and there is no justification for performing famous compositions in natural tuning for no particular reason.[25]

Thus, Planck turns to the autonomous realm of music, "ever-changing" though it be, as a new locus of authority: "Academic rules must regulate themselves according to art, not turn it upside down."[26] In the process, he had to reconcile himself to historicity, despite his overarching preference for the timeless absolute. For Hui, "this conception of physics as oscillating between the worlds of sense and reality but also asymptotically approaching unity with the real world is a rather delightful solution that allowed Planck room to historicize scientific thought while maintaining an antipositivist stance."[27] To her formulation I would like to add several new aspects in which Planck's return to the quest for the absolute was affected by his musical encounter with historical relativity.

At the most basic level, Planck's musical findings required an implicit confrontation with the old authorities, particularly his revered mentor Helmholtz, who had maintained the fundamental, absolute status of natural tuning. But in October 1893, when Planck delivered his paper on tuning, Helmholtz suffered a serious fall, after which his health was never the same until his death the following year. Indeed, Planck called 1894 "the black year of German physics," during which Heinrich Hertz and August Kundt also died prematurely.[28] Thus, Planck's struggle with these musical-physical questions notably overlapped with the deaths of three of his most important colleagues, leaving him as the sole remaining professor of physics in Berlin by the end of 1894. Arguably, Planck's return to issues of thermodynamics and black-body radiation in that year was affected by the sudden loss of many of his Berlin colleagues, which required him to take up these other physical investigations after their demise. Helmholtz's death effectively forestalled any awkward controversy between teacher and student on the musical issues that mattered so much to both of them.

This was, indeed, the first occasion on which Planck could have experienced what he much later described in his autobiography as a generality that has come to be called "Planck's principle": "A new scientific truth does not triumph by convincing its opponents and making them see the light, but rather because its opponents eventually die, and a new generation grows up that is familiar with it."[29] Though Planck remarks this in the immediate context of the struggle between Boltzmann and Wilhelm Ostwald about the reality of atoms, his comment immediately follows his description of his musical experiments. We lack documentary evidence about the conversations between Helmholtz and Planck in the aftermath of his 1893 work on tuning, but they may have been strained and awkward, given Planck's well-known deference to authority, and to Helmholtz in particular. Helmholtz's passing gave way to new voices expressing contrary views, as had Planck. Those onerous encounters with death in the "black year" of 1894 provided a forceful education in dimensions of historicity, which Planck received first of all during his musical struggle with tuning and convention.

Planck certainly confronted these issues as he participated in the controversy about the reality of atoms, during which he changed his views toward those of Boltzmann, a passionate advocate of real (rather than purely theoretical) atoms and the great champion of their statistical nature. During the years 1894 through 1900, Planck's research on the equilibrium radiation of black bodies turned on the application of Boltzmann's methods. As with his work on the Eitz harmonium, Planck combined his underlying theoretical preoccupations with the external demands of practicality: an electric lightbulb manufacturer had requested help determining the optimal radiance for their product, which in turn devolved on thermodynamic and electrodynamic questions that had emerged already by 1859, when Gustav Kirchhoff had devised and named the idealization of heating a perfectly black oven to a given temperature, allowing some of its radiant light to escape through a small hole for observation. Kirchhoff had established that this "black-body radiation" was universal in terms of its distribution of energy over frequency, depending only on the oven's temperature, and not its size, shape, or materials. Because of this absolute quality characterizing the energy distribution, Planck was interested in finding what determined the exact shape of that distribution, which was becoming ever better known through the efforts of his experimental colleagues in Berlin.

Approaching this problem, Planck brought to bear the thermodynamics of entropy that he had developed in his earlier work. But two elements of his treatment had distinctly musical antecedents. To sample the distribution of energies inside the oven, Planck introduced an imaginary resonator as a probe. This resonator he assumed to be an idealized pendulum or harmonic oscillator, and thus an idealization of a vibrating string. This theoretical contrivance allowed him to express the relation between the radiant energy in the cavity of the oven at a certain frequency (chosen to be the resonant frequency of the imaginary resonator itself) and the entropy of that ambient radiation. Then Planck could apply the second law of thermodynamics to the entropy by calculating the number of possible ways in which the available energy might be distributed over a whole ensemble of imaginary resonators, each probing the oven's radiation at its own particular resonant frequency.

In the previous chapter, we saw the prehistory of this hypothesis in Maxwell's assumption that molecules could be treated as vibrating bodies whose modes of resonant oscillation are then visible as the spectrum of the substance formed by those molecules. Under that assumption, the walls of the cavity itself would be formed of an immense number of those vibrating molecules. Thinking of their vibrations in terms of sound, the oven (or indeed any chunk of matter) is essentially a chorus, an assemblage of many individual "singers," each sounding its own particular note or resonant frequency. The light inside the heated oven, thus, would be like the chord formed by all those individual voices, sounding a continuum of pitches, according to classical theory.

Planck at times calls these "Hertzian oscillators," referring to Heinrich Hertz's 1886 experiments that used a spark gap and capacitor to induce a rapidly oscillating electric spark capable of producing a detectable amount of electromagnetic radiation, as predicted by Maxwell's equations (figure 17.5a). To produce the rapid interruption of the electric spark, Hertz used a vibrating contact that, when operating, generates an audible pitch (♪ sound example 17.6); his electromagnetic vibrator is also in fact a source of ordinary sound vibrations. Indeed, the physical structure of these mechanical vibrators is essentially identical to the reed of a harmonium (figure 17.5b), though Hertz used a metal vibrator that could conduct the electricity for the spark. Thus, Hertz's physical contrivance has a continuous lineage with the other imaginary and real oscillators we have already discussed, including those envisaged by Stoney and Balmer. As they had used a vibrator to model light-emitting atoms and molecules, Planck uses his oscillator to model an idealized probe of the ambient radiation emitted by the excited atoms in the oven cavity.

Far more often than "oscillator," Planck uses the term "resonator," the term Helmholtz also used for his instrument tuned to resonate sympathetically at a certain pitch (see figure 14.6). Planck's resonator is likewise coupled with the vibrations it both registers and retransmits.[30] As Helmholtz used many resonators to map out the frequency spectrum, each tuned to a pitch of interest given by a harmonium, Planck too envisages a whole ensemble of resonators, each capable in principle of being tuned to any given pitch.[31] In that sense, his thought experiment envisages an imaginary harmonium, an instrument capable of

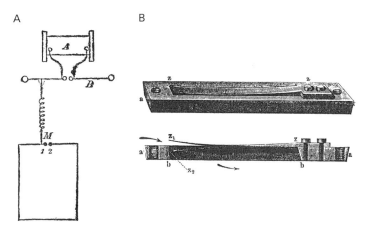

Figure 17.5
(a) Heinrich Hertz's illustration of his oscillator, which produces a rapidly oscillating electric spark across the gap near B (♪ sound example 17.6). (b) Harmonium reeds, as illustrated by Helmholtz, which have an essentially identical structure to the vibrators Hertz used to interrupt his electric spark.

sounding (by thermal excitation) and resounding (by sympathetic vibration) a whole spectrum of pitches. By preferring the Helmholtzian and musical term "resonator" over the more neutral "oscillator," Planck indicates the acoustic origins of his thought experiment: he is doing for electromagnetic vibrations conceptually what Helmholtz did for sound experimentally.

The climax of this extended comparison between sound and electromagnetic vibrations connects Planck's central innovation and his involvement in tuning controversies. Unlike the purely continuous possibility of reeds or tubes resonating at any arbitrary length, Planck tunes his resonators in an analogous way to the Eitz harmonium he had spent so long studying. That is, at a crucial point in his calculations he allows the resonators to have only discrete possible energies, just as the harmonium allows only discrete multiples of the fundamental unit of pitch. For the Eitz harmonium, this unit was $\frac{1}{104}$ of an octave; for ordinary "tempered tuning," it was the equal-tempered semitone, $\frac{1}{12}$ of an octave. In contrast, the "natural tuning" had no obvious "fundamental unit," for it envisaged different-sized versions of semitones (major or minor, for instance) and was inherently an unequal temperament. Planck's 1893 argument had led him to accept the "unnatural" equal-tempered scale, even though its number of divisions depended on convention. In his work on black-body radiation, using his resonator, Planck essentially argued for the "equal temperament" of electromagnetic radiation, that is, the equally spaced quanta of his expression $E = h\nu$, where E is their energy, ν their frequency, and h the constant that sets their spacing or "temperament."

To be sure, this analogy is not made explicit in any of Planck's own writings and must be read judiciously: the equally spaced *energy* quanta need to be distinguished from equally spaced semitones in *frequency*. But the implicit analogy is strong and, in retrospect, less surprising, given Planck's intensive work on the Eitz harmonium and on the issue of natural versus tempered tuning in the year just preceding his return to the thermodynamics of electromagnetic radiation. Comparing these two consecutive projects, Planck's black-body "harmonium" is set up so that, in terms of a "tuning pitch" ν, every "note" on that instrument sounds integer multiples of $E = h\nu$. In that sense, the analogy is quite exact between harmonium and black body: parallel to Eitz's 104 keys per octave, Planck's constant h sets the minimum spacing between adjacent "notes" on the atomic harmonium.

Planck also kept before him the question of the arbitrariness of the "standard pitch" for his electromagnetic "harmonium." In 1899, he realized that his assumption of the "tuning constant" h led, along with the Newtonian gravitational constant G and the speed of light c, to a fundamental and universal set of units that "necessarily retain their meaning for all times and for all civilizations, even extraterrestrial and non-human ones, and therefore [should] be designated as 'natural units.'"[32] The exalted universality of this realization moved him at the time to tell his young son Erwin that he had made a great discovery, comparable to those made by Newton or Copernicus.[33] By this, Planck seemed to mean the disclosure of these "natural," universal units, more than his success in fitting the experimental data for black-body radiation per se. From the point of view of the extended analogy with a cosmic harmonium, he had discovered the natural wavelength of that instrument, the "Planck length" he calculated to be 4.13×10^{-35} m. Corresponding to this universal length, Planck also calculated a universal time unit, 5.391×10^{-44} sec. Though he did not mention it, its inverse should then be a universal "Planck frequency," 1.855×10^{43} Hz. If we return to Young's attempt to state the "pitch" of a light vibration in musical terms, Planck's cosmic harmonium is tuned to a quite low A (426 Hz) 135 octaves above middle C, according to contemporary equal temperament (♪ sound example 17.7).[34] Alternatively, we can understand Planck's equation $E = h\nu$ as translating any energy E into a frequency $\nu = E/h$, so that energy corresponds to *pitch*, with Planck's constant h as the conversion factor.

Planck's term "natural" brings to mind his foregoing struggle with the problem of natural versus equal-tempered tuning. There, he had to yield to conventionality and the human feeling (which he shared) for the greater sweetness (as well as familiarity) of tempered versus natural intervals. But when he transferred the problem of tuning to the black body, Planck was able both to have complete equality of "temperament" through the equally spaced "notes" of the quantum harmonium as well as naturalness, through the fundamental "pitch" implied by the very temperament itself, the Planck frequency. The tuning dilemmas he had faced while playing the Eitz harmonium turned out to be solvable in the case of the black-body resonators.

Even if one hesitates to take this analogy as definitive, Planck's experience with musical issues arguably had a significant effect on his whole approach to the black-body question. Though an avowed champion of what he conceived as the universal, transhuman project of physics, he confronted and came to terms with the human aspects of tuning, in untrained hearers, skilled musicians, and even in himself, bearing out Heilbron's general observation that "his ear gradually lost its absolutism and allowed him greater satisfaction."[35] At the same time, though, his investigation of the musical dilemmas of tuning allowed him to return to physics and find a new absolute there: the paradoxes of musical temperament gave him new flexibility of mind that helped him take the next step past the impasses of black-body theory, while at the same time enabling him to hold on to the absolute at an even deeper level, namely his "natural units" as expressions of the universal constant h. He later described his postulation of the quantum as "an act of desperation," undertaken in the midst of the greatest crisis of his professional life as a physicist, wrestling with the contingent and seeking the absolute.[36] It seems right to look for the sources of the strength that saw him through not only in his general Wilhelmian or religious steadfastness (as informed his chosen motto, "Ye Must Have Faith") but in the particular details of his own sensibility, for which music loomed so large in general and for which his musical interlude of 1893–1894 was so significant.[37]

However intriguing may be the clues and resemblances that link Planck's musical and theoretical work, these musical formations very quickly became embedded in the sedimented strata of mathematical formalism. Whatever unconscious similarities might have moved him as he turned from tuning to black bodies, his theoretical formulation soon accumulated a growing mathematical vesture that amounted to a kind of "body." Hertz had observed that, in the end, "Maxwell's theory is Maxwell's system of equations," rather than the various mechanical models that had moved Maxwell during the process of discovery.[38] Likewise, Planck's theory became his equations of quantization and the attendant black-body radiation spectrum. As such, their musical prehistory was for the most part embedded deep inside that formalism or even underneath it, in the sense of Husserl's geological strata of meaning and intention. So difficult were the conceptual issues and the downright paradoxes of the emerging quantum theory that at many points it seemed that only formalism could see it through, leaving intuition and visualizability behind as mere anthropomorphic illusions.

18 Unheard Harmonies

Many of the physicists in the generation after Planck continued to be enthusiastic *Kulturträger*, devoted especially to music. Einstein was famously loyal to his violin and to Mozart, yet wrote that "music does not influence research work, but both are nourished by the same sort of longing, and they complement each other in the satisfaction they offer."[1] Indeed, there is no evidence of direct involvement of music in his work such as we have considered earlier in this book. This was not because music was insignificant for him as an intellectual or a scientist; as his sister noted,

> music served as his only distraction. He could already play Mozart and Beethoven sonatas on the violin, accompanied by his mother on the piano. He would also sit down at the piano and, mainly in arpeggios full of tender feeling, constantly search for new harmonies and transitions of his own invention. And yet it is really incorrect to say that these musical reveries served as a distraction. Rather, they put him in a peaceful state of mind, which facilitated his reflection. For later on, when great problems preoccupied him, he often suddenly stood up and declared: "There, now I've got it." A solution had suddenly appeared to him.[2]

His musicality was so deeply embedded in his larger *Weltanschauung* that it could no longer be distinguished from his general views. Einstein's philosophical praise of harmony and beauty in physical theory may be sublimated expressions of his underlying musical feeling, but now so generalized and universalized that only the presence of charged terms like "harmony" bears witness to their underlying origins.

We can, though, use this insight to decode what Einstein and others meant by "beauty" in mathematics or physical theory—namely, a kind of architectonic proportion and interrelation, an intensity of content, coherence, and significance that makes equations "elegant" rather than cumbersome, at least in their eyes. Though it is frustratingly difficult to go further than this generality, the elusiveness of the physicists' and mathematicians' idea of beauty or harmony is a result of the sedimentation to which Husserl drew attention. Consider, for example, the case of Werner Heisenberg, devoted to the piano from his youth onward, whose autobiography enshrines a musical moment that he considered definitive of his path in life. Near Munich, during the turbulent revolutionary days of 1919, he was attending a youth meeting at a castle, where his anguish over the lack of any "unifying center" to his life "was brought home to me with increasingly painful intensity the longer I listened. I was suffering almost physically, but I was quite unable to discover a way toward the center through the thicket of conflicting opinions." As he felt more and more upset and as the evening shadows grew,

> quite suddenly, a young violinist appeared on a balcony above the courtyard. There was a hush as, high above us, he struck up the first great D minor chords of Bach's Chaconne [♪ sound example 18.1]. All at once, and with utter certainty, I had found my link with the center. The moonlit Altmühl Valley below would have been reason enough for a romantic transfiguration, but that was not it. The clear phrases of the Chaconne touched me like a cool wind, breaking through the mist and revealing the towering structures beyond. There had always been a path to the central order in the language of music, in philosophy and in religion, today no less than in Plato's day and in Bach's. That I now knew from my own experience.[3]

Plato's *Timaeus* figured largely in Heisenberg's subsequent reflections, which led from this musical epiphany to his ultimate decision to study physics. The inner turmoil he experienced on the way to finding a new quantum theory also reflected his acute sense of the cognitive dissonances of that confusing period, felt as intensely as the tension of musical dissonance. Yet his published papers show no recognizable trace of these musically toned inner experiences precisely because they were so thoroughly transformed and embedded into the structure of matrix mechanics that they can no longer be distinctly perceived.

For Einstein and Heisenberg, as for many others of their and the next generation, the musical groundwork had now become part of the mathematical and theoretical structures that they tended to take for granted as their point of departure. To revert to our earlier metaphor, the instrument of theoretical physics had been built and its tuning already worked out in the course of the episodes studied earlier in this book; the newer generations played it in their several styles, more or less taking it for granted, often without reflecting on that instrument's origins. Even so, the seemingly stratified musical content sometimes resurfaced in surprising cases, such as that of Erwin Schrödinger, who was distinctly unmusical, if not antimusical. As his biographer observed, "almost uniquely among theoretical physicists, Erwin not only did not play any instrument himself, but even displayed an active dislike for most kinds of music, except the occasional love song. He once ascribed this antipathy to the fact that his mother died from a cancer of the breast, which he thought was caused by mechanical trauma from her violin. More likely he learned this distaste for music as a child, echoing his father's lack of response to his mother's art."[4]

On the other hand, Schrödinger was deeply interested in color theory.[5] In the 1920s, he was recognized as the world authority in this field, the successor to Helmholtz, and was asked to write the authoritative monograph on the subject. Schrödinger kept publishing papers on aspects of color theory through 1925, on the eve of his famous work on the quantum-mechanical wave equation that bears his name. Thus, we might compare the color-theoretic episode in his work with Planck's musical investigations of 1893–94, though these occupied proportionately much less of Planck's career than Schrödinger's work on color occupied him.

Schrödinger essentially carried forward the Young–Helmholtz three-color theory by bringing to completion Helmholtz's comparison of the manifolds of color and geometric space. Helmholtz was moving toward determining the true geometry of color space, which Schrödinger showed was a non-Euclidean manifold and even determined the exact form of its distance function (its "metric," in geometric terminology). The curved quality of the lines of shortest length in this color space, implicit in Helmholtz's work, came forward explicitly in Schrödinger's theory (figure 18.1). Helmholtz and Riemann, in their complementary ways, had gone from the complexities of color vision to a vision of a dynamical space, which Riemann posited as curved. Their dialogue contrasted sound and color perception to bring out the implicit geometry behind those modes of perception, namely the manifolds involved in seeing and hearing. Riemann in particular envisaged a new kind of geometric physics that would involve curved, multidimensional manifolds. Einstein accomplished exactly that in his 1916 general theory of relativity, treating gravitation and acceleration as manifestations of curved space-time. In 1920, directly influenced by Einstein's work, Schrödinger took the same curved manifolds Einstein had drawn from Riemann and went back to the problem of color perception with which the whole story had started. Schrödinger played in color theory the role Einstein played in space-time geometry and gravitation: each applied Riemannian geometry to fulfill the intuition of Helmholtz. Einstein showed that space-time, in the presence of matter, was non-Euclidean; Schrödinger showed that human color perception also obeyed non-Euclidean geometry.

In this dénouement, it seems at first glance that these developments concerned only visual perception, not sound. Yet despite Schrödinger's disinterest, music in a particularly Pythagorean vein shaped his greatest achievement, his wave equation. His work on color theory prepared him to apply Riemannian, curved higher-dimensional manifolds to other physical problems. And so when in 1926, just after his color theory work, he took up the problem of the behavior of atoms and electrons, he was ready with the tools that had been so successful with gravitation and now also with color perception.[6] What happened next showed the enduring power of the musical shaping of physical law. Heisenberg's matrix mechanics had given numerically satisfactory predictions of atomic processes, but without disclosing any intuitable—specifically, visualizable—mode of understanding these results. Louis de Broglie had argued, by analogy, that, as light had particle as well as wave qualities, so too should particles have analogous wave properties. In 1926, Schrödinger was then a young theorist in Zurich; according to an eye-witness report by Felix Bloch, then a graduate student, the senior professor Peter Debye said: "Schrödinger, you are not working right now on very important problems anyway. Why don't you tell us some time about that thesis of de Broglie, which seems to have attracted some attention."[7] At one of

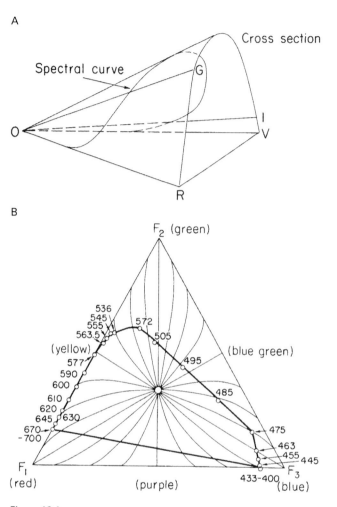

Figure 18.1
Two figures from Schrödinger's "Outline of a Theory of Color Measurement for Daylight Vision" (1920). (a) The color cone, showing the three-dimensional manifold of color vision with the curve of the spectrum. (b) The curved geodesics (lines of shortest distance in color space) shown on a chromaticity diagram; the inner area corresponds to the range of human daylight vision.

the next colloquia, Bloch recalled that Schrödinger presented a clear account of de Broglie's reasoning and how it led to an explanation of Bohr and Sommerfeld's quantization rules, capable of accounting for the discrete spectral lines whose puzzling spacing we considered in the last chapter.

When he had finished, Debye casually remarked that he thought this way of talking was rather childish. As a student of Sommerfeld he had learned that, to deal properly with waves, one had to have a wave equation. It sounded quite trivial and did not seem to make a great impression, but Schrödinger evidently thought a bit more about the idea afterwards. Just a few weeks later he gave another talk in the colloquium which he started by saying: "My colleague Debye suggested that one should have a wave equation: well, I have found one!"[8]

Debye's casual intervention led Schrödinger to seek a wave equation for de Broglie waves. Essentially, Debye underlined the force of the implicit analogy: what de Broglie called waves must require some relation to the wave phenomena of mechanics, acoustics, and optics; therefore mathematically it should be possible to express this analogy through a wave equation of the sort known to apply in those fields. Schrödinger simply took the general form of de Broglie's wave and worked backward to see what partial differential equation it could satisfy, following the pattern of sound or water waves. He found that he was led rather directly to what now is called Schrödinger's equation.[9]

Further, once in possession of that equation, Schrödinger began to investigate its general properties—even though he admitted he had no idea what was "waving." Above all, he knew that Bohr's atomic theory, expressed more generally by Sommerfeld's quantization condition, yielded the hydrogen spectral lines, though here again neither Bohr nor anyone else knew *why* these quantization conditions held, only that they "worked" to explain the spectra. The furthest Bohr and Sommerfeld could go was a simple picture in which quantization of atomic energy levels corresponded to an integral number of "electron waves" fitting around the orbit so as to join back on themselves smoothly (figure 18.2). But now, with a wave equation in hand, Schrödinger realized that these quantized energy levels corresponded exactly with the overtone series of the waves, its "eigenvalues" or "proper values." As he put it at the beginning of his first seminal paper (1926) on the hydrogen atom: "The customary quantum conditions can be replaced by another postulate, in which the notion of 'whole numbers,' merely as such, is not introduced. Rather, when integralness does appear, it arises in the same natural way as it does in the case of the *node-numbers* of a vibrating string."[10] Despite his aversion to music, Schrödinger found himself taking the mathematical step from a wave equation to its eigenvalues *in terms of a vibrating string*. Given de Broglie's proposed wavelength λ for a particle of momentum p ($\lambda = h/p$), Schrödinger was essentially taking the reverse step of Stoney and Balmer, who went from the spectral line to the overtone: Schrödinger interpreted his overtones as the stationary states that (according to Bohr) were the starting and ending points of atomic transitions.

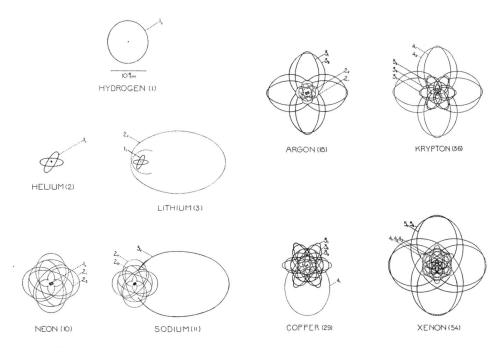

Figure 18.2
An illustration of the orbits of electrons in the Bohr–Sommerfeld theory for various elements, from H. A. Kramers and Helge Holst, *The Atom and the Bohr Theory of Its Structure* (1923).

In his later presentations of his argument (beginning with the second installment of his 1926 papers), Schrödinger characteristically brought forward the optical version of his analogy, in which he connected acoustics and optics with mechanics through an analogy with the Hamilton–Jacobi equation of dynamics. Given his visual bent, it is not surprising that, in those later presentations, he tended to describe his equation in optical language, using concepts like rays: lecturing on his theory in 1928, he opined that "Hamilton's wave-picture, worked out in the way discussed above, contains *something* that corresponds to ordinary mechanics, namely, the *rays* correspond to the mechanical *paths*, and *signals* move like *mass-points*."[11] Nevertheless, in his very first 1926 presentation of the nascent theory, Schrödinger did not use this optical language but instead spoke only of vibrating strings. He even found himself in the curious situation of trying to express optics in terms of sound, not just by analogy (as had Euler and Young) but somehow more literally and directly, invoking the sonic difference tones introduced by Helmholtz:

The emission frequencies [of the hydrogen spectral lines] appear therefore as deep "difference tones" of the proper vibrations themselves. It is quite conceivable that on the transition of energy from one to another of the normal vibrations, *something*—I mean the light wave—with a *frequency* allied to each frequency *difference*, should make its appearance. One only needs to imagine that the light wave is causally related to the *beats*, which necessarily arise at each point of space during the transition; and that the frequency of the light is defined by the number of times per second the intensity maximum of the beat-process repeats itself.[12]

This extraordinary passage, with its tumultuous language and jumpy italics, expresses Schrödinger's struggle to yoke together the aspects of light and sound that provide the mathematical analogies he used to describe the atomic events underlying spectral emission. Above all, he seems forced to these verbal contortions because he is trying to picture processes that resolutely defy any visualization. His allusions to sound emerge under the pressure of trying to visualize the unvisualizable: in the atomic realm, the extended analogy with sound helped him interpret his equation, lacking any other means to connect it to visual reality.

To be sure, after his initial insight in terms of sound vibrations, Schrödinger's visual orientation and even his color theory came into play when he considered how an atomic field of force might affect the resulting "overtones"; in that case, the difference between relative values of potential and kinetic energies acts to "curve" the manifold in which Schrödinger's waves act, which one might compare to the way that the human visual system "curves" the manifold of color perception. Thus, Schrödinger's equation represents a kind of synthesis that embeds overtones within a "curved," non-Euclidean environment dependent on the energy of the system and the forces at work there. In that sense, his equation combines the idiomatically musical element of overtones with the visual component of rays traversing a curved manifold (figure 18.3). Using this generalized wave description, Schrödinger was able to put forward an extended analogy: the new quantum mechanics is to ordinary mechanics as diffractive wave optics is to the geometrical optics of rays. Yet despite his free use of optical or visual metaphor, Schrödinger frankly acknowledged he had no idea what these "waves" might be, only that his wave equation plus the boundary conditions typical for vibration problems led to overtones corresponding to the observed hydrogen spectra. Nor was it clear in what "space" his wave equation operated; at first, he hoped the waves (whatever they were) were physical in the sense of occupying ordinary space, but when he began to consider more complex atoms with N electrons, he realized that instead the wave equation would have to be formulated within a $3N$-dimensional "configuration space."

So great was the power of the generalized mathematics of the wave equation that, even without any concept of the nature of the waves involved, Schrödinger could deduce their overtones and experimental implications. Though proud of his optical analogies, Schrödinger was left mainly with negative conclusions about the intuitive or visual meaning of his wave equation: "no special meaning is to be attached to the electronic path itself,"

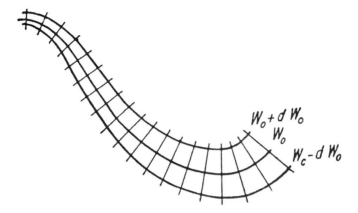

Figure 18.3
Schrödinger's diagram of curving families of the action function W, from "Quantization as a Problem of Proper Values II" (1926), showing the non-Euclidean effect of atomic potentials.

so we must relinquish "the ideas of 'place of the electron' and 'path of the electron.' If these are not given up, contradictions remain. This contradiction has been so strongly felt that it has even been doubted whether what goes on in the atom could ever be described within the scheme of space and time"—though Schrödinger continued to hope that his comparison between geometrical and physical optics could provide some kind of intelligibility.[13] But what could not be seen might be heard, so to speak; though Schrödinger could not give a visual picture of his waves, their overtones were physically manifest in atomic transitions and spectral lines.

In the years that followed, the increasingly sophisticated mathematical formalism of quantum mechanics subsumed such reminiscences of acoustics or optics into abstract vectors in a many-dimensional Hilbert space, a mathematical manifold on which the machinery of the theory operated and which could be mined for observable predictions. Schrödinger's sense of the failure of intuition was increasingly borne out; as quantum theory became more and more powerful in its predictive ability, it became less and less visualizable, until physicists like P. A. M. Dirac and Richard Feynman abandoned any pretense of trying to "understand," in the sense that we believe we can understand ordinary phenomena.[14] All that mattered was that quantum theory "works": calculations gave experimentally verifiable predictions, however statistical in nature.

Yet at the same time, Dirac asserted that "it is more important to have beauty in one's equations than to have them fit experiment," a criterion he drew from the mathematician Hermann Weyl and which led him to formulate the relativistic quantum equation now named after him. This aesthetic criterion arguably is a veiled form of the search for harmony that, as we have seen, was the heir of Pythagorean longings for a harmonious and coherent mathematical theory that would connect with observable phenomena. Among the themata that Gerald Holton identified as the enduring, recurrent motifs of science, the theme of harmony may be the most pervasive and perhaps the deepest.[15] We have seen this harmony go from the almost audible music of the spheres, to the soaring (but inaudible) polyphony Kepler found in the planetary songs, to the imaginary (yet visible) strains of Planck's cosmic harmonium. In between, Chladni, Wheatstone, Faraday, and Helmholtz sought harmony in the gritty actualities of human hearing; we are only at the beginning of exploring how complex data (such as from stock markets, galaxy surveys, climatic studies, and seismometers) might be grasped through *hearing* them, via "sonification" rather than visualization.[16] Beyond overtly musical meanings, "harmony" became a touchstone for scientists, a way of stating their deepest, largest goals, their shared (or contested) senses of the explanatory order they sought. Still, the elusive sense of world-harmony has become a rather ghostly apparition, now so dispersed throughout the mathematical structure as no longer to be recognizable as music. In that sense, the Pythagorean dream succeeded to such an extent that it subsumed and even evaporated its own musical content. If, as John Keats wrote, "heard melodies are sweet, but those unheard / are sweeter," can unhearable harmonies be the sweetest of all?[17] Yet what is left of the archaic quest for world-harmony if its connection with music becomes purely metaphorical?

For example, consider how twentieth-century string theory began with a mathematical relation between physical processes, the Veneziano amplitude (1968), which unified several important theoretical features of the behavior of high-energy particles. Many physicists were struck with the beauty of that expression, whose few symbols related several different kinds of physical processes and satisfied a surprising number of theoretical desiderata. Even more amazing, this amplitude was a function Euler had devised long before in an utterly different context.[18] It seemed miraculous that this relic from "classical" mathematical physics could be the key to understanding high-energy behavior in the extreme quantum realm, two and a half centuries later. A number of theorists extended the Veneziano amplitude into the dual resonance model, hoping that it would develop into a theory of the strongly interacting particles. Then in 1969–70, Yoichiro Nambu, Holger Nielsen, and Leonard Susskind independently suggested that the dual resonance model could be understood as describing the behavior of relativistic strings. Each found his way to this idea through mathematical analogies.

For his part, Nambu reexpressed the Euler beta function in an alternative mathematical form and then realized that its spectrum of possible values "immediately suggested a one-dimensional harmonic oscillator system, like an oscillating string of some length moving in four dimensions." He began hypothetically and analogically, using such phrases as "suggested," "could be labeled," and "could be interpreted." Gradually, as the analogy cohered more compellingly, he shifted to a more direct mode of expression: the number of states "reminds" him of the Hagedorn model of hadrons, in which the rising mass spectrum of these strongly interacting particles recalls the rising spectrum of a vibrating string. Where Hagedorn thought of the interacting hadrons as "fireballs" having a common "melting temperature," Nambu thought of them as strings. Then, "since the external hadrons *should also be* strings, I *formed a pictu*re that the scattering *is* a process of two incoming strings joining ends and separating again."[19] In his description, the strings seem to come into *physical* existence as the analogy with simple harmonic oscillators coheres with more and more facets of the argument. Nambu moved between a hypothetical, mathematical view and emergent physical insights about strings joining and separating.

Thus, the relativistic string began as an analogy of an analogy of an analogy, first to the quantum-mechanical harmonic oscillator (via Planck, Heisenberg, and Schrödinger), which is an analogical extension of Newtonian mechanics, which itself was a bold (and in certain respects counterintuitive) mathematical representation of ordinary physical strings. Yet in Nambu's thinking, and in many developments since then, this analogic hybrid has increasingly been treated as a candidate model of physical reality. The striking accord between the features of the Veneziano amplitude and the dynamics of a hypothetical string led to physical pictures of such strings inhabiting 26-dimensional worlds whose degree of "reality" remains hotly debated.[20]

Though several degrees removed from everyday experience, these analogies can only operate by disclosing perceptible resonances between the terms they connect. For instance, the experimental detection of a high-energy particle requires the observation of what is still called a *resonance*, namely finding in particle data the same bell-shaped response curve first derived from a glass resonating at its natural frequency (see figure 18.4, ♪ sound example 18.2).[21] Thus, music continues to link vibrating bodies and particle physics, for resonance is the hallmark of musical tone. Every effort of quantum field theory, string theory, or loop quantum gravity (different as they are) ultimately may be traceable, however distantly, to vibrating bodies and their sonorous mathematics. We must weigh the continuity of that connection as well as how far it has been stretched. As with Wheatstone's Enchanted Lyre (figure 13.3), the faintness of the sound transmitted from its hidden source is far less significant than the wonder of hearing it at all, however faintly. Indeed, the mysterious faintness of that sound augments its wonder and its beauty; distance lends enchantment to a sound heard from afar, no less than to a distant view. So too, I think, the stretch to connect vibrating bodies and resonant particles intensifies the felt power of an analogy that can sustain itself so far.

Long ago, Pythagoras's younger contemporary (and critic) Heraclitus expressed this important yet deeply surprising aspect of harmony: "The unapparent harmony is stronger than the apparent one" (*Harmoníē aphanēs phanerēs kreíssōn*).[22] His words compress several senses: hidden (or invisible, *a-phanēs*, un-apparent) harmony is stronger (*kreíssōn*), more powerful, better than what is visible (*phanerēs*) or apparent. The ever more hidden harmonies that science seeks indeed have ever greater power (with all its promise and danger) but are also may be more excellent, more beautiful, if Heraclitus is right.[23]

For this and many other reasons the harmonic presuppositions of modern physics remain subject to question. Kepler's harmonic astronomy excluded the diverse, unrelated solar

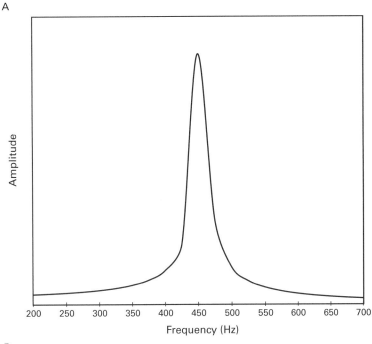

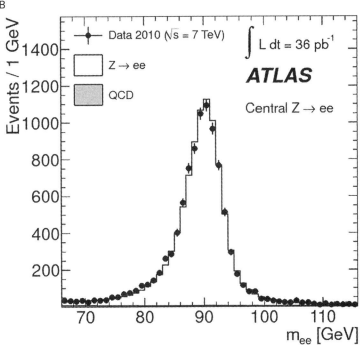

Figure 18.4
Two examples of resonance: (a) resonance of a wine glass at its natural frequency $f_0 = 454.73$ Hz, plotting the amplitude of the glass's motion versus the frequency of the sound exciting it (♪ sound example 18.2); (b) resonance showing the detection by the ATLAS collaboration at CERN of Z neutral intermediate bosons (mass 90 GeV) by their decay into an electron-positron pair, plotting the number of detected events versus the invariant mass-energy of the observed electron-positron pair.

systems that would populate Descartes's infinite space. For a long time, the universality of Newtonian law, general relativity, and quantum theory specified the overarching harmony of all observed phenomena. But various zones of the universe may have utterly different and unrelated local physical "laws," as has emerged in the consideration of possible string theory models. Some have embraced this "landscape" of different, even divergent or contradictory, universes within a larger "multiverse"; others have insisted that finally the universe must be a unity, with a single set of universal laws.[24] This choice between competing antimusical and musical themes mirrors the contrast between the sonic worlds of John Cage and Josquin des Prez. Randomness may finally rule, perhaps according to an "anthropic principle" that explains our local laws as merely the coincidence that in this region sentient observers happen to be physically possible. Others find this purely contingent (even ad hoc) reasoning disturbingly incoherent and opportunistic. Yet harmony emanating from a transcendent source seems unified to a troubling degree, requiring a single universal pattern, such as Pythagoras heard in the smithy "by the favor of a god."

Even Plato's mighty demiurge did not create the eternal Forms but merely used them as patterns, with some discretion.[25] Must cosmic harmony have a unique, essentially divine source? Even those open to such ideas may worry about a single power capable of imposing itself on the whole cosmos. Physics is on the verge of its own Darwinian turn: perhaps each aspiring universe must struggle with its own inner tensions, as well as with the other universes; arguably only a limited number might survive. Could that number be only one? If not, do global principles limit or fix the number of universes within the multiverse? For instance, consider the principles that constrain the formation of soap bubbles and whose analogues might constrain universes considered as cosmic "bubbles" (figure 18.5).[26] Generalized considerations of harmony and dissonance may be the decisive factors determining whether any given universe survives or indeed whether any incoherent assemblage of unharmonized universes can endure. In the end, music may be the final and deepest—perhaps the only—raison d'être.

In the beginning, music provided the middle ground, the epistemic interface through which natural philosophy could connect with mathematics, bridging the invisible realm of number and visible phenomena.[27] In the sisterhood of the quadrivium, "music" meant the dispassionate "music of the spheres" (*musica mundana*) rather than the expressive art of moving the passions, both known since antiquity. Their intertwining histories would require another volume.[28] In the examples considered in this book, that passion-laden music informed the episodes surrounding the mathematical innovation of "irrational numbers," Kepler's doleful song of the Earth, Euler's "modern" mathematics of sadness. But for the most part, the impassive, transcendent music of the spheres has remained the touchstone of harmony even into the quantum period and beyond: for instance, Planck's critique of "anthropomorphism" and his quest for a truly "natural" tuning sought an unchanging cosmic standard. To a striking extent, the project of natural philosophy, modern as well as ancient, has remained faithful to the search for cosmic harmony

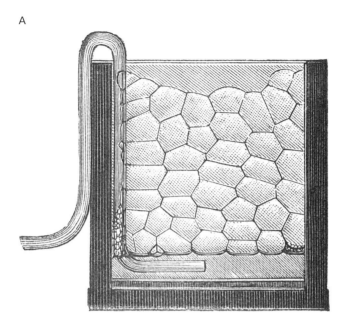

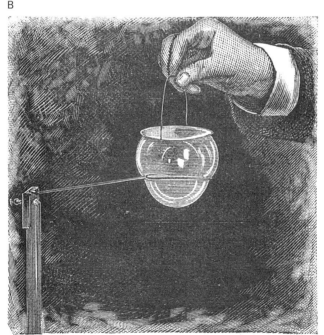

Figure 18.5
(a) Soap bubbles formed between two plates of glass separated about one centimeter; nowhere do more than three films meet one another and all the meeting points have equal angles. (b) A bubble blown within another bubble. From C. V. Boys, *Soap-Bubbles and the Forces Which Mould Them* (1890).

governed by mathematics. String theory and loop quantum gravity are only the latest "grand unified theories" based on the synthetic power of mathematics. Despite the enormous historical changes across this vastly diverse range of time and theories, this cosmic sense of music and quest for universal harmony remains a recurrent theme. Indeed, contemporary theoretical physics at every point takes its bearings from the Pythagorean project to explain how *all is number*, so that even "chaos" has its mathematical parameters and universal exponents.

From its beginnings until now, science has followed music, which first connected the senses to the invisible realm of mathematical theory. In our treatment, even the "scientific revolution" has been an episode in the larger story of how music, mathematics, and experimental natural philosophy decisively came together. Music, after all, first brought forward experiment as well as the mathematical approach to physics, bridging the ancient divide between number and magnitude: music harmonized experience with mathematics. Ironically (or perhaps with poetic justice), the very success of this hybrid enterprise tended to bury the musical traces under ever denser, more powerful mathematical formalism. One might legitimately wonder whether the ancient quadrivium gave up the ghost, exhausted after giving birth to modern science. The language of "sedimentation" suggests a decorous form of burial, with dignified geological overtones; even intentional "desedimentation" suggests exhuming graves, however piously. In contrast, though, the examples considered in this book show many points at which these ancient musical concepts return to life as persistent themes and continuing questions. For music, mathematics, and science, the rest is not silence.

Notes

Introduction

1. Whitehead 1967, 20.

2. In the late twentieth century, many historians of science turned toward a more "externalist" sociological understanding of science, in preference to "internalist" studies of the evolution of certain concepts within the confines of its environing discipline. Turning away from the presumption of objectivity or impersonality, the sociological approach treats scientific research as all-too-human, never "pure," driven by such social forces as affect other human endeavors (thus, e.g., Shapin 1994, 2010). Precisely through its intermediary character, music bridges these "inside" and "outside" views of natural philosophy.

3. Such an endeavor would consult, revise, and transcend Edmund Husserl's initiatives, applied in Klein 1985, 65–84. Among recent efforts to reengage these larger connections, see Daston and Galison 2007. Concerning the synthetic vocation of the history of science, see Holton 2009.

4. The story about his operatic experience is related by his biographer William Stukeley, cited in Gouk 1999, 224.

5. In the vast literature on "scientific revolution(s)," note especially the thoughtful work of Cohen 1994, 2010, which complements his classic treatment of music in Cohen 1984. Note also the thoughtful reinterpretation of "presentism" in Oreskes 2013, which seems to me to make important distinctions.

6. Thus, I leave aside those (such as Christiaan Huygens) mainly occupied with making music into a science, which Cohen (1984, 205–230) has treated so well, and as Kassler (1995; 2001, 83–124) and Gouk (1999, 193–223) have done for Robert Hooke. I will only add a few thoughts on Francis Bacon to Gouk's (1999, 157–170) pioneering treatment. I barely mention Robert Fludd and Athanasius Kircher, both deeply involved in music and natural philosophy, but whose relation to the new philosophy was antagonistic or tangential. In Kircher's case, much remains to be done to explore the whole range of his thought and clarify music's place in it. For some helpful beginnings, see Findlen 2004; Engelhardt and Heinemann 2007; Pangrazi 2009; Kircher 2011; McKay 2012.

7. For treatments of the relation of music and sound to the biological and medical sciences, see Kassler 1995; Horden 2000; Volmar 2012, 2013b.

8. See, e.g., Galison 1997; Daston and Lunbeck 2011. Ford (2004, 314–315) notes that Aristotle mostly used the term *mousikē* to denote what we call music, rather than poetry.

9. Walker 1978; Cohen 1984; Gouk 1982; Gouk 1999; Kassler 1995, 2001; Wardhaugh 2008.

10. Thompson 2002; Sterne 2003; Jackson 2006; Hui 2013a,b. Hui, Kursell, and Jackson 2013 includes a rich variety of approaches.

11. Schwartz 2011; Smith 1999 treats an early modern soundscape in rich detail. Jay 1994 investigates the modern denigration of vision as the primary modality of knowledge.

12. See Van Wymeersch 1999; Gozza 2000; Moreno 2004; Kittler 1999, 2006; Erlmann 2010, 2004; Ziemer 2008.

13. Panofsky 1954. See also Palisca 1961.

14. See the exchange on Panofsky's thesis, expressed (with some revisions) in Panofsky 1956a, followed by Rosen 1956 and Panofsky 1956b.

15. See Drake 1970a, 43–62; 1970b, 1992.

16. Drake and Galilei 2000, 100–108 (143–150); see also Walker 1978, 27–33; Cohen 1984, 85–97.

17. Drake 1970a, 53.

1 Music and the Origins of Ancient Science

1. I will restrict myself to Greek developments, but the foundations of Chinese and Indian natural philosophy were also deeply involved with music. Indeed, evidence suggests influence flowing from India to Greece; though the interrelations with China need further study, the centrality of music in all three traditions is no accident, whatever may be the variety of historical influence and parallel invention. See Burkert 1972, 471, 477; Sedlar 1980; Doshi 1985; Shankman and Durrant 2002.

2. Barker 1984, 2:5, 28. Martínez 2012 critiques many myths about Pythagoras.

3. Ibid., 2:29. See also Hagel 2010, 143–151.

4. Philolaus quoted from Barker 1984, 2:36; *Odyssey* 5.248; *Iliad* 22.255. Spitzer 1963, 7–19, surveys Greek treatments of cosmic harmony.

5. Barker 1984, 2:38, 36. For "unlimited" and "limiter," see also Burkert 1972, 251–269.

6. Barker 1984, 2:256–257; see also the helpful commentary in Nicomachus 1994, 13–27, 83–97.

7. Boethius 1989, I.10.

8. For a sensitive treatment of the significance of the fifth hammer, see Heller-Roazen 2011, who judges that "one can only speculate as to the reasons" for its discord (17).

9. See Raasted 1979; Lloyd 1987, 277–278. In Nicomachus's version, Pythagoras went home to experiment with strings tensed with different weights (Barker 1984, 2:257–258), which would also prove problematic. For another account of sounding "discs," see Barker 1984, 2:30–31.

10. Creese 2010 gives the most complete account of this instrument and its theory, including a valuable discussion of the smithy story (81–93).

11. For Ptolemy (*Harmonics* 16.32), see Barker 1984, 2:291; Burkert 1972, 375–378. About 1580, Vincenzo Galilei adduced experiments showing that the pitch of a string depends linearly on the square root of its tension (rather linearly), but did not test the hammer story, which Mersenne (1972, 166–171) finally rejected as "very false" (1634); see Galilei 2003, 326, 329, 339; Chua 2001. On the other hand, Walker (1978, 23–26) argued that Galilei himself could not have done the experiments he claimed to have performed.

12. Here following Porphyry's *Commentary*, in Barker 1984, 2:34–35.

13. Philolaus and Aristotle cited in Barker 1984, 2:38, 34; for further context, see Heath 1966, 47–48, 94–97, 115–116, 187–189.

14. *Timaeus* 37a.

15. Laertius 1972, sec. 8:25.

16. *Phaedrus* 253b: "convincing the boy they love and training [*rhythmizontes*] him to follow their god's pattern." Translations not otherwise identified are my own.

17. *Republic* 522c (Plato 2006, 237).

18. The first ten Greek letters, followed by the accent ´, denoted the first ten numbers: 1, 2, 3, 4, … , 9 were α´, β´, γ´, δ´, … , θ´, while 10, 11, 12, … were ι´, ια´, ιβ´, … and 20 was κ´. Thus, Greek numerals were more alphabetic (and harder to use) than Roman.

19. *Republic* 526a (Plato 2006, 242).

20. For *arithmos*, see Klein 1992, 46–60; for *rhythmos*, see Pollitt 1972, 56–60.

21. *Iliad* 23.239.

22. This possibility goes beyond the suggestions of Tannery (1902) and Szabó (1978, 99–184) that "all the important terms of the theory of proportions have their origins in the theory of music" (Szabó 1978, 170). See also Borzacchini 2007.

23. See Pesic 2003, 5–21.

24. Laertius 1972, sec. 8:25. For Plato and the concept of zero, see Pesic 2004.

25. Most interpretations of the Platonic dyad stem from the testimony of Aristotle, *Metaphysics* 987b23–35 (Aristotle 1984, 1561–1562); see Watson 1973.

26. *Republic* 526d–e (Plato 2006, 243); the identification of the Idea of the Good as the One has also been put forward as a central tenet of the so-called unwritten teachings of Plato, as an esoteric continuation of the exoteric discourse contained in his extant dialogues; see Klein 1992; Watson 1973.

27. *Republic* 527e (Plato 2006, 244).

28. Ibid., 534d; 525d–e (Plato 2006, 242).

29. For surreal numbers, see Knuth 1974; Conway 2001.

30. See Pesic 2003, 20. Adrastus (second century C. E.) is quoted from Barker 1984, 2:214; see Creese 2010, 5, 243. *Psophos* (noise) ironically inverts *sophos* (wise): the unwise merely make noise, as when Plato makes fun of those who imitate "the sound [*psophos*] of wind" (*Republic* 327a).

31. Successive musical intervals combine through multiplication because each successive interval acts to shorten (or lengthen) the string representing the initial pitch. Thus, five successive rising 9:8 tones will lengthen the initial string by $(9:8)^5 \approx 1.802 \ldots < 2:1$, less than an octave; six tones overshoot the octave because $(9:8)^6 \approx 2.027 \ldots > 2:1$. For an overview of the problem of temperament, see Bibby 2004.

32. *Republic* 529a–d (Plato 2006, 246–247).

33. Cited in Barker 1984, 2:150. For a lively reconsideration and defense of Aristoxenus, see Levin 2009.

34. See Barker 2000.

35. See Ptolemy, *Harmonics* 99.1–111.15 (the end of the surviving text), in Barker 1984, 2:378–391, and the commentary in Barker 2000, 158–191.

36. For the "New Music," see West 1992, 356–372; Csapo 2004. See Hagel 2010, 3–10, 17–28, 53–60, for the common use of modulation.

37. See Jaeger 1969. Spitzer (1963, 10), argues that Democritus first joined together the four studies of the quadrivium.

2 The Dream of Oresme

1. For an excellent treatment of the whole problem of the transmission of ancient music theory, see Mathiesen 1999; regarding the transmission of Boethius, see Mellon 2011. Concerning the period considered in this chapter and following, see Carpenter 1955, 24–27, 115–118, 313–315. For the earlier history of music in the quadrivium see Libre 1969, 175–191; Moyer 1992, 11–35; and Vendrix 2008.

2. Zoubov 1961; Abdounar 2008.

3. The earliest printed editions of Euclid in Latin date from 1482; vernacular translations were published in the 1560s and 1570s; see Grafton 1991, 23–46.

4. Euclid himself had proved that there were an unlimited number of *kinds* (genera) of irrational magnitudes, compared to the single genus of rational quantities; see Pesic 2003, 18–21.

5. See *Timaeus* 39d and Campion 1994, 243–247. Hipparchus (second century B. C. E.) is credited with the discovery that the location of the spring equinox would slowly precess around the zodiac from its initial position, returning to its starting point after 26,000 solar years, the "Great Year." Observationally, the "north star" and the whole field of stars would change over that time as the north celestial pole traces out a slow circle: the present pole star, Polaris, will give way to others but return to its place after 26,000 years. Plato's "Great Year" seems to have been a more general concept of cosmic recurrence based on the completion of the known cycles of the planets.

6. Oresme 1968a, 514–515, 340–341 For *chorea* and planets, see especially Wright 2001, 129–158; for the religious context of *chorea* in the church, see also Mews 2009. For the *carole*, see Mullaly 2011, 47–48.

7. Oresme 1968a, 522–523; Oresme attributes this view to "*La Perpective* de Witelo."

8. Ibid., 530–531. For Oresme's relation to the Copernican view, see Blumenberg 1987, 158–168.

9. Oresme 1968a, 536–539.

10. Oresme 1968b, 12.

11. Oresme 1971, 284–289

12. Ibid., 288–311, at 295–297.

13. Ibid., 310–321, at 310–311, 316–317. See also Zoubov 1961, 96–98.

14. Oresme 1971, 312–313.

15. Ibid., 320–323.

16. Oresme 1968a, 480–481. Zoubov 1961, 102, also thinks that the verdict goes to Geometry; see also Kassler 2001, 26–35.

17. Oresme 1968a, 480–481.

18. Apoc. 5:9, 14:3; cf. Ps. 39:4, 143:9, 149:1.

19. Oresme 1968a, 480–483.

20. Pesic in preparation-a.

21. The term *ars nova* as a historical epoch was only introduced by Johannes Wolf in 1904, so we cannot assume that Oresme would have responded to this term as we would; on the other hand, the very title of de Vitry's treatise would suffice to make the connection I am putting forward here. For the *ars nova* and de Vitry, see Hoppin 1978, 353–357.

22. Grant 1965, 328. De Vitry had become bishop of Meaux, hence dating this work during his bishopric, 1351–1361; see Oresme 1971, 328; 1968b, 122–123, 447, 471–472, 477; 1966, 12–13. For the history of Pythagoreanism in the Middle Ages, see Joost-Gaugier 2006, 116–133.

23. De Vitry and Plantinga 1961. See also Werner 1956, 132, who notes that "this new, mathematically grounded theory of musical measurement proved serviceable to the hitherto blocked development of musical notation."

24. For Oresme's references to de Muris, see Oresme 1968b, 450; 1966, 58n, 125–126, 299; 1971, 78–79, 97–103, 86–97.

25. For Gersonides, see Werner 1956, which mistakenly refers to Gersonides's work as *Sefer ha-Mispar* (*Book of Number*), an earlier and less sophisticated work by Rabbi Abraham ben Meir ibn Ezra (1090–1167).

26. For the rhythmic issues, see Hoppin 1978, 354–357, 362–367.

27. Oresme 1971, 212–215.

28. Ibid., 294–295.

29. Ibid., 304–305.

30. Ibid., 316–317.

31. Oresme 1968b, 222–225, 450, which speculates that he may have learned Archimedes's ideas from de Muris. In the following century, Nicholas of Cusa argued (incorrectly but ingeniously) that the circle can be squared (Boyer 1991, 271–272). For Oresme's general approach to magnitudes and intensities, see Taschow 1999; 2003, 59–199; and Heller-Roazen 2011, 49–59.

32. Oresme 1968a, 482–483.

3 Moving the Immovable

1. Tinctoris 1961, 77. Regarding the fate of the spheres, see Donahue 1981.

2. For the relation between theory and practice in Gaffurius, see Westman 2011, 41–42.

3. For its history, see M. Lundberg 2011.

4. Glarean 1965, 87; he discussed this adage with his friend Erasmus, who had included it in his *Adagia*.

5. Aristotle, *Politics* 1342b9–11 (1984, 2129); note that the poet, Philoxenus, is one of the avant-garde practitioners of the New Music. See Csapo 2004, 233–234.

6. Glareanus 1965, 87. Zarlino lists this motet under mode 4, which "accommodates itself marvelously to lamentful words which contain sadness or supplicant lamentation"; he does not remark on the mode altering in this motet. See Zarlino 1983, chap. 21. Glarean's attribution of this motet to Josquin is now considered dubious; Macey (2009) argues in favor of Nicolas Champion being its composer. Glarean's arguments are not affected, however, so my text continues to follow him in referring to the composer as Josquin. Thomas (2009) surveys the evidence that *Absalon fili mi* is by Pierre La Rue.

7. The octave species (to use the modern terminology) gives the pattern of intervals constituting the lower pentachord (intervals over a perfect fifth) of the mode; for both Ionian and Aeolian, the octave species is T T S T T, compared to Dorian T S T T T and Phyrigan S T T S T.

8. Aristotle, *Physics* 224b34–35 (1984, 380).

9. See Aristotle, *Physics* 254b12–26 (1984, 425) and Pesic 2014a.

10. Oresme 1968b.

11. Bacon 1968, 4:216, 2:342; see Pesic 2014a.

12. For the case of Albert the Great, see Partington 2004; for the theory of transmutation, see Principe 2013, 25–26, 37–38, 125–127.

13. For a helpful explanation of degrees of impossibility or improbability involved in such "absurdities," see Funkenstein 1975. For his knowledge of Greek, see Copernicus 1985, 3–19; for his relation to the question of the ordering of the spheres, see Westman 2011, 48–55.

14. Knoll 1975, 143.

15. Dyer 2007, 2009; Ward 2013.

16. Heilbron 2010, 218. For the relation of Oresme and Buridan to Copernican thought, see Blumenberg 1987, 152–168.

17. Copernicus 1985, 90, 126n327 (Martianus).

18. Copernicus 1992, 22. For the aesthetic issues, see Gingerich 1993, 193–204.

19. Ptolemy 1998, 35, 37.

20. Ibid., 45.

21. Regarding the meanings and resonances of *symmetria*, see Westman 2011, 135–137, 187–190.

22. His published text refers to the Pythagoreans at Copernicus 1992, 3. His original text (Copernicus 1992, 25–26, 361) expanding these references was deleted before publication (whether by his own or another hand). For his relation to Aristarchus, see Gingerich 1993, 185–192.

23. Both Galileo and the church authorities referred to "Pythagorean" heliocentric cosmology; see G. Galilei 1890, 2:198; Heilbron 2010, 110–111; and Copernicus and Rhäticus 1959, 138–139. For the general history of Pythagoreanism during this period, see Joost-Gaugier 2009.

24. Copernicus and Rhäticus 1959, 138–139.

25. For his final stance on heliocentrism, see Westman 1975, 299–305.

26. Gilbert 1958, 215.

27. Quoted from V. Galilei 2003, xvii. See Drake 1992; R. Lundberg 1992; Palisca 1992; cf. the critique of Vincenzo's originality and importance in Pirrotta 1984, 219–222.

28. See Walker 1978, 14–26.

29. I will try to address the larger dimensions and context of this project in Pesic in preparation-b.

30. Referring to the choral antistrophe and epode; V. Galilei, de'Bardi, and Mei 1960, 133.

31. V. Galilei 2003, 77. As confirmed by Palisca's note, the Italian term *stelle* clearly signifies planets rather than fixed stars both because of context (the sentence makes no sense if it were to refer to the stars, which even for Copernicus and Kepler were not considered to have any ordering of distance) and also because of the musical metaphor described in the text.

32. See Gingerich 2002.

33. Zarlino 1579, 1588, vol. 4. For Zarlino's theoretical and mathematical views, see Mambella 2008; Heller-Roazen 2011, 61–73.

34. I thank Owen Gingerich for drawing my attention both to Zarlino's calendrical writings and to his ownership of a copy of Copernicus's book, as cataloged in Gingerich 2002, 133.

35. G. Galilei 1890, 10:68. Heilbron (2010, 112) argues that Galileo had adopted this opinion "five or six?" years before 1595.

36. The work in question is Clavius 1999, first published 1581; for the dating of Galileo's notebook and its sources, see G. Galilei 1977, 22–23, 264–265.

37. G. Galilei 1977, 71–77.

38. G. Galilei 1989, 74–75.

39. Ibid., 196.

40. Quoted from Westman 2011, 489; for Hooke's attempts to provide experimental verification of the Copernican view, see 504–510.

4 Hearing the Irrational

1. Recorde 1557, sigs. Aiir, Sir, Siv; Neal 2002, 49–55, at 50. See also Van Wymeersch 2008.

2. See Boyer 1991, 304; Dijksterhuis 1970, 16–19, 21–22, 38–39; Bos 2001, 119–143; Rasch 2008. In 1585, the Flemish mathematician and engineer Simon Stevin also advocated decimal notation; see Klein 1992, 186–197.

3. The seminal work on Viète in relation to Greek mathematics is Klein 1992, 150–185, 321–322n10, which contains a translation of Viète's *Isagoge* (313–353). For cryptographic parallels, see Pesic 1997a,b; 2000a, 59–83; 2000b, 73–83; and Panza 2006.

4. Field 1997, 67; 2005, 24–31, 282–284, 312–316. See also Moyer 2008; Peterson 2011, 106–124.

5. Boyer 1991, 281–282; Aubel 2008; Neal 2002, 49.

6. Stifel 1544, fol. 7v, 55r–58r.

7. Boethius 1989, 4.11; Stifel 1544, fols. 70r–75v; for the sources, see Euclid and Porphyry 1991, props. 3, 16; Barker 1984, 2:190–208 and Barbera 1984; Knorr 1975, chap. 7; Field 2011; Moyer 2011.

8. Modern convention has an octave comprise 1,200 cents, of which an equal-tempered semitone would be 100 cents, a "major semitone" ($2{,}187{:}2{,}048 = 3^7{:}2^{11}$) 114.7 cents, a "minor semitone" ($256{:}243 = 2^8{:}3^5$) 90.2 cents, and a Pythagorean comma ($3^{12}/2^{19} = 531{,}441/524{,}288$) 24.5 cents. See Fauvel, Flood, and Wilson 2004, 13–27.

9. Stifel 1544, fol. 76r; for the earlier theorists, see de Muris 1992, 292–301, at 294; Hentschel 1998, 39–60, at 41.

10. As in its 1482 Latin translation: Busard 2005, 1:160–161.

11. Oresme 1966, 60–65, 304–309; 1971, 78–161, 296–305, at 297; see also Abdounar 2008.

12. Lefèvre d'Étaples and Jordanus 1496, fol. g6v, cited in Stifel 1544, fol. 76v.

13. For details of Stifel's computations, see Pesic 2010.

14. Stifel 1544, fol. 79v. I thank William Donahue for his kind help with the translations from Stifel's Latin.

15. Ibid., fol. 103r.

16. Cardano 1967, 10:222, as translated in Cardano 2007, 1.

17. Cardano 1967, 2:337, cited in Cardano 1973, 22n36, which also includes the quotation from Miller.

18. Cardano 1973, 45, cites Lefèvre; see also Pesic 2010; Barbour 1972, 7. For a lively introduction to questions of temperament, see Duffin 2007.

19. Cardano 1967, 4:281; 2007, 204.

20. Boethius 1989, 1.21. Boethius's semitone 243:256 is not an exact equal division of the tone. The trihemitone is 294.1 cents, slightly smaller than the modern minor third, 300 cents.

21. Ibid. The ditone is 407.8 cents, slightly larger than the modern major third, 400 cents.

22. Palisca 1985, 88–110; he comments on Vincenzo on 10 n 35. For Vicentino, see Berger 1980; Cordes 2007.

23. Palisca 1985, 119; see also Kaufmann 1966.

24. Vicentino 1996, 302–314, on 304, discussed in Moyer 1992, 168–184. Regarding the presence of Lasso, see Kaufmann 1966, 24n5, though Hell and Leuchtmann (1982, 112) think he may have arrived later. In either case, Lasso's chromatic *Prophetiae Sibyllarum* dates from 1550 to 1552 and may well show the influence of Vicentino.

25. Vicentino 1996, 313–314; see Boncella 1988; McKinney 2005.

26. For the Jesuits' reliance on the certainty of mathematics, see Gatto 1994, 17–64; Consentino, Homann, and Lukács 1999, 50–51; Dear 1995, 39. In 1591, Robert Parsons denounced the English algebraist Thomas Harriott as an Epicurean atheist and conjurer because of his mathematical atomism; see Neal 2002, 25–27; Kargon 1966, 27.

27. See Tiella 1975; Rasch 2002; Wright 2002; Barbieri 2002. The split keys of Vicentino's instrument were not unique; other keyboards had earlier used this device. For sound examples, hear the CD attached to Cordes 2007.

28. Cardano (1973, 194–195) described Vicentino's new instruments.

29. Vicentino's contemporaries were most familiar with just intonation (see box 4.2), in which the "just diatonic semitone" ($16:15 \approx 1.067$, 111.7 cents) was a common solution, but there were other competing possibilities, such as the "just chromatic semitone," sometimes called the "minor semitone of the minor tone" ($25:24 \approx 1.042$, 70.7 cents).

30. Vicentino defined a minor diesis as "one-half of the minor semitone" and the major diesis as "identical" to a minor semitone, but then we still have to divide that minor semitone exactly in half. See Vicentino 1996, 59–62 (fol. 17r–18v); Berger 1980, 7–18.

31. Boethius cites "Philolaus, a Pythagorean," who divided the tone unequally. See Boethius 1989, 96–97; West 1992, 135–136. Boethius argues that the Pythagorean comma is the "ultimate interval heard which can really be perceived" (Boethius 1989, 96–97).

32. See Plato, *Republic* 531a; Aristotle, *Posterior Analytics* 84b37–39; and Aristotle, *Metaphysics* 1016b18–24 (Aristotle 1984, 1:138, 2:1605), collected in Barker 1984, 2:55, 70, 72, 135. See also Quintilianus 1983, 81, 95.

33. Quintilianus 1983, 84; Strunk and Treitler 1998, 57.

34. The word *diesis* originally meant a "letting through," suggesting the performance practices of wind instruments. See Aristotle, *Politics* 134a21; and Longinus, *On the Sublime* 39.2, in West 1992, 81–107, esp. 105–106.

35. For Aristoxenus's references to the diesis, see Barker 1984, 2:135, 140, 143, 145, 154, 165–166, 182, 184.

36. Ibid.,137.

37. See Levin 2009 for a similar account of Aristoxenus. In Ptolemy's view, Aristoxenus constructed the enharmonic genus by essentially assigning the unit of 6 to each diesis, in units where the tone is 24 units; see Barker 1984, 2:384–391, 270.

38. Barker 1984, 2:170n1. Socrates draws a square on the diagonal of a unit square in his conversation with the slave boy in Plato's *Meno* 84d–85c.

39. Vicentino 1996, 6, 12 (fol. 3r, 4v).

40. For the role of Lodovico Fogliano's geometric construction to divide intervals (on the model of figure 4.1), see Pesic 2010.

41. Vicentino 1996, 207 (fol. 66v).

42. Zarlino 2011; Strunk and Treitler 1998, 299. See also Mambella 2008.

43. See Berger 1980, 15–16.

44. Vicentino 1996, 207 (f.66v).

45. See Oettinger 2003.

46. These compositions can be found in Cardano 1973, 139, 154–171.

47. Vicentino 1996, 33 (fol. 10v). For commentary on the musical details of his motets vis-à-vis his critics, see li–lviii; recordings of this and his other motets are available on the CD accompanying Cordes 2007.

48. On the other hand, Vincenzo Galilei noted that enharmonic music "was never sung without the instrument named above [archicembalo] and if by misfortune one of the singers lost his way while singing, it was impossible to put him back on the right track." After Vicentino's death, "it was practiced neither by his students nor by anybody else"; Berger 1980, 73.

49. Cardano was critical of Vicentino's scheme for tuning, which he found "not unserviceable, but … not entirely accurate"; see Cardano 1973, 194–195. Regarding Zarlino, see also Moyer 1992, 202–225.

50. See Berger 1980, 44–56, 88–95; he notes that Artusi, like Vincenzo Galilei, found the use of equal temperament for the harpsichord "strange" (92). The quotations from Artusi 1934 are taken from Lindley 1982, 400–404.

51. Klein (1992, 147–148) emphasized the connection between Viète's mathematical and astronomical works but did not recognize the significance of music as a consequential meeting ground between mathematics and perception.

52. Gosselin 1577, fol. 2r, quoted and discussed in Klein 1992, 262n225. Gosselin translated Tartaglia 1578 into French, which also experimented with the division of a tone into equal semitones (see Moyer 1992, 126–134).

5 Kepler and the Song of the Earth

1. Strictly, proportional to the semimajor axis of its elliptical orbit. The standard modern edition is Kepler 1937, to be cited hereafter as KGW, followed by volume number and page, here 6:302. I will cite the translation Kepler 1997 as HW 411, in this case. Regarding the "third law," see Field 1988, 142–163; Gingerich 1993, 348–356, 388–406. The term "law" is due to later scholars, who thereby emphasized Kepler's anticipation of crucial Newtonian results (Wilson 1989).

2. See Dickreiter 1973, who emphasizes Kepler as theorist. For the details of Kepler's arguments, see Stephenson 1994a; Field 1988, 96–166; Martens 2000, 112–141. See also Cohen 1984, 13–34; Field 2004. I am particularly indebted to Walker 1978, 34–62; Holton 1988; Werner 1978; Harburger 1980; Gingerich 1993, 388–406; Koyré 1992.

3. HW 505 (KGW 6:374). Heller-Roazen (2011, 112–140) emphasizes Kepler's self-identification with Pythagoras.

4. Ibid. For Fludd, see Ammann 1967.

5. See Dickreiter 1973, 123–138.

6. KGW 6:141, 158–159, 162; 15:238, 15:397.

7. Dickreiter (1973, 124) lists some of the polyphonic music used in Württemberg.

8. Ibid., 125. For a helpful study of Kepler's milieu during this period, see Methuen 1998.

9. Dickreiter 1973, 126.

10. Ibid., 164.

11. Stephenson 1994b,a; Pesic 2000a, 87–112.

12. Dickreiter 1973, 129.

13. Ibid.

14. Ibid., 130; KGW 14:13. Kepler already refers to Lasso's motets in a letter of 1599 (KGW 14:9).

15. Letter to Herwart von Hohenburg of August 6, 1599 (KGW 14:29). For Kepler on tuning, see Bühler 2013, 53–69.

16. Dickreiter 1973, 131, citation from Chytil 1904, preface.

17. Among recent studies, see Comberiati 1987; Lindell 1994; Kmetz 1994; Saunders 1995.

18. Maier 1989, discussed by Liessem 1969; Meinel 1986.

19. Dickreiter 1973, 132–133.

20. Evans 1984, 190–193; see Yates 1991, 78–83; Tomlinson 1993, 45–46.

21. Caspar 1993, 262, from Kepler's letter to Philipp Muller after September 13, 1622, KGW 18:78–79, discussed also in Pesic 2000b. For Kepler's attitude toward the occult arts, see Rosen 1984; Vickers 1984.

22. KGW 17:80, translated by H. Floris Cohen.

23. KGW 6:397.

24. Dickreiter 1973, 134.

25. Ibid. Though this was a very early foray into ethnomusicology, already in 1578, the Swiss theologian Jean de Léry had transcribed some Brazilian songs (Léry 1990).

26. HW 217 (KGW 6:158).

27. Dickreiter 1973, 135.

28. Ibid., 137.

29. Caspar 1993, 248; KGW 17:254. For Kepler's reading of Vincenzo, see Gingerich 1993, 396–398.

30. Dickreiter (1973, 138–139) argues that Kepler's knowledge of music theory was "not very many-sided." Indeed, Kepler manuscripts in the Pulkowa library reveal that he had read only three-quarters of Galilei's book.

31. Caspar 1993, 266.

32. Where Boethius and Macrobius defined the major third as 81:64 (two whole tones, each 9:8) and the minor third as 32:27, Kepler advocated the simpler intervals of just intonation (5:4 and 6:5, respectively).

33. HW 137 (KGW 6:99). Though praising Ptolemy for including just intonation, Kepler still criticized him for not finally advocating it; HW 138 (KGW 6:99). HW 192 (KGW 6:139) is the unique mention of Zarlino in Kepler's text.

34. Walker 1978, 35–53. Here I go beyond Walker's comparison of mathematics and music as "parallels" (39–40) drawn from a single archetype.

35. See Pesic 2000b, 57–59.

36. HW 138 (KGW 6:99).

37. Kepler goes past the commonplaces dating back to medieval theorists about allowed melodic intervals; see Hucbald et al. 1978.

38. HW 217 (KGW 6:158).

39. Here he refers to "Euclid" for a vocabulary of melodic devices, by which he means the *Introductio harmonica* now attributed to Cleonides, a student of Aristoxenus. See HW 218n125; there is a parallel passage defining this terminology in Aristides Quintilianus (Barker 2004, 2:430–431).

40. HW 218 (KGW 6:158). In the case of *Victimae paschali*, Kepler shows how the direct motion of *agogē* (as in the setting of the words "paschali laudes" or of "immolent") sets off the continuous intonation (*tonē*) of "-demit oves Christus in-" and the "playing" alternations (*petteia*) of "-cens … re- … li- … peccat-." In contrast, the Turkish chant uses "a pure *plokē*, although not a natural one," throughout its course, meaning the continuous twisting or twining of the melodic line.

41. Walker 1978, 38–40; see also Tomlinson 1993, 76–84.

42. *Sacrae cantiones quinque vocem* (Lasso 1894, 9:49–52), cited at HW 221, 234, 239 (KGW 6:161, 171, 174). Kepler mentions Lasso's *Ubi est Abel* and *Tristis est anima mea* at HW 253 (KGW 6:184).

43. HW 243 (KGW 6:177). For the fame of this motet, see Braun et al. 1994, 2:139–142. See also Boetticher 1954, cited by Lossius 1570, book 1, chapter 7, a book in Kepler's school library in Linz (Dickreiter 1973, 145).

44. *Musica autoschediastikē* (1601), expanded in his *Musica poetica* (1606); see Burmeister 1993, 205–206; see the translation and commentary Palisca 1972. For a discussion of this motet and its performance practice, see Smith 2011, 111–124. Concerning Burmeister, see Ruhnke 1955, 130–135, 162–165. That we possess no specific reference to this motet might be explained by the disappearance of some of Kepler's letters to Seth Calvisius in which Burmeister might well have been discussed; see Dickreiter 1973, 60–61.

45. HW 221 (KGW 6:161). Kepler writes the same slightly incorrect rhythm both times he cites this passage, probably quoting from memory and indicating how familiar this motet is to him, as noted by Dickreiter 1973, 175–176; Braun et al. 1994, 141. For the authentic text, see Lasso 1894, 9:49. Even Kepler's mistake is revealing; by incorrectly citing the opening e' as dotted, he has the expressive minor sixth e'–c' arrive on the downbeat in the cantus, an accented dissonance, whereas the authentic text lacks his dot and consequently arrives on the offbeat, resolving by suspension. Thus, Kepler's rhythmic mistake throws the expressive semitonal descent c'–b'–a' into higher relief.

46. HW 239 (KGW 6:174).

47. HW 238 (KGW 6:173).

48. HW 441 (KGW 6:323).

49. HW 449–450 (KGW 6:329).

50. HW 430 (KGW 6:316). For further discussion of planetary songs, see Tomlinson 1993, 63–100.

51. HW 423 (KGW 6:311).

52. HW 439 (KGW 6:322).

53. Walker 1978, 59–60.

54. HW 441 (KGW 6:323).

55. For Kepler's conservative musical tastes, see Field 1988, 118; *In me transierunt* sets Ps. 88:16; 38:10, 17, 21. This motet includes Kepler's "planetary" chords E *mollis* and C *durus* (in modern terms, E minor first inversion and C major second inversion) in mm. 15, 28, 30, 31, 58.

56. HW 440 (KGW 6:322). Kepler assigns Earth the Phrygian mode whose final note is *mi*, "because its motions revolve within a semitone [16:15]."

57. Though Gouk (1999, 129) asserts that the modern tonic sol–fa system was transmitted to English scholars via Kepler, there is no evidence of this modern system to be found anywhere in his work. The present sol–fa system began to be used only about 1600 in France. In earlier solmization, each pitch derives its name from its place in a hexachord (a group of six sequential pitches, beginning either on C, G, or F); hence, *In me transierunt* begins e la *mi*, c sol *fa*, b fa *mi*, whose terminal syllables are *mi fa mi*. In contrast, if the motet began with a semitone (as in e–f–e: e la *mi*, f fa *ut*, e la *mi*), the solmization would be changed (in this case, to *mi ut mi*).

58. To be sure, other examples of "*mi fa mi*" would have worked as well, such as Josquin's "Miserere mei Deus," probably the most famous example in the sixteenth century, but Kepler nowhere mentions Josquin. Lasso's *Locutus sum* also begins with a prominent "mi fa mi," but only after an initial leap of a fifth.

59. Several ancient stories connect erotic excitement with music; see West 1992, 31. Plato's "nuptial number" (*Republic* 546b–547a) probably inspired Kepler's discussion of the "progeny" of geometrical figures (HW 253; KGW 6:184). In his *Harmonie Universelle* (1636), Mersenne remarked on this passage, though omitting the detailed sexual imagery (Mersenne 1963, 3:188); in 1577, Salinas (1958, 56) used much milder sexual imagery about music (Cohen 1984, 64).

60. May 12, 1608, to Joachim Tanckius, a Leipzig physician (KGW 16:154–165, at 157).

61. HW 241 (KGW 6:175).

62. HW 242 (KGW 6:176).

63. Ibid.

64. Although *ekphusis* can mean "bursting out," as HW glosses, its far more direct meaning here is ejaculation as the act of begetting.

65. HW 354 (KGW 6:265); for the cube and octahedron as "spouses," see HW 407 (KGW 6:299).

66. Walker 1978, 53–57, at 57.

67. HW 360 (KGW 6:266); cf. *Georgics* 2:326.

68. See Pesic 2000a, 108–112.

69. HW 444–445 (KGW 6:325–326).

70. HW 442–46 (KGW 6:324–328). Stephenson (1994a, 170–185) emphasizes this point.

71. HW 446 (KGW 6:328).

72. Ibid.

73. Zarlino 1968, 151. Kepler refers to Artusi, Zarlino's student, at HW 254 (KGW 6:185), though never to Zarlino directly.

74. Palisca (1972, 42–46) also brings in Francis Bacon's description (1605) of what Thomas Morley called the "false close."

75. HW 447–48 (KGW 6:328).

76. HW 442–43 (KGW 6:324).

77. HW 449–450 (KGW 6:329).

78. HW 417 (KGW 6:306).

79. Kepler 1981, 223 (KGW 1:79).

80. Ibid. (KGW 8:127).

81. For Kepler's relations with different Christian denominations, see Caspar 1993, 77–85, 111–115, 146–148; for detailed discussion of his beliefs, see Hübner 1975.

82. HW 491 (KGW 6:363).

83. HW 488–489 (KGW 6:360–361).

84. KGW 7:330, cited in Panofsky 1954, 29–31 as translated in Kepler 1995, 932.

85. Kepler 1981, 223 (KGW 1:79); discussed in Pesic 2000a, 112.

6 Descartes's Musical Apprenticeship

1. Descartes 1961, 53.

2. The exemplary treatment of Beeckman, to which I am much indebted, is Cohen 1984, 116–161.

3. Koyré 1978, 117; Gaukroger, Schuster, and Sutton 2000, 4–59.

4. Beeckman 1939, 4:62.

5. Van Berkel 2000 also discusses Descartes's debt to Beeckman.

6. Brown 1991 and Dear 1995, 32–62, treating the status of music at 39.

7. Descartes 1996, 1:21.

8. Cohen 1984, 163; see also the commentary of Erlmann 2010, 37–47.

9. Descartes 1996, 2:23; Clark and Rehding 2001, 6.

10. Descartes 1961, 11–12. Augst (1965, 125) connects the *Compendium* with the beginnings of Descartes's new science. Van Wymeersch (1999) judges the *Compendium* "the first field of application of the new epistemological approach which Descartes systematized ten years later" (163); she too connects Descartes's language of "clear and distinct ideas" with the *Compendium* (101–108).

11. Descartes 1961, 12–13.

12. Cohen 1984, 163.

13. Descartes 1961, 14.

14. Ibid., 15.

15. Ibid.

16. Ibid., 21; he also notes octave overblowing in pipes (18). Cf. the Aristotelian discussion of how a note "contains" the sound of note an octave higher, as when boys and men sing together: *Problems* 19:8, 918a19–21 (Aristotle 1984, 2:1430).

17. For the philosophical consequences, see Van Wymeersch 1999; for the effect on the Cartesian concept of the self, see Moreno 2004.

18. Descartes 1996, 1:21. The addressee is this letter is not explicit, but the editors infer Mersenne. All translations from Descartes's letters are mine.

19. Ibid., 1:142.

20. For the larger context of Descartes's attitude toward "wonders," see Daston and Park 2001, 292–331.

21. Descartes 1996, 2:23–29, 31.

22. Ibid., 1:70.

23. Ibid., 1:74.

24. Ibid., 1:85–86.

25. Ibid., 1:83, 85–86.

26. Ibid., 1:88.

27. Ibid., 1:87.

28. Ibid., 1:100, and elsewhere Descartes also refers Mersenne to the *Compendium*.

29. Descartes 1996, 1:101–102.

30. Ibid., 1:102; on sunspots, see Schuster and Brody 2013.

31. Cohen 1984, 166–169.

32. Descartes 1996, 1:142; cf. Pirro (1973), who judged Descartes to be "held back by ignorance of [musical] practice," uncertain in his musical judgments.

33. Descartes 1996, 1:223–224.

34. Descartes 1979, 1–7. Augst (1965, 120) describes the *Compendium* as giving "the first in a series of geometrical models leading to the design of the cosmic machine" in *Le Monde*.

35. Gaukroger 2000, 1995, 234–237.

36. Descartes 1996, 1:270–271.

37. For his later presentation of the arguments against the void, see Descartes 1983, 46–49.

38. Walker 1978, 81–110, at 102; cf. Pirro 1973, 100–120, which is more critical of Descartes's musical discernment.

39. For Kepler on the infinite universe, see Koyré 1957, 58–87; Harrison 1987, 46–49; Heller-Roazen 2011, 131–140.

7 Mersenne's Universal Harmony

1. Beaulieu 1995, 87.

2. Quoted in Beaulieu 1995, 25.

3. For the larger context of the Republic of Letters, see Grafton 1991 and Bots and Waquet 1997. For treatments of Mersenne, see Cohen 1984, 97–114, 191–201; Gouk 1999, 170–178; Dear 1988. For his role as intelligencer, see Grosslight 2013.

4. Beaulieu 1995, 173–185; for the development of the Académie, see Brown 1967; Cohen 1981, 3–5.

5. Mersenne 1623, question 9, article 1, column 869. See Hine 1973; Lewis 2006, 113–140.

6. Egan 1962, 63–64, 70.

7. Ibid., 94.

8. Ibid., 103.

9. Ibid., 105–107. Mersenne consistently capitalizes *Soleil*, which I have followed only when he refers to God.

10. Quoted in Beaulieu 1995, 256–257.

11. Egan 1962, 154.

12. Ibid., 202.

13. For the French reaction to Galileo, see Lewis 2006. In 1634, Mersenne published his own description of Galileo's mechanical propositions, *Les méchaniques de Galilée*, and even planned writing a defense of Galileo, though he later gave this up.

14. Beaulieu 1995, 103–104.

15. On the musicality of sloths, see Clark and Rehding 2001, 2–4.

16. He cites Psalms 83:8. Mersenne includes a concluding "Livre de l'utilité de l'harmonie" as a kind of appendix to the whole work; see Mersenne 1963, vol. 3 (following the "Livre des instrumens de percussion").

17. Mersenne 1963, 2:103, 107.

18. Ibid., 1:169. Note that he uses French feet (0.325 m = 12.8 in).

19. Mersenne 1963, 3:208; Mersenne had read Vincenzo Galilei's experiments relating string pitch to the square root of the tension.

20. Regarding the accuracy of this determination, see Dostrovsky 1975, 197–198.

21. Mersenne 1963, 3:251–252.

22. Sauveur 1987; Maxham 1976.

23. Mersenne 1963, 3:209.

24. Ibid., 1:213.

25. Ibid., 3:211.

26. For Huygen's work on music, see Cohen 1984, 209–230. For the controversy about Huygen's role in the discovery of the pendulum watch, see Iliffe 1992, 39–52.

27. Dear 1988, 139.

28. Mersenne 1963, 3:208.

29. Ibid. Descartes (1996, 1:88, 102) mentions his "little treatise" (clearly the *Compendium*) in two 1629 letters to Mersenne.

30. See, e.g., Christensen 2002, 249–252; 2011; Van Wymeersch 2011. For the perspectives of Hugo Riemann and Heinrich Schenker, see Rehding 2003; Jonas 1982, 18.

31. Mersenne 1963, 3 ("Livre des instrumens de percussion"):36–37.

32. Ibid., 3:209.

33. Ibid., 3:211; Dear 1988, 187–188. Cf. Darrigol 2012, vi, who judges that "the basic concept of sound as a compression wave was not available until the last third of the seventeenth century."

34. Mersenne 1963, 3 ("Livre des instrumens de percussion"):37.

35. For his work on organs, see Gauvin 2013. Cf. Darrigol 2012, 39–49, who considers that Descartes's mechanical medium theory of light "needed neither sound nor waves" (30).

36. Mersenne 1963, 3 ("Livre des instrumens de percussion"):37–38; 1957, 535–536.

37. Mersenne 1963, 3 ("Livre des instrumens de percussion"):36–37.

38. Ibid., 37–38; Mersenne 1957, 535–536. See also Beaulieu 1995, 70–81. Maury 2003, 179–238, treats the issue of the vacuum.

39. Mersenne 1957, 3 ("Livre des instrumens de percussion"):8–39; Mersenne 1957, 536–537.

40. Mersenne 1957, 3 ("Livre des instrumens de percussion"):8–39; Mersenne 1957, 536–537.

41. Mersenne 1963, 3:211.

8 Newton and the Mystery of the Major Sixth

1. Newton 1983, 388–391.

2. For these notebooks, see Westfall 1980, 83n52; for a transcription and discussion of Newton 1665, see Pesic 2006. Newton's only other writings on music are brief comments in Add. Ms. 39582, fol. 30^v and "Questiones quaedam Philosophiae," Add. MS. 3996, fol. 33, transcribed in Newton 1983, 388–391.

3. See Newton 1665, fol. 104–113, especially the circular diagrams on fol. 109r–109v that strongly resemble those in Descartes 1961, 34, 36; see also Bühler 2013, 73–79. For Newton's study of Descartes, see Westfall 1980, 88–105; Wardhaugh 2008, 53–56. For Newton's circular tonometer and colorimeter, see Adams 2013.

4. Gouk 1999, 224–257, is the pioneering study of Newton's "Of Musick." For a detailed treatment of Newton's study of scales, see Bühler 2013, 97–121.

5. Newton 1665, fol. 138; I here regularize the spelling, compared to the original orthography in Pesic 2006.

6. For Hooke's 1672 letter to Oldenburg, see Newton 1959, 1:111; relying on it, Gouk (1999, 241) claims that Hooke "first suggested to Newton the analogy between colour and musical tone," though in this letter Hooke does not discuss color explicitly. Of course, the whole analogy between light and sound has a much longer history, excellently surveyed in Darrigol 2009; 2010; 2012, 58–66, 86–93, 144–152.

7. See *Republic* 516d. In *De sensu* (*Sense and Sensibilia*), Aristotle compares ratios in music with colors at 439b20–440b25 (Aristotle 1984, 698). For alchemical comparisons, see Gouk 1999, 231.

8. Pesic 2006, fol. 139. For Newton's use of square brackets, see Shapiro 2002, 253n24. Though he made preliminary calculations about equal temperament, Newton does not refer further to them in "Of Musick" but (like Kepler) restricts himself to just intonation; see Gouk 1999, 252.

9. In his reading of Kepler, Newton would have found references to Aristoxenus (e.g., HW 197).

10. Add. Ms. 3996, fol. 33, in Newton 1983, 388–391.

11. For Newton's association of idolatry and corruption with geocentric cosmology, see Gouk 1999, 252.

12. See Newton 1978, 177–235; 1959, 2:418–419; Guilford and Kassler 2004, 175–178. For Newton's application of these arguments to the color of the sky, see Pesic 2005, 45–52.

13. Newton, second letter on light and colors for the Royal Society (1675), quoted in Pesic 2005, 192–193.

14. For Newton's "analogy of nature," see Armstrong 1972; Newton 1984, 1:537–549, especially 545, 550n10; McLaren 1985; Hall 1993, 56–57, 112–113; Sepper 1994, 95–99, 205–270; Shapiro 1994.

15. See Newton 1979, Book III, 317–339, which ends inconclusively with Newton's remark that he was "interrupted, and cannot now think of taking these things into farther Consideration," followed by his long series of "Queries, in order to a farther search to be made by others."

16. See Kepler 1997, 138 and above. Gouk (1999, 232–237) suggests that Newton's choice of the Dorian mode reflects his preference for a symmetric, palindromic sequence of ratios (namely T S T T T S T), which she discerns in his "ideal" scale; see also Pesic 2006 and Bühler 2013, 121–129.

17. For his extensive experiments following "Grimaldo," see Newton 1979, Book III, 317–339. See also Shapiro 2001. For Grimaldi, Pardies, and Ango, see Darrigol 2012, 58–64.

18. For Newton's "fits" and his concept of "bigness," see Sabra 1963; 1981, 231–250, 273–342; Westfall 1967; Hall 1993, 163–179; Shapiro 1993, 2001, 2002.

19. "An Hypothesis hinted at for explicating all the aforesaid properties of light," Add. Ms. 3970, fol. 528^v.

20. Ibid., fol. 521^v. Newton 1984, 1:546–547n28 contains a helpful account of the development of Newton's thinking on this issue.

21. Newton 1979, 200–201.

22. Certain insects apparently have a full octave (or more) of color perception and may find what seems a behaviorally similar identity in two wavelengths of frequency ratio 2:1 to that identity our ear perceives in D and d. For instance, Frisch 1971, 10, notes that the spectral sensitivity of honey bees is 300–650 nm, which implies the possibility of their experiencing a "color octave" denied to human vision (400–700 nm).

23. This is also the view of Sepper 1994, 123–127, at 123. Cf. Darrigol 2012, 80–85, 89–108, which does not comment on the problem of the color octave.

24. For musical analogies in Newton's cosmology, see McGuire and Rattansi 1966; Gouk 1999, 251–257; Tonietti 2000.

25. Malebranche 1997, 689, 716–717.

26. Voltaire 1967, 149–158.

27. McGuire and Rattansi 1966; Westfall 1980, 510n136.

9 Euler: The Mathematics of Musical Sadness

1. Yushkevich, Bogolyubov, and Mikhaïlov 2007, 375.

2. See Tserlyuk-Askadskaya 2007. For a reproduction of Euler's notebook, see Bredekamp and Velminski 2010, 39–64.

3. "Dissertatio physico de sono," E2, III.1.183–196. The original text of this and other works by Euler may also be found in Euler 1911. For convenience, I will cite them by the standard Eneström number of each item, here E2, and its place in the *Opera omnia* by series, volume, and pages, here III.1.183–296. These works (along with helpful listing of translations and secondary literature) can be found at the online Euler Archive at http://www.math.dartmouth.edu/~euler/. Euler's very first published paper, "Constructio linearum isochronarum in medio quocunque resistente," E1, II.6.1–3, concerned the brachistochrone problem, finding a curve along which a particle falls with the shortest time; see Sandifer 2007a, 3–5.

4. All my citations of E2 are from the translation by Bruce 2013.

5. In his time, the flute was notably different in shape (with a U-shaped bend near the mouthpiece) than Mersenne's straight (transverse) flute. Though he only begins to consider the effect of temperature (whose full treatment came with Laplace and the beginnings of thermodynamics at the end of the century), Euler includes the relative effects of varying air pressure and density, following Newton.

6. Ronald 1996, 147–148. On Euler's work in naval science, see Sandifer 2007b; 2007c, 219–222.

7. See Busch 1970; Muzzulini 1994; Gertsman 2007; Warusfel 2009, 165–185; Velminski 2009b, 150–171. For the context of Euler's other early Petersburg works, see Calinger 1996.

8. Smith 1960, 24 (preface).

9. Ibid., 37.

10. Ibid., 42.

11. Pelseneer 1951, 480–482, at 482. Recall that superparticular ratios have the form $n:(n+1)$; see above, 32.

12. Smith 1960, 119–122 (IV.35–39).

13. Among the very few other attempts, note Birkhoff 1933. For a brief summary, see Newman 1956, 4:2185–2208. Birkhoff's basic equation, $M = \dfrac{O}{C}$ (where M is the aesthetic measure, O the order, and C the complexity), is consistent with Euler's approach.

14. Smith 1960, 27–28 (E33, III.1.197–427).

15. Helmholtz 1954, 229–233.

16. As pointed out by Jeans (1968, 155–156).

17. Smith 1960, 68 (II.7).

18. Ibid., 71 (II.12).

19. Ibid., 71–72 (II.13).

20. Ibid., 72 (II.14).

21. Aristotle, *Poetics* 1453b10–12 (Aristotle 1984, 2326); his terms are *tragikē hedonē* and *katharsis*.

22. Smith 1960, 73 (II.15–16).

23. Ibid., 23. See also Tserlyuk-Askadskaya 2007.

24. For instance, we learn that the standard musical pitch he knew was a full major second lower than the present standard (A440); Smith 1960, 42.

25. Ibid., 119–122 (IV.35–39). Euler seems unaware of earlier work on logarithms in music; see Wardhaugh 2008, 43–56; Bühler 2013, 39–41.

26. Recall (box 4.1) that equal temperament divides the octave into twelve equal semitones, each given by the irrational factor $\sqrt[12]{2}$. For instance, J. S. Bach's *Well-Tempered Keyboard* (1722) required a temperament capable of playing in all twenty-four major and minor keys, though not necessarily equally; see Duffin 2007. Euler discusses equal temperament in his early "Adversaria mathematica" (1726, fol. 45r), cited in Bühler 2013, 225, and reproduced in Bredekamp and Velminski 2010, 53. Euler's *Tentamen* mentions equal temperament briefly at Smith 1960, 204–205 (IX.17); the rest of the book uses just intonation.

27. Ibid., 121 (IV.38).

28. John Wallis devised its name in 1695; see Gowers 2008, 192–193, 315–317.

29. Euler 1985, 302–305 (E71, I.14.187–216).

30. For his argument, see Euler 1985 and Sandifer 2007a, 234–248; 2007c, 185–190.

31. Cited in Weil 1984, 172; Dunham 1999, 7. Calinger 1996, 130–133, argues that the Bernoullis were a more important influence on Euler's turn to number theory than Goldbach.

32. See Leibniz 1989, 212; Tserlyuk-Askadskaya 2007.

33. Weil 1984, 267. For Leibniz's work on music, see Bühler 2010; 2013, 130–175; for the Euler/Leibniz connection, see also Downs 2012.

34. The *proper* divisors of a number exclude that number itself.

35. See E152, I.2.86–162 (Dunham 1999, 7–12); Sandifer 2007c, 49–62.

36. For Euler's work on harmonic progressions, see McKinzie 2007. For the general history of the harmonic series, see Green 1969. See also Bullynck 2010.

37. Weil 1984, 267.

38. Cited in Hakfoort 1995, 60–65, at 61. See also Sachs, Stiebitz, and Wilson 1988.

39. "The Seven Bridges of Königsberg," in Newman 1956, 1:573–580 (E53, I.7.1–10).

40. Euler's 1736 paper is generally regarded as the origin of graph theory, a field introduced by J. J. Sylvester in 1878 whose terminology was codified by George Pólya and others about 1936; see Biggs, Lloyd, and Wilson 1986.

41. "Elementa doctrina solida," E230, I.26.71–93; "Demonstratio nonnullarum insignium proprietatum, quibus solida hedris planis inclusa sunt praedita" E231, I.26.94–109. For commentary, see Sandifer 2007c, 9–18. Velminski 2009a collects these papers, highlighting their connection with Euler's work on the Königsberg problem, though Sachs, Stiebitz, and Wilson 1988 point out that Euler did not note this connection. See also Mahr and Velminski 2010. Later proofs of Euler's formula exploited the analogy with an Euler walk; see Fajtlowicz and Mathew 2012.

42. For an excellent presentation of the details of both arguments and their connections, see Richeson 2008; see also Debnath 2010, 153–173.

43. For a closed, orientable surface of genus g, $\chi = 2 - 2g$; for example, Möbius strips are nonorientable because their "inside" is not distinct from their "outside." See also Blatter and Ziegler 2010.

44. For instance, comparing superparticular ratios [$n:(n + 1)$] or multiples [$1:n$] to other classes of ratios.

45. Thus, a quadratic equation has degree 2 because it contains no power of x higher than x^2; a cubic equation has degree 3 and no power higher than x^3; and so on.

10 Euler: From Sound to Light

1. For an excellent overview of the reception of Newton's theory, see Hakfoort 1995, 11–71.

2. Ibid., 79–80.

3. For the connection with de Mairan's work in 1717, see Hakfoort 1995, 63, 37–42. For Malebranche's use of the analogy with sound, see 56. See also the excellent treatment in Darrigol 2009, 115–185; 2012, 152–161, treating Malebranche, de Mairan, and Bernoulli at 136–152.

4. Hakfoort 1995, 80–82.

5. Quoted in Hakfoort 1995, 72; regarding Euler's theory, see also Home 1988; Pedersen 2008.

6. My translation, from the manuscript cited in Hakfoort 1995, 79–80.

7. The assessment of the response to the 1744 lecture follows Hakfoort 1995, 80–82; Pedersen (2008, 393) considers the analogy between sound and light to be "the hard core of Euler's optical research program." Euler's extended 1746 presentation of his theory was "Nova theoria lucis et colorum" (E88, III.5.1–45).

8. Hakfoort 1995, 98.

9. Quoted in Jean Formay's summary of Euler's 1744 "Pensées" in Hakfoort 1995, 90–91.

10. See Pesic 2005, 42–52.

11. Hakfoort 1995, 111.

12. Ibid., 113. Note that here f has been substituted for Euler's original notation of α for frequency.

13. This quote from Rameau's *Génération harmonique* (1737) is cited in Cohen 2001, 68–92, at 73–74.

14. See Rehding 2003, 15–35.

15. "Du véritable caractere de la musique moderne" (E315, III.1.516–539). See Knobloch 2008.

16. I thank Noam Elkies for pointing out to me these problems in Euler's voice-leading.

17. "De harmoniae veris principiis per speculum musicum repraesentatis" (E457, III.1.568–587). In 1840, Fétis (1994, 97) noted the priority of Euler's analysis of the dominant seventh, but was in general very critical of Euler's approach (69–84).

18. Euler 1837, 2:71, 1:112, 1:109. See also Welsh 2010.

11 Young's Musical Optics

1. Peacock 1855, 128.

2. Ibid., 12, and Gurney 1831, which both rely on Young's autobiographical notes; see also Hilts 1978. The most complete modern biography is Wood and Oldham 1954, which notes that Young's father and grandmother "were not merely nominal Quakers, but active members of the Society" and adduces "a certain affinity between the Quaker pursuit of truth, with its emphasis on verification in personal experience, and the scientific method" (3). More recently, see Robinson 2006. Regarding the Quaker background, see Isichei 1970; Cantor 2004; 2005, 64, 82–83, 111; Mathieson 2007.

3. Peacock 1855, 7, 22–23.

4. Ibid., 35–41; Robinson 2006, 36–40.

5. Cantor 2004, 147–149.

6. Regarding a Quaker doctor of the generation before Young, it was noted that "music, dancing, the theatre, the opera, wine, women and song, gambling, attendance at cock-fights, bull-baitings, race meetings, all the rough hearty joys of the Englishman of the time were incompatible with the Quaker costume he wore" (Wood and Oldham 1954, 35).

7. Darrigol 2009, 188n140; 2012, 167–168.

8. Wood and Oldham 1954, 49–50.

9. Ibid., 49.

10. Peacock 1855, 114.

11. Ibid., 115–120, at 118, 120.

12. Ibid., 121. Young himself attributed "the ultimate extent of his uncle's protection" to Burke's "friendly and indulgent" interest and his "good offices" (Hilts 1978, 251).

13. Wood and Oldham 1954, 50.

14. Peacock 1855, 129.

15. Wood and Oldham 1954, 65.

16. Young 2002, 4:613–631.

17. See Euler 1837, 1:34–56, 83–87, which first appeared in English in 1795, and Cantor 1983, 117–123.

18. The superb accounts in Darrigol 2009; 2012, 166–187, place Young in the larger context of this analogy.

19. Young 2002, 4:543.

20. Ibid. For Euler's statement of the analogy between sound and light, see Euler 1837, 1:85.

21. Newton's rings appear even with incoherent light, thus allowing Young's analogy with coherent musical tones to go forward, whereas other optical setups would depend on the issue of coherence. The centrality of coherence in Young's thought is particularly emphasized in Kipnis 1991.

22. Young 2002, 4:565; he discusses the history of the organ at 1:404.

23. Ibid., 4:627.

24. Ibid., 4:544. For Tartini, see Polzonetti 2001; for his combination tones, see Helmholtz 1954, 152–159.

25. Ibid., 4:627.

26. Ibid., 4:546–547.

27. Jackson 2006, 172–176.

28. See Young 2002, 4:562–572, here quoted at 562, 565–567; Pesic 2013c gives a detailed discussion of Young's treatment of this issue.

29. Young 2002, 4:633.

30. See Pesic 2005, 167–169.

31. Young 2002, 4:633; for his optometer, see 575–577.

32. For his "Letter to Mr. Nicholson … Respecting Sound and Light," see Young 2002, 4:607–612; for "On the Theory of Light and Colours," see Young 2002, 4:613–631.

33. "On the Theory of Light and Colours" (Young 2002, 4:618–620); see also Cantor 1970a.

34. Young 2002, 4:617. In his next paper, "An Account of Some Cases of the Production of Colours," Young will change these three primaries to red, green, and violet, whose ratios are as 7, 6, and 5, to meet Wollaston's corrections of the spectral ratios.

35. The Newton quote about "the analogy of nature" is cited at Young 2002, 4:617; the following quotes come from 618.

36. Young 2002, 4:624–626 (emphasis in original). For the development of the technology of these gratings, see Jackson 2000.

37. See Hilts 1978, 252.

38. See Pesic 2006; see also Shapiro 1980.

39. Young 2002, 4:627.

40. Ibid., 4:624–626.

41. See Pesic 2006.

42. Young 2002, 4:633–638.

43. Ibid., 4:624–626.

44. Ibid., 4:633. He also adduces "coloured atmospherical halos" and supernumerary rainbows as meteorological examples of his colored fringes, writ large in the heavens; ibid., 4:634–635, 643–645.

45. "Experiments and Calculations Relative to Physical Optics" (Young 2002, 4:639–648, at 639). See also Mollon 2002 and Kipnis 1991.

46. Young 2002, 4:624–626.

47. Oddly, Young does not calculate the value of the incident wavelength of light for any of these cases, as he had done in his 1801 paper for Newton's rings and for the diffraction grating. Though some have therefore questioned whether he really performed the measurements, the table shown seems perfectly definite, unless one doubts that the numbers listed there really were observed by Young (rather than cooked up after the fact). See Worrall 1976; cf. Kipnis 1991, 118–124. Young may have thought it sufficient to show the consistency of his new experiment with those of Newton, relying on his 1801 determination of wavelength from Newton's rings and diffraction gratings to establish that number's value.

48. All quotes in this paragraph from Young 2002, 4:645.

49. See Jones 1975.

50. Cited in Hilts 1978, 252.

51. Regarding Young's work at the Royal Institution, see Peacock 1855, 134–137, Cantor 1970b, and Robinson 2006, 85–94. For details of his exposition of music and light in these lectures, see Pesic 2013c.

52. Young's penchant for encyclopedism led him to contribute articles not just on optics but also on Egypt (a seminal work in the beginnings of Egyptology), bridges, and tides, among many others; see Robinson 2006, 179–188, which discusses the reception of Young by the French school at 165–178. See also Arago 1832; Frankel 1976; James 1984.

53. For Malus and polarization, see Pesic 2005, 84–89. See also Park 1997, 252–253, 273–274, and especially Darrigol 2012, 187–224.

54. Young 1855, 383.

55. Young 2002, 380.

56. Young 1855, 383.

57. For detailed discussion, including the work of Fresnel and Arago, see Buchwald 1989, 205–232.

58. Ibid., 203–214.

59. See Wood and Oldham 1954, 186, quoted and echoed by Robinson 2006, 173.

60. Young 1855, 1:412–417, at 414, 415.

61. For Fresnel's final understanding of transversality, see Buchwald 1989, 228–231.

62. See Gordon 1982, 27–30; Buchwald and Josefowicz 2010, 316–327.

63. Cited in Hilts 1978, 254.

64. For further references and details on topics discussed throughout this chapter, see Pesic 2013c.

12   Electric Sounds

1. Cited in Mautner and Miller 1952, 225.

2. Ibid., 228.

3. Lichtenberg 2000, 180.

4. Lichtenberg 1997, 151.

5. Carlson 1965; Takahashi 1979; Schiffer 2003, 244.

6. Chladni 1809, v–vii. For Hooke's earlier discovery of similar figures, see Gouk 1999, 219–221; Wardhaugh 2008, 106–110.

7. For an outstanding treatment of these and other aspects of Chladni's works, see Jackson 2006, 13–44; see also Schwartz 2011, 177–180.

8. Koertge 2008, sec. on Chladni.

9. The committee's report is included in Savart 1819, 114–188, at 115.

10. Ørsted 1998, 180. In 1802, Ørsted had noted that Chladni had discovered that "the notes a flute sound far higher in hydrogen gas than in oxygen gas" (130). See also Christensen 2013, 207–213; for Ørsted's relation to Ritter, see 108–112, 147–155, 242–244.

11. For Ritter as Romantic, see Wetzels 1990.

12. Ørsted 1998, 183.

13. Strickland 1998, 457.

14. Ritter 1806, 3:115–116; Strickland 1998, 458.

15. Ritter 1806, 2:124; Strickland 1998, 458.

16. Ørsted 1998, 183–184.

17. For Ørsted's writings on music, see Christensen 2013, 214–220. Benjamin cited in Ritter 2010, 4–5, 470–507, at 477, 479.

18. Rosen 1995, 58–78, at 59; see also Ritter 2010, 301–303, 330–331, 335, 444–447, 477–485.

19. Ritter 1806, 1:160, 2:232; Strässle 2004, 32–33; my translation.

20. Christensen 1995, 167; for Benjamin's reading of this passage, see Strässle 2004 and the brilliant commentary of Erlmann 2010, 185–202.

21. Cited in Rosen 1995, 59.

22. Ørsted 2011, 136–210, which describes many visits to the Parisian theaters and several encounters with Biot (163, 184, 209).

23. Ørsted 1998, 185–191, at 191.

24. Ibid., 279.

25. Ibid., 274.

26. Ibid., 280.

27. Ibid., 420. See Christensen 2013, 336–349.

28. Ørsted 1998, 417–418; see also Snelders 1990.

29. Biot and Savart 1820; Hashimoto 1983.

13 Hearing the Field

1. Faraday 1991, 1:53; for his love of music, see Williams 1971, 10. A "serpent" here denotes a bass wind instrument of curving shape, the ancestor of a tuba.

2. Hirshfeld 2006, 4; on Davy, see Lawrence 1990.

3. Ampère 1936, 2:562.

4. Faraday 1821; for his misunderstanding, see Williams 1971, 151–152.

5. Faraday 1818. "Singing flames" continued to fascinate many savants; see Tyndall 1898, 244–286; Jones 1945.

6. Faraday 1991, 1:221–223.

7. Faraday 1821, 196.

8. Ibid., 197.

9. Faraday 1932, 1:279–280. For a helpful survey of the history, see Williams 1971, 169–183.

10. Williams 1971, 175.
11. Ibid., 176.
12. Cited in Bowers 2001, 7.
13. Wheatstone 1879, 1–12; see also Bowers 2001, 15–31.
14. Wheatstone 1879, 5. For the meeting with Ørsted, see Christensen 2013, 212.
15. Ibid., 8.
16. Ibid., 21–29.
17. Ibid., 21.
18. Ibid., 36–46.
19. Faraday 1991, 1:448.
20. Wheatstone 1879, 42; emphasis in the original.
21. Ibid., 44.
22. Proceedings of the Royal Institution 1829.
23. Williams 1971, 177.
24. Bowers 2001, 35–41, at 37; Atlas 1996, 28–34; Chladni 1821.
25. As per his 1834 account, Wheatstone 1879, 84–96, at 84; see also Bowers 2001, 57–68; Canales 2009, 138, 151, 159.
26. Wheatstone 1879, 58–59.
27. Williams 1971, 178.
28. Ibid., 179.
29. Faraday 1932, 329–359; see also Faraday 1831.
30. Faraday 1831, 309. See the helpful discussion in Tweney 1992a,b.
31. Faraday 1831, 336.
32. Faraday 1991, 1:556–557.
33. Faraday 1831, 337–338. See also Coleridge's poem "The Eolian Harp" (1795) and Abrahms 1957; Bidney 1985.
34. Faraday 1831, 328.
35. Faraday 1932, 1:353.
36. Ibid., 1:358–359.
37. Faraday 1965, 1:3.
38. Faraday 1932, 1:368; 1965, 1:4.
39. Faraday 1965, 1:16, 18, 19.
40. From a manuscript in the Royal Institution included in Williams 1971, 181, which notes the analogy between sound and electricity, as does Hirshfeld 2006, 117–118.
41. See Hubbard 1968, 27–67; Bowers 2001, 117–138.
42. Tyndall 1961, 149.
43. Wheatstone 1879, 138–140, at 138.
44. For this development, see Galison 2003, which mentions Wheatstone on 30.
45. Wheatstone 1879, 141–142.
46. Ibid., 143–151.
47. On the different values for this speed, see Faraday 1965, 3:515–516, 575–579.
48. Though Faraday was well aware of Young's work, he shows no awareness of the private letter in which Young had recorded his misgivings about the ether.
49. Faraday 1965, 3:449–451, 161–168.

14 Helmholtz and the Sirens

1. Helmholtz 1882, 1:12–75; see also Bevilacqua 1993.
2. Helmholtz 1882, 2:760–839.
3. Ibid., 2:858–880; see also Debru 2001; see Meulders 2010, 89–106.
4. Helmholtz 1882, 2:229–260.
5. Helmholtz 1971, 144–22, at 213–222; Hatfield 1990; Lenoir 1993, 124–126; Hui 2008, 77–79.
6. For the relation between his physiological, mathematical, and philosophical work, see Richards 1977.
7. Helmholtz 1882, 2:45–70; on the relation between visual color and sound-color (*Klangfarbe*), see Kursell 2013.
8. Helmholtz 1867, 282–288, 293, which incorporated the work of Grassmann 1854; MacAdam 1970, 53–60.
9. Helmholtz 1962a, 2:64; see also Lenoir 1993.
10. Helmholtz 1962a, 2:76.
11. Ibid., 2:66.
12. Ibid., 2:77, originally published in Helmholtz 1882, 2:78–82.
13. Helmholtz 1962a, 2:117.
14. Ibid., 2:77.
15. For excellent discussions of Helmholtz within his larger musical and cultural milieus, see Hiebert and Hiebert 1994; Jackson 2006, 146–148; Hui 2008, 39–82; Erlmann 2010, 217–270; Hui 2013a.
16. Koenigsberger 1965, 14; Helmholtz 1993, 43n2, 68.
17. Helmholtz 1962b, 250–286. Regarding the relation between painting and visual science in Helmholtz, see Hatfield 1993, 522–588, at 535–540. On the significance of the *Kulturträger* in science, see Sonnert 2005.
18. Helmholtz 1882, 1:256–302, 420–423, 424–426.
19. For helpful overviews of his project, see Vogel 1993; Meulders 2010, 153–199; Fowler 2004.
20. For Helmholtz's use of "instruments as agents of change," see Pantalony 2009, 20–36. For Weber's anticipation of Helmholtz's recording technique, see Kittler 1999, 26.
21. Helmholtz 1954, 229–233.
22. For the larger context of this instrument, see Rehding 2014. See also Kursell 2013.
23. Helmholtz 1954, 8, 11, 13.
24. Ibid., 155–159, at 157, with mathematical details at 411–413, 418–420. See also Jackson 2006, 178–179.
25. Helmholtz 1954, 158.
26. Ibid., 170, 173.
27. Helmholtz 1903, 1:265–365, at 309; Helmholtz 1962b, 93–185, at 130–131.
28. Helmholtz 1903, 1:120–155, at 122; Helmholtz 1962b, 22–58, at 23.
29. Helmholtz 1954, 369–370. In the second edition, following this paragraph the text then cuts to the final paragraph on 371. Regarding the importance of invariance in his thinking, see Hatfield 1993, 552–553.
30. Meulders 2010, 71.

15 Riemann and the Sound of Space

1. Koenigsberger 1965, 207–208.
2. Gauss 2005.
3. Riemann 1990, 304–319; quoted in Pesic 2007, 23–45.
4. See Pesic 2007, 2; see also Scholz 1982; Nowak 1989.
5. Riemann 1990, 318; quoted in Pesic 2007, 33; see also Grattan-Guinness and Cooke 2005, 506–520.
6. Riemann 1990, 304, 306; quoted in Pesic 2007, 23, 24.

7. Helmholtz 1867, 288–297. Young 2002, 4:617.

8. Riemann 1990, 309–310; quoted in Pesic 2007, 26–27. In modern notation, Riemann generalized the quadratic Euclidean line element $ds^2 = dx_1^2 + dx_2^2 + dx_3^2$ (in terms of three spatial coordinates x_1, x_2, x_3) to a general quadratic form $ds^2 = g_{11} dx_1^2 + g_{12} dx_1 dx_2 + g_{22} dx_2^2 + \ldots = \Sigma_{\mu\nu} g_{\mu\nu} dx^\mu dx^\nu$ (summed over all n dimensions, $\mu, \nu = 1, \ldots, n$), where $g_{\mu\nu}$ is the "metric tensor."

9. Riemann 1990, 320–325; see also Ionescu-Pallas and Sofonea 1986; Archibald 1991; Laugwitz 1999, 254–263, 269–272.

10. Maxwell 1868; 1890, 2:125–143.

11. Riemann 1990, 168–176.

12. See Pesic 2007, 41–45.

13. Riemann 1990, 370–382; 1984; Laugwitz 1999, 281–287.

14. Riemann 1990, 587–589; Laugwitz 1999, 2–3.

15. Riemann 1990, 373; 1984, 32–33.

16. Riemann 1990, 375; 1984, 35.

17. See Ritchey 1991; Tonietti 2011.

18. Riemann 1990, 376; 1984, 35.

19. Helmholtz 1971, 366–408, at 370–377; Hui 2008, 78–79. For Ernst Mach's sign theory of hearing, see Hui 2013a.

20. Riemann 1990, 380; 1984, 37. For the developing concept of attention, see Crary 1999, 30, 64, 104–105; Steege 2012, 82–122.

21. Riemann 1990, 374; 1984, 34.

22. Jacob Henle, the friend and editor who published Riemann's paper posthumously, noted that "Riemann thought that the mathematical problem to be solved was in fact hydraulic [*hydraulisches*]"; quoted in Riemann 1990, 370n. See also Riemann 1990, 807–810, and Gallagher 1984.

23. Helmholtz 1882, 2:503–514, 515–581, collected in Helmholtz 1873.

24. Koenigsberger 1965, 267.

25. Eventually, Helmholtz's detailed description of the ear was superseded by later anatomical findings, particularly because the larger context of the processing of hearing became understood as involving the auditory system of the brain as well. Rather than being a kind of "nerve piano," its separate cilia sympathetically responding to incoming pitches, the cochlea currently is considered to comprise a series of chambers of variable resonant frequency, in which the cilia respond to the local amplitude of vibration, rather than its frequency. As Riemann surmised, the overall functioning of hearing may be described in terms of inputs and outputs of a complex electrical network. Cf. Sterne 2003, 62–67; Erlmann 2010, 312–314.

26. Riemann 1867; Dedekind 1990.

27. Koenigsberger 1965, 254–255. "Monodromous" means that two congruent bodies remain congruent after one of them has undergone a complete rotation around any axis.

28. For d'Alembert's and Lagrange's remarks, see Archibald 1914; Bork 1964.

29. Helmholtz 1868; 1882, 2:610–617, at 610–611, as translated in Pesic 2007, 47–52, at 47, though the date of publication of this paper should be listed as 1868, as shown by Volkert (1993).

30. Helmholtz 1868; 1882, 2:611; Pesic 2007, 47–48.

31. See also Helmholtz's longer 1868 paper on this subject: Helmholtz 1977, 39–58; 1882, 2:618–639. For a helpful modern account of Helmholtz's argument, see Adler, Bazin, and Schiffer 1975, 7–16; also Rosenfeld 1987, 333–338; Wahsner 1994; Darrigol 2003; Darrigol 2007.

32. Goethe, *Faust*, Part I, line 1237. Helmholtz devoted two major essays to Goethe: Helmholtz 1995, 1–17, 393–412.

33. Helmholtz 1868; 1882, 2:617; Pesic 2007, 51.

34. For Beltrami, see Gray 1989, 147–154.

35. Torretti 1978, 155–179; Schüller 1994; Volkert 1996.

36. For the non-Euclidean character of color space, see chapter 18.
37. Helmholtz 1868; 1882, 2:616–617; Pesic 2007, 50–51; see also Heinzmann 2001.
38. The second edition (1885) of the *Handbuch* mentions Riemann and describes color perception as a three-dimensional manifold comparable to space; Helmholtz 1896, 336.
39. For the role of Heinrich Grassmann, see Torretti 1978, 109; Scholz 1980.
40. Helmholtz 1954, 370, based on the fourth German edition (1877); cf. Helmholtz 1865, 560; Helmholtz 1870, 576. See also Vogel 1993, 273.
41. Helmholtz 1954, 370.
42. See Helmholtz 1995, 76–95.
43. Quoted in Pesic 2007, 53–70, at 53.
44. Ibid., 64.
45. Ibid., 61.
46. Ibid., 59.
47. For Immanuel Kant on space as a manifold, see *Critique of Pure Reason* B50; see also Lenoir 2006.
48. Quoted in Pesic 2007, 59.
49. See Helmholtz 1977, 159–160; Hyder 1999.
50. Quoted in Pesic 2007, 68.
51. See, e.g., Fullinwider 1990.
52. Helmholtz writes his four-dimensional line element as $ds^2 = dx^2 + dy^2 + dz^2 + dt^2$, in which he then allows t to become imaginary ($t = i\tau$), so that the four-dimensional manifold is now pseudospherical and hence $ds^2 = dx^2 + dy^2 + dz^2 - d\tau^2$, exactly the form of the Lorentzian line-element used by Einstein and Minkowski, if $\tau = ct$, where c is the speed of light.
53. Abbott 1992; Pesic 2007, 54–55; Reichenbach 1970, 308.
54. See Hawkins 1994; 2000, 124–130.
55. See Hawkins 2000, 34–42; Rowe 1992. For Clifford's response, see Pesic 2007, 71–87.
56. Quoted in Pesic 2007, 109–116, at 110.
57. Heinzmann 1992, 2001.
58. See Pesic 2007, 100; for the Poincaré and Klein models, see Rosenfeld 1987, 236–246.
59. Einstein 1987, 1:220.
60. Solovine and Einstein 1956, viii; Holton 1996, 205; Cahan 2000, 59–74.
61. See Stein 1977; Carrier 1994; Friedman 2002; Ferreiros 2006.
62. Quoted in Howard 2005, 34.
63. Einstein as quoted in Pesic 2007, 190. The mathematical details not given in his 1854 lecture are provided in Riemann 1990, 423–436; Farwell and Knee 1990.
64. Einstein 1987, 6:569; second Einstein quote from Pesic 2007, 161.

16 Tuning the Atoms

1. Sommerfeld 1934, v.
2. Ångström 1855, 327.
3. See Pais 1986, 166–205.
4. H_α 6562.1 Å, H_β 4860.7 Å, H_γ 4340.1 Å, H_δ 4101.2 Å.
5. Stoney 1871, 291.
6. Helmholtz 1954, 23–24.

7. Stoney here instances the Fourier series, so that "the nth of these lines is represented by the term $C_n \sin(nx + \alpha_n)$, in which C_n is the amplitude of the vibration; and consequently C_n^2 represents the brightness of the line" (Stoney 1871, 293).

8. Ibid.; on the clarinet, see Helmholtz 1954, 98–99.

9. Stoney 1871, 295.

10. Ibid., 296.

11. See Stoney and Reynolds 1871, which explicitly mentions Helmholtz and the violin string on 47.

12. There is only a brief mention of this result in Stoney 1880.

13. Balmer 1885a, 551–552.

14. Ibid. McGucken (1969, 131) notes that "certainly Balmer knew of Stoney's earlier work."

15. Balmer 1885a, 551–552. Pais 1986, 172, though generally useful as a summary, misleadingly translates *Grundton* as "keynote."

16. Balmer 1885a, 553.

17. Helmholtz 1954, 40–41. Balmer would have known these as "cylinder functions [*Zylinderfunktionen*]."

18. The frequency goes as $C(m+2n)^p$, where C is a coefficient depending on the plate, m and n are integers, and p is roughly 2 for a circular plate. The coefficient p can vary between 1.4 and 2.4 for other, more complicated shapes, such as cymbals, hand bells, or church bells. As a mathematician, Balmer would also have known the approximate expression of Bessel functions involving squares; see Rayleigh 1945; Airey 1910.

19. His second publication (Balmer 1885b) extends his first results to several more recently discovered hydrogen lines, which, compared with his formula for $n = 2$, $m = 5$–16, he finds "agreement that must surprise to the highest degree"; his fourth and final paper (Balmer 1897) addresses a number of elemental spectra and, in an addendum, discusses the work of Johannes Rydberg (1890). Pais (1986, 173) implies misleadingly that Balmer did not address other elements besides hydrogen until this publication, whereas he does at least mention his thoughts in Balmer 1885a, 559–560.

20. Balmer 1885c.

21. Rayleigh 1945. In the whole work, the word "music" is only used explicitly five times. For the distinction between discovery and justification, see Reichenbach 2006, 382.

22. Rayleigh 1945, vi–vii.

23. This is true, for example, of Rydberg 1890.

24. Husserl 1970, 52–53.

25. Ibid., 360–361. For discussions of this process, see Klein 1985, 65–84; Derrida 1989, 98–107.

26. I have discussed this in Pesic 2000a, 2–3, and applied it throughout that work.

17 Planck's Cosmic Harmonium

1. Husserl 1939, 212. The phrase is not included in the standard edition and translation (Husserl 1970); Klein (1985, 372) concludes that "this sentence is based on Husserl's own words, uttered in conversation with Fink," the editor who first published the essay in 1939.

2. Planck 1998, 7.

3. For his comments on Helmholtz and energy, see Planck 1998, 19–20, 99–107.

4. This and the following general information about Planck's musical life come from the excellent account in Heilbron 1986, 3, 34. For a superb account of the "singing savants" in the earlier part of the century, see Jackson 2006, 45–74.

5. For Planck's personal recollections of Helmholtz, see Planck 1949, 15, 24–25.

6. Planck 1893, 428; Pesic 2014c.

7. Heilbron 1986, 34.

8. Hui 2013b; Hiebert 2003.

9. Planck 1949, 26.

10. Helmholtz 1954, 316.

11. Ibid., 314.

12. See Swafford 1997, 509–510.

13. See, e.g., Duffin 2007.

14. Planck 1949, 26–27.

15. Planck 1893, translated in Pesic 2014c. Note that this paper does not appear in Planck's collected physics papers (Planck 1958), indicating the disciplinary divide as perceived by his editors.

16. See the discussion of this issue in Duffin 2007.

17. Planck 1893, 418; Pesic 2014c.

18. The passage in question comes from the eighth volume of Schütz's *Sämmtliche Werke*, edited by Spitta and published in 1889, only four years before.

19. For an explanation of how exactly these five commas result, see Pesic 2014c.

20. Planck 1893, 439–440; Pesic 2014c.

21. Planck 1949, 27.

22. Planck later became the first champion of Einstein's relativity theory, on the grounds that it revealed a deeper, less anthropomorphized absolute; see Planck 1998, 112–130, and Gorham 1991.

23. See Heilbron 1986, 4.

24. Hui 2013b, 141–142.

25. Planck 1893, 438; Pesic 2014c.

26. Planck 1893, 439.

27. Hui 2013b, 142.

28. Planck wrote this in 1935, during the even blacker years of the Nazi period; see Mulligan 1994.

29. Planck 1949, 33–34. For Planck's principle and its historical ramifications, see Gorham 1991.

30. For a thorough discussion of Planck's use of the resonator concept, see Kangro 1976, 132–148.

31. Helmholtz mentions his use of the harmonium to generate exact pitches for his resonator experiments in Helmholtz 1954, 56.

32. Planck 1899, 479–480.

33. Hermann 1973, 39.

34. The Planck length is $\sqrt{\frac{hG}{2\pi c^3}}$, the only quantity with units of length that can be formed from Planck's constant h, the Newtonian gravitational constant G, and the speed of light c (the factor 2π generally divides h as unit of angular momentum, though Planck himself did not include this factor). In contemporary physics, this length is often taken to give the scale at which quantum effects enter into the fundamental structure of space-time itself, which may no longer be a continuum at that scale. Likewise, the Planck time $\sqrt{\frac{hG}{2\pi c^5}} = 5.391 \times 10^{-44}$ sec, corresponds to a "Planck frequency" $\sqrt{\frac{2\pi c^5}{hG}} = 1.855 \times 10^{43}$ Hz. My calculation of the corresponding "pitch" assumes concert A as 440 Hz.

35. Heilbron 1986, 34.

36. Hermann 1971, 21.

37. Planck 1981, 214.

38. Hertz 1962, 21.

18 Unheard Harmonies

1. Letter to Paul Plaut, October 23, 1928, quoted in Einstein 1979, 78.

2. Einstein 1987, 1:lxiii (translation: xxi).

3. Heisenberg 1971, 10–11.

4. Moore 1989, 18.

5. Ibid., 120–129. Schrödinger's writings on color are collected in MacAdam 1970, 134–193.

6. In 1922, Schrödinger also was influenced by Hermann Weyl's work on extending general relativity to a "gauge theory" including electromagnetism; see O'Raifeartaigh 1997, 77–106; Moore 1989, 146–148.

7. Bloch 1976, 23.

8. Ibid., 23–24.

9. He initially used relativistic expressions and found an equation (now called the Klein–Gordon equation) that he could not immediately interpret; in its place, he then put forward a nonrelativistic expression, now called the Schrödinger equation. See Pais 1986, 288–289.

10. Schrödinger 1982, 1.

11. Schrödinger 1928, 7.

12. Schrödinger 1982, 10.

13. Ibid., 26.

14. Regarding the problem of visualization, see Pesic 2002, 97–99. A *Hilbert space* comprises vectors (which may be real or complex) having a positive measure of distance (given by their "inner product") and is *complete*, meaning that any Cauchy sequence (in which the successive terms become arbitrarily close to each other as the sequence progresses) converges to a limit within the space.

15. Dirac 1963, 47; Weyl 2009, 11–12; Holton 1988.

16. Kramer 1994; Kramer et al. 2010; Hermann, Hunt, and Neuhoff 2011; Volmar 2012, 2013a,b; Supper 2012.

17. "Ode to a Grecian Urn," lines 11–14. For a later physicist's expression of such longings, see Wilczek and Devine 1988.

18. For an overview, see Vecchia 2008; Cappelli et al. 2012, 221–235. In a 1730 letter to Goldbach, Euler first defined what was only named the "beta function" by Jacques Binet in 1839; see Cajori 2011, 271–272.

19 Nambu 2012, 279; emphasis added. Susskind (2012) also relied on the analogy with the harmonic oscillator. Nielsen (2012) traversed a different, but finally convergent, metaphorical path, realizing that the Feynman diagrams for the processes described by the Veneziano amplitude could be understood in terms of the behavior of an elastic sheet.

20. In popular writings, Smolin 2006 and Woit 2006 challenge the rhapsodic string encomium of Greene 2000.

21. Erlmann 2010 places this resonance in the larger context of the history of reason.

22. Brann 2011, 42–43.

23. I address science and the hiddenness of nature in Pesic 2000.

24. See Susskind 2005; Carr 2007.

25. The demiurge even "outsourced" to the young gods the creation of human beings; see *Timaeus* 42d–e.

26. A growing literature considers cosmology in terms of bubbles; see Kleban 2011, Salem 2011.

27. The valuable accounts in Daston and Galison 2007 and Daston and Lunbeck 2011 remain centered on visual and material modalities and evidence. The present work may complement these accounts and help situate issues of "objectivity" and "observation" (as post-Kantian concerns) within the larger historical and metaphysical framework that aural issues require.

28. Pesic in preparation-b.

References

Abbott, Edwin A. 1992. *Flatland: A Romance of Many Dimensions*. New York: Dover.

Abdounar, Oscar João. 2008. Ratios and music in the Late Middle Ages: A preliminary survey. In *Music and Mathematics: In Late Medieval and Early Modern Europe*, ed. Philippe Vendrix, 23–69. Turnhout: Brepols.

Abrahms, M. H. 1957. The correspondent breeze: A romantic metaphor. *Kenyon Review* 19:113–130.

Adams, Charles R. 2013. "*In Experiments Where Sense Is Judge*": Isaac Newton's tonometer and colorimeter. *Journal of the Oughtred Society* 22:41–45.

Adler, Ronald, Maurice Bazin, and Menahem Schiffer. 1975. *Introduction to General Relativity*, 2nd ed. New York: McGraw-Hill.

Airey, John R. 1910. The vibrations of circular plates and their relation to bessel functions. *Proceedings of the Physical Society of London* 23:225–232.

Ammann, Peter J. 1967. The musical theory and philosophy of Robert Fludd. *Journal of the Warburg and Courtauld Institutes* 30:198–227.

Ampère, André-Marie. 1936. *Correspondance du Grand Ampère*. Paris: Gauthier-Villars.

Ångström, Anders Jonas. 1855. Optical researches. *Philosophical Magazine* 9:327–342.

Arago, F. 1832. *Éloge historique du Docteur Young*. Paris.

Archibald, R. C. 1914. Time as a fourth dimension. *Bulletin (New Series) of the American Mathematical Society* 20:409–412.

Archibald, Thomas. 1991. Riemann and the theory of electrical phenomena: Nobili's rings. *Centaurus* 34:247–271.

Aristotle. 1984. *The Complete Works of Aristotle*. Ed. Jonathan Barnes. Princeton, NJ: Princeton University Press.

Armstrong, H. L. 1972. Comment on Newton's inclusion of indigo in the spectrum. *American Journal of Physics* 40:1709.

Artusi, Giovanni Maria. 1934. *Discorso secondo mvsicale di Antonio Braccino [pseud.] Da Todi. Per la dichiaratione della lettera posta ne' Scherzi Musicali del Sig. Claudio Monteuerde*. Venice: G. Vincenti.

Atlas, Allan W. 1996. *The Wheatstone English Concertina in Victorian England*. Oxford: Clarendon Press.

Aubel, Matthias. 2008. *Michael Stifel: Ein Mathematiker im Zeitalter des Humanismus und der Reformation*. Augsburg: Rauner.

Augst, Bertrand. 1965. Descartes's compendium on music. *Journal of the History of Ideas* 26:119–132.

Bacon, Francis. 1968. *The Works of Francis Bacon*. Ed. James Spedding, Robert Leslie Ellis, and Douglas Denon Heath. New York: Garrett.

Balmer, J. J. 1885a. Notiz über die Spectrallinien des Wasserstoffs. *Verhandlungen der naturforschenden Gesellschaft in Basel* 7:548–560.

Balmer, J. J. 1885b. Zweite Notiz über die Spectrallinien des Wasserstoffs. *Verhandlungen der naturforschenden Gesellschaft in Basel* 7:750–752.

Balmer, J. J. 1885c. Notiz über die Spectrallinien des Wasserstoffs. *Annalen der Physik* 261:80–87.

Balmer, J. J. 1897. Eine neue Formel für Spektralwellen. *Verhandlungen der naturforschenden Gesellschaft in Basel* 11:448–463.

Barbera, André. 1984. Placing *Sectio Canonis* in historical and philosophical contexts. *Journal of Hellenic Studies* 104:157–161.

Barbieri, Patrizio. 2002. The evolution of open-chain enharmonic keyboards c. 1480–1650. *Schweizer Jahrbuch für Musikwissenschaft* 22:145–184.

Barbour, J. Murray. 1972. *Tuning and Temperament: A Historical Survey.* New York: Da Capo Press.

Barker, Andrew, ed. 1984. *Greek Musical Writings.* Cambridge: Cambridge University Press.

Barker, Andrew. 2000. *Scientific Method in Ptolemy's Harmonics.* Cambridge: Cambridge University Press.

Beaulieu, Armand. 1995. *Mersenne: Le grand Minime.* Brussels: Fondation Nicolas-Claude Fabri de Peiresc.

Beeckman, Isaac. 1939. *Journal tenu par Isaac Beeckman, de 1604 à 1634.* Ed. Cornelis De Waard. La Haye: M. Nijhoff.

Berger, Karol. 1980. *Theories of Chromatic and Enharmonic Music in Late 16th Century Italy.* Ann Arbor, MI: UMI Research Press.

Bevilacqua, Fabio. 1993. Helmholtz's *Ueber die Erhaltung der Kraft*: The emergence of a theoretical physicist. In *Hermann von Helmholtz and the Foundations of Nineteenth-Century Science*, ed. David Cahan, 291–333. Berkeley, CA: University of California Press.

Bibby, Neal. 2004. Tuning and temperament: Closing the spiral. In *Music and Mathematics: From Pythagoras to Fractals*, ed. John Fauvel, Raymond Flood, and Robin Wilson, 13–27. New York: Oxford University Press.

Bidney, Martin. 1985. The Aeolian harp reconsidered: Music of unfulfilled longing in Tjutchev, Mörike, Thoreau, and others. *Comparative Literature Studies* 22:329–343.

Biggs, Norman E., Keith Lloyd, and Robin J. Wilson. 1986. *Graph Theory, 1736–1936.* Oxford: Clarendon Press.

Biot, Jean-Baptiste, and Félix Savart. 1820. Note sur le magnetisme de la pile de Volta. *Annales de Chimie et de Physique* 15:222–223.

Birkhoff, George David. 1933. *Aesthetic Measure.* Cambridge, MA: Harvard University Press.

Blatter, Christian, and Günter M. Ziegler. 2010. Eulers Polyederformel und die Arithmetisierung der Gestalt. In *Mathesis & Graphé: Leonhard Euler und die Entfaltung der Wissenssystems*, ed. Horst Bredekamp and Wladimir Velminski, 243–256. Berlin: Akademie Verlag.

Bloch, Felix. 1976. Heisenberg and the early days of quantum mechanics. *Physics Today* 29(12):23–27.

Blumenberg, Hans. 1987. *The Genesis of the Copernican World.* Cambridge, MA: MIT Press.

Boethius. 1989. *Fundamentals of Music.* Ed. Claude V. Palisca, trans. Calvin M. Bower. New Haven: Yale University Press.

Boetticher, W. 1954. Orlando di Lasso als Demonstrationsobject in der Kompositionslehre des 16. und 17. Jahrhunderts. In *Bericht über den Internationalen Musikwissenschaftlichen Kongress, Bamberg, 1953*, ed. Wilfried Brennecke, Willi Kahl, Rudolf Steglich, 124–127. Kassel: Bärenreiter-Verlag.

Boncella, Paul Anthony Luke. 1988. Denying ancient music's power: Ghiselin Danckerts' essays in the "Generi Inusitati." *Tijdschrift van de Vereniging voor Nederlandse Muziekgeschiedenis* 38:59–80.

Bork, Alfred M. 1964. The fourth dimension in nineteenth-century physics. *Isis* 55:326–338.

Borzacchini, Luigi. 2007. Incommensurability, music, and continuum: A cognitive approach. *Archive for History of Exact Sciences* 61:273–302.

Bos, H. J. M. 2001. *Redefining Geometrical Exactness: Descartes' Transformation of the Early Modern Concept of Construction.* New York: Springer.

Bots, Hans, and Françoise Waquet. 1997. *La République des lettres.* Paris: De Boeck.

Bowers, Brian. 2001. *Sir Charles Wheatstone FRS: 1802–1875*, 2nd ed. Stevenage: Institution of Electrical Engineers in association with the Science Museum.

Boyer, Carl B. 1991. *A History of Mathematics.* Ed. Uta C. Merzbach. New York: Wiley.

References

Brain, R. M., R. S. Cohen, and O. Knudsen, eds. 2007. *Hans Christian Ørsted and the Romantic Legacy in Science: Ideas, Disciplines, Practices.* Dordrecht: Springer.

Brann, Eva T. H. 2011. *The Logos of Heraclitus: The First Philosopher of the West on Its Most Interesting Term.* Philadelphia: Paul Dry Books.

Braun, Werner, Theodor Göllner, Heinz von Loesch, and Klaus Wolfgang Niemöller. 1994. *Deutsche Musiktheorie des 15. bis 17. Jahrhunderts.* Darmstadt: Wissenschaftliche Buchgesellschaft.

Bredekamp, Horst, and Wladimir Velminski, eds. 2010. *Mathesis und Graphé: Leonhard Euler und die Entfaltung der Wissenssysteme.* Berlin: Akademie Verlag.

Brown, Gary I. 1991. The evolution of the term "mixed mathematics." *Journal of the History of Ideas* 52:81–102.

Brown, Harcourt. 1967. *Scientific Organizations in Seventeenth Century France.* New York: Russell & Russell.

Bruce, Ian, ed. 2013. Euler's dissertation *De sono.* http://www.17centurymaths.com/contents/euler/e002tr.pdf.

Buchwald, Jed Z. 1989. *The Rise of the Wave Theory of Light: Optical Theory and Experiment in the Early Nineteenth Century.* Chicago: University of Chicago Press.

Buchwald, Jed Z., and Diane Greco Josefowicz. 2010. *The Zodiac of Paris: How an Improbable Controversy Over an Ancient Egyptian Artifact Provoked a Modern Debate between Religion and Science.* Princeton, NJ: Princeton University Press.

Bühler, Walter. 2010. Musikalische Skalen und Intervalle bei Leibniz unter Einbeziehung bisher nicht veröffentlichter Texte. I. *Studia Leibnitiana* 42:129–161.

Bühler, Walter. 2013. *Musikalische Skalen bei Naturwissenschaftlern der frühen Neuzeit: Eine elementarmathematische Analyse.* Frankfurt: Peter Lang.

Bullynck, Maarten. 2010. Leonhard Eulers Weg zur Zahlentheorie. In *Mathesis & Graphé: Leonhard Euler und die Entfaltung der Wissensystems*, ed. Horst Bredekamp and Wladimir Velminski, 157–175. Berlin: Akademie Verlag.

Burkert, Walter. 1972. *Lore and Science in Ancient Pythagoreanism.* Cambridge, MA: Harvard University Press.

Burmeister, Joachim. 1993. *Musical Poetics.* Trans. Benito V. Rivera. New Haven: Yale University Press.

Busard, H. L. L., ed. 2005. *Campanus of Novara and Euclid's Elements.* Stuttgart: Steiner.

Busch, Hermann Richard. 1970. *Leonhard Eulers Beitrag zur Musiktheorie.* Regensburg: G. Bosse.

Cahan, David, ed. 1993. *Helmholtz and the Foundations of Nineteenth-Century Science.* Berkeley, CA: University of California Press.

Cahan, David. 2000. The Young Einstein's physics education: H. F. Weber, Hermann von Helmholtz, and the Zurich Polytechnic Physics Institute. In *Einstein: The Formative Years, 1879–1909*, ed. Don Howard and John Stachel, 43–82. Basel: Birkhäuser.

Cajori, Florian. 2011. *A History of Mathematical Notations.* Mineola, NY: Dover.

Calinger, Ronald. 1996. Leonhard Euler: The first St. Petersburg years (1727–1741). *Historia Mathematica* 23:121–166.

Campion, Nicholas. 1994. *The Great Year: Astrology, Millenarianism, and History in the Western Tradition.* London: Arkana.

Canales, Jimena. 2009. *A Tenth of a Second: A History.* Chicago: University of Chicago Press.

Cantor, G. N. 1970a. The changing role of Young's ether. *British Journal for the History of Science* 5:44–62.

Cantor, G. N. 1970b. Thomas Young's lectures at the Royal Institution. *Notes and Records of the Royal Society of London* 25:87–112.

Cantor, G. N. 1983. *Optics After Newton: Theories of Light in Britain and Ireland, 1704–1840.* Manchester: Manchester University Press.

Cantor, G. N. 2004. Real disabilities? Quaker schools as "nurseries" of science. In *Science and Dissent in England, 1688–1945*, ed. Paul Wood, 147–166. Aldeshot: Ashgate.

Cantor, G. N. 2005. *Quakers, Jews, and Science: Religious Responses to Modernity and the Sciences in Britain, 1650–1900.* Oxford: Oxford University Press.

Cappelli, Andrea, Elena Castellani, Filippo Colomo, and Paolo Di Vecchia, eds. 2012. *The Birth of String Theory*. Cambridge: Cambridge University Press.

Cardano, Girolamo. 1967. *Opera omnia*. New York: Johnson Reprint.

Cardano, Girolamo. 1973. *Writings on Music*. Trans. Clement A. Miller. Rome: American Institute of Musicology.

Cardano, Girolamo. 2007. *The Great Art; or, The Rules of Algebra*. Trans. T. Richard Witmer. Mineola, NY: Dover.

Carlson, Chester F. 1965. History of electrostatic recording. In *Xerography and Related Processes*, ed. John H. Dessauer and Harold E. Clark, 15–49. New York: Focal Press.

Carpenter, Nan Cooke. 1955. Music in the medieval universities. *Journal of Research in Music Education* 3:136–144.

Carr, Bernard, ed. 2007. *Universe or Multiverse?* Cambridge: Cambridge University Press.

Carrier, Martin. 1994. Geometric facts and geometric theory: Helmholtz and 20th-century philosophy of physical geometry. In *Universalgenie Helmholtz: Rückblich nach 100 Jahren*, ed. Lorenz Krüger, 276–291. Berlin: Akademie Verlag.

Caspar, Max. 1993. *Kepler*. Trans. Clarisse Doris Hellman. New York: Dover.

Chladni, Ernst Florens Friedrich. 1809. *Traité d'acoustique*. Paris: Courcier.

Chladni, Ernst Florens Friedrich. 1821. Weitere Nachrichten von dem neulich in der musikalischen Zeitung erwähnten Chinesischen Blasinstrumente Tscheng oder Tschiang. *Allgemeine Musikalische Zeitung* 23(22):369–374.

Christensen, Dan C. H. 1995. The Orsted–Ritter partnership and the birth of Romantic natural philosophy. *Annals of Science* 52:153.

Christensen, Dan Charly. 2013. *Hans Christian Ørsted: Reading Nature's Mind*. Oxford: Oxford University Press.

Christensen, Thomas Street, ed. 2002. *The Cambridge History of Western Music Theory*. Cambridge: Cambridge University Press.

Christensen, Thomas Street. 2011. Mersenne and the mechanics of musical proportion. In *Proportions: Science, musique, peinture & architecture*, ed. Sabine Rommevaux, Philippe Vendrix, and Vasco Zara, 247–260. Turnhout: Brepols.

Chua, Daniel K. L. 2001. Vincenzo Galilei, modernity, and the division of nature. In *Music Theory and Natural Order from the Renaissance to the Early Twentieth Century*, ed. Suzannah Clark and Alexander Rehding, 17–29. Cambridge: Cambridge University Press.

Chytil, Karel. 1904. *Die Kunst in Prag zur Zeit Rudolf II*. Prague: Verlage des Kunstgewerblichen Museums der Handels- und Gewerbe-Kammer.

Clark, Suzannah, and Alexander Rehding, eds. 2001. *Music Theory and Natural Order from the Renaissance to the Early Twentieth Century*. Cambridge: Cambridge University Press.

Clavius, Christoph. 1999. *In Sphaeram Ioannis de Sacro Bosco commentarius*. Hildesheim: Olms-Weidmann.

Cohen, Albert. 1981. *Music in the French Royal Academy of Sciences: A Study in the Evolution of Musical Thought*. Princeton, NJ: Princeton University Press.

Cohen, David E. 2001. The "Gift of Nature": Musical "instinct" and musical cognition in Rameau. In *Music Theory and Natural Order from the Renaissance to the Early Twentieth Century*, ed. Suzannah Clark and Alexander Rehding, 68–92. Cambridge: Cambridge University Press.

Cohen, H. Floris. 1984. *Quantifying Music: The Science of Music at the First Stage of the Scientific Revolution, 1580–1650*. Dordrecht: Reidel.

Cohen, H. Floris. 1994. *The Scientific Revolution: A Historiographical Inquiry*. Chicago: University of Chicago Press.

Cohen, H. Floris. 2010. *How Modern Science Came into the World: Four Civilizations, One 17th-Century Breakthrough*. Amsterdam: Amsterdam University Press.

References

Comberiati, Carmelo Peter. 1987. *Late Renaissance Music at the Habsburg Court: Polyphonic Settings of the Mass Ordinary at the Court of Rudolf II (1576–1612)*. New York: Gordon and Breach Science Publishers.

Consentino, Giuseppe, Frederick A. Homann, and Ladislaus Lukács. 1999. *Church, Culture, and Curriculum: Theology and Mathematics in the Jesuit Ratio Studiorum*. Philadelphia: Saint Joseph's University Press.

Conway, John Horton. 2001. *On Numbers and Games*, 2nd ed. Natick, MA: A. K. Peters.

Copernicus, Nicolaus. 1985. *Minor Works*. Trans. Edward Rosen. Baltimore, MD: Johns Hopkins University Press.

Copernicus, Nicolaus. 1992. *On the Revolutions*. Ed. Jerzy Dobrzycki, trans. Edward Rosen. Baltimore, MD: Johns Hopkins University Press.

Copernicus, Nicolaus, and Georg Joachim Rhäticus. 1959. *Three Copernican Treatises*. Trans. Edward Rosen. Mineola, NY: Dover.

Cordes, Manfred. 2007. *Nicola Vicentinos Enharmonik: Musik mit 31 Tönen*. Graz: Akademische Druck- und Verlags-Anstalt.

Crary, Jonathan. 1999. *Suspensions of Perception: Attention, Spectacle, and Modern Culture*. Cambridge, MA: MIT Press.

Creese, David E. 2010. *The Monochord in Ancient Greek Harmonic Science*. Cambridge: Cambridge University Press.

Csapo, Eric. 2004. The politics of the new music. In *Music and the Muses: The Culture of "Mousikē" in the Classical Athenian City*, ed. Penelope Murray and Peter Wilson, 207–248. Oxford: Oxford University Press.

Cunningham, Andrew, and Nicholas Jardine, eds. 1990. *Romanticism and the Sciences*. Cambridge: Cambridge University Press.

Darrigol, Olivier. 2003. Number and measure: Hermann von Helmholtz at the crossroads of mathematics, physics, and psychology. *Studies in History and Philosophy of Science* 34:515–573.

Darrigol, Olivier. 2007. A Helmholtzian approach to space and time. *Studies in History and Philosophy of Science* 38:528–542.

Darrigol, Olivier. 2009. The analogy between light and sound in the history of optics from Malebranche to Thomas Young. *Physis* 46:111–217.

Darrigol, Olivier. 2010. The analogy between light and sound in the history of optics from the Ancient Greeks to Isaac Newton. Part 1. *Centaurus* 52:117–155.

Darrigol, Olivier. 2012. *A History of Optics from Greek Antiquity to the Nineteenth Century*. Oxford: Oxford University Press.

Daston, Lorraine, and Peter Galison. 2007. *Objectivity*. New York: Zone Books.

Daston, Lorraine, and Elizabeth Lunbeck, eds. 2011. *Histories of Scientific Observation*. Chicago: University of Chicago Press.

Daston, Lorraine, and Katharine Park. 2001. *Wonders and the Order of Nature, 1150–1750*. New York: Zone Books.

Dear, Peter. 1988. *Mersenne and the Learning of the Schools*. Ithaca: Cornell University Press.

Dear, Peter. 1995. *Discipline and Experience: The Mathematical Way in the Scientific Revolution*. Chicago: University of Chicago Press.

Debnath, Lokenath. 2010. *The Legacy of Leonhard Euler: A Tricentennial Tribute*. London: Imperial College Press.

Debru, Claude. 2001. Helmholtz and the psychophysiology of time. *Science in Context* 14:471–492.

Dedekind, Richard. 1990. Analytische Untersuchungen zu Bernhard Riemann's Abhandlungen über die Hypothesen, welche der Geometrie zu Grunde liegen. *Revue d'histoire des sciences* 43:237–294.

de Muris, Johannes. 1992. *Die Musica Speculativa des Johannes de Muris: Kommentar zur Überlieferung und kritische Edition*. Ed. Christoph Falkenroth. Stuttgart: F. Steiner.

Derrida, Jacques. 1989. *Edmund Husserl's Origin of Geometry: An Introduction*. Lincoln: University of Nebraska Press.

Descartes, René. 1961. *Compendium of Music*. Trans. Walter Robert. Rome: American Institute of Musicology.

Descartes, René. 1979. *Le Monde, ou, Traité de La Lumière*. Trans. Michael Sean Mahoney. New York: Abaris Books.

Descartes, René. 1983. *Principles of Philosophy*. Trans. Reese P. Miller and Valentine Rodger Miller. Dordrecht: D. Reidel.

Descartes, René. 1996. *Oeuvres de Descartes*. Ed. Charles Adam and Paul Tannery. Paris: Vrin.

de Vitry, Philippe, and Leo Plantinga. 1961. Philippe de Vitry's "Ars Nova": A translation. *Journal of Music Therapy* 5:204–223.

Dickreiter, Michael. 1973. *Der Musiktheoretiker Johannes Kepler*. Bern: Francke.

Dijksterhuis, E. J. 1970. *Simon Stevin: Science in the Netherlands around 1600*. The Hague: Martinus Nijhoff.

Dirac, Paul A. M. 1963. The evolution of the physicist's picture of nature. *Scientific American* 208(5):45–53.

Donahue, William H. 1981. *The Dissolution of the Celestial Spheres 1595–1650*. New York: Arno Press.

Doshi, Saryu, ed. 1985. *India and Greece: Connections and Parallels*. Bombay: Marg Publications.

Dostrovsky, Sigalia. 1975. Early vibration theory: Physics and music in the seventeenth century. *Archive for History of Exact Sciences* 14:169–218.

Downs, Benjamin. 2012. Sensible pleasures, rational perfection: Leonhard Euler and the German rationalist tradition. *Mosaic: Journal of Music Research* 2. http://mosaicjournal.org/index.php/mosaic/article/viewFile/41/51.

Drake, Stillman. 1970a. *Galileo Studies: Personality, Tradition, and Revolution*. Ann Arbor: University of Michigan Press.

Drake, Stillman. 1970b. Renaissance music and experimental science. *Journal of the History of Ideas* 31:483–500.

Drake, Stillman. 1992. Music and philosophy in early modern science. In *Music and Science in the Age of Galileo*, ed. Victor Coelho, 3–16. Dordrecht: Kluwer Academic.

Drake, Stillman, and Galileo Galilei. 2000. *Two New Sciences, Including Centers of Gravity and Force of Percussion*. Toronto: Wall & Emerson.

Duffin, Ross W. 2007. *How Equal Temperament Ruined Harmony (and Why You Should Care)*. New York: W. W. Norton.

Dunham, William. 1999. *Euler: The Master of Us All*. Washington, DC: Mathematical Association of America.

Dyer, Joseph. 2007. The place of *musica* in medieval classifications of knowledge. *Journal of Musicology* 24:3–71.

Dyer, Joseph. 2009. Speculative "*musica*" and the medieval university of Paris. *Music & Letters* 90:177–204.

Egan, John Bernard. 1962. Marin Mersenne: *Traite de l'Harmonie universelle*: Critical translation of the second book. PhD diss., Indiana University.

Einstein, Albert. 1979. *Albert Einstein: The Human Side*. Ed. Helen Dukas and Banesh Hoffmann. Princeton, NJ: Princeton University Press.

Einstein, Albert. 1987. *The Collected Papers of Albert Einstein*. Ed. John J. Stachel et al. Princeton, NJ: Princeton University Press.

Engelhardt, Markus, and Michael Heinemann, eds. 2007. *Ars Magna musices—Athanasius Kircher und die Universalität der Musik*. Laaber: Laaber-Verlag.

Erlmann, Veit, ed. 2004. *Hearing Cultures: Essays on Sound, Listening, and Modernity*. Oxford: Berg.

Erlmann, Veit. 2010. *Reason and Resonance: A History of Modern Aurality*. New York: Zone Books.

Euclid and Porphyry. 1991. *The Euclidean Division of the Canon*. Trans. André Barbera. Lincoln: University of Nebraska Press.

Euler, Leonhard. 1837. *Letters on Different Subjects in Natural Philosophy: Addressed to a German Princess*. New York.

Euler, Leonhard. 1911. *Opera omnia*. Leipzig: B. G. Teubner.

Euler, Leonhard. 1985. An essay on continued fractions. Trans. Myra F. Wyman and Bostwick F. Wyman. *Theory of Computing Systems* 18:295–328.

Euler Archive. n.d. http://www.math.dartmouth.edu/~euler/.

Evans, Robert John Weston. 1984. *Rudolf II and His World: A Study in Intellectual History, 1576–1612*. Oxford: Clarendon Press.

Fajtlowicz, Siemion, and Stephanie Mathew. 2012. Three new proofs of Euler's Characteristic Formula. *Congressus Numerantium* 212:165–170.

Faraday, Michael. 1818. On the sounds produced by flame in tubes, &c. *Quarterly Journal of Science* 5:274–280.

Faraday, Michael. 1821. Historical sketch of electro-magnetism. *Annals of Philosophy* 18:195–200, 274–290.

Faraday, Michael. 1831. On a peculiar class of acoustical figures; and on certain forms assumed by groups of particles upon vibrating elastic surfaces. *Philosophical Transactions of the Royal Society of London* 121:299–340.

Faraday, Michael. 1932. *Faraday's Diary*. Ed. Thomas Martin. London: Bell.

Faraday, Michael. 1965. *Experimental Researches in Electricity*. New York: Dover.

Faraday, Michael. 1991. *The Correspondence of Michael Faraday*. London: Institution of Electrical Engineers.

Farwell, Ruth, and Christopher Knee. 1990. The missing link: Riemann's "Commentatio," differential geometry, and tensor analysis. *Historia Mathematica* 17:223–255.

Fauvel, John, Raymond Flood, and Robin Wilson, eds. 2004. *Music and Mathematics: From Pythagoras to Fractals*. New York: Oxford University Press.

Ferreiros, Jose. 2006. Riemann's Habilitationsvortrag at the crossroads of mathematics, physics, and philosophy. In *The Architecture of Modern Mathematics: Essays in History and Philosophy*, ed. Jose Ferreiros and Jeremy Gray, 67–96. Oxford: Oxford University Press.

Fétis, François-Joseph. 1994. *Esquisse de l'histoire de l'harmonie*. Trans. Mary I. Arlin. Stuyvesant, NY: Pendragon Press.

Field, J. V. 1988. *Kepler's Geometrical Cosmology*. Chicago: University of Chicago Press.

Field, J. V. 1997. *The Invention of Infinity: Mathematics and Art in the Renaissance*. Oxford: Oxford University Press.

Field, J. V. 2004. Musical cosmology: Kepler and his readers. In *Music and Mathematics: From Pythagoras to Fractals*, ed. John Fauvel, Raymond Flood, and Robin Wilson, 29–44. New York: Oxford University Press.

Field, J. V. 2005. *Piero Della Francesca: A Mathematician's Art*. New Haven: Yale University Press.

Field, J. V. 2011. Ratio and proportion in the Renaissance. In *Proportions: Science, musique, peinture & architecture*, ed. Sabine Rommevaux, Philippe Vendrix, and Vasco Zara, 29–50. Turnhout: Brepols.

Findlen, Paula, ed. 2004. *Athanasius Kircher: The Last Man Who Knew Everything*. New York: Routledge.

Finocchiaro, Maurice A., ed. 1989. *The Galileo Affair: A Documentary History*. Berkeley, CA: University of California Press.

Ford, Andrew. 2004. Catharsis: The Power of Music in Aristotle's Poetics. In *Music and the Muses: The Culture of "Mousikē" in the Classical Athenian City*, ed. Penelope Murray and Peter Wilson, 309–336. Oxford: Oxford University Press.

Fowler, David. 2004. Helmholtz: Combinational tones and consonance. In *Music and Mathematics: From Pythagoras to Fractals*, ed. John Fauvel, Raymond Flood, and Robin Wilson, 77–88. New York: Oxford University Press.

Frankel, Eugene. 1976. Corpuscular optics and the wave theory of light: The science and politics of a revolution in physics. *Social Studies of Science* 6:141–184.

Friedman, Michael. 2002. Geometry as a branch of physics: Background and context for Einstein's "Geometry and Experience." In *Reading Natural Philosophy*, ed. David B. Malament, 193–229. Chicago: Open Court.

von Frish, Karl. 1971. *Bees: Their Vision, Chemical Senses, and Language*. Ithaca, NY: Cornell University Press.

Fullinwider, S. P. 1990. Hermann von Helmholtz: The problem of Kantian influence. *Studies in History and Philosophy of Science* 21:41–55.

Funkenstein, Amos. 1975. The dialectical preparation for scientific revolutions: On the role of hypothetical reasoning in the emergence of Copernican astronomy and Galilean mechanics. In *The Copernican Achievement*, ed. Robert S. Westman, 165–203. Berkeley, CA: University of California Press.

Galilei, Galileo. 1890. *Le opere di Galileo Galilei*. Ed. Antonio Favaro. Florence: G. Barbèra.

Galilei, Galileo. 1977. *Galileo's Early Notebooks: The Physical Questions*. Trans. William A. Wallace. Notre Dame, IN: University of Notre Dame Press.

Galilei, Vincenzo. 2003. *Dialogue on Ancient and Modern Music*. Trans. Claude V. Palisca. New Haven: Yale University Press.

Galilei, Vincenzo, Giovanni de'Bardi, and Girolamo Mei. 1960. *Letters on Ancient and Modern Music to Vincenzo Galilei and Giovanni Bardi: A Study with Annotated Texts*. Ed. Claude V. Palisca. Rome: American Institute of Musicology.

Galison, Peter. 1997. *Image and Logic: A Material Culture of Microphysics*. Chicago: University of Chicago Press.

Galison, Peter. 2003. *Einstein's Clocks, Poincaré's Maps: Empires of Time*. New York: W. W. Norton.

Gallagher, Robert. 1984. Riemann and the Göttingen School of Physiology. *Fusion* 6(3):24–30.

Gatto, Romano. 1994. *Tra scienza e immaginazione: Le matematiche presso il Collegio Gesuitico Napoletano (1552–1670 Ca.)*. Florence: L. S. Olschki.

Gaukroger, Stephen. 1995. *Descartes: An Intellectual Biography*. Oxford: Clarendon Press.

Gaukroger, Stephen. 2000. The foundational role of hydrostatics and statics in Descartes' natural philosophy. In *Descartes' Natural Philosophy*, ed. Stephen Gaukroger, John Schuster, and John Sutton, 60–80. London: Routledge.

Gaukroger, Stephen, John Schuster, and John Sutton, eds. 2000. *Descartes' Natural Philosophy*. London: Routledge.

Gauss, Carl Friedrich. 2005. *General Investigations of Curved Surfaces*. Ed. Peter Pesic. Mineola, NY: Dover.

Gauvin, Jean-François. 2013. Organ making and natural philosophical knowledge in Mersenne's *Harmonie Universelle*. Unpublished manuscript.

Gertsman, E. V. 2007. Euler and the history of a certain musical-mathematical idea. In *Euler and Modern Science*, ed. A. P. Yushkevich, N. N. Bogolyubov, and G. K. Mikhaĭlov, trans. Robert Burns, 335–347. Washington, DC: Mathematical Association of America.

Gilbert, William. 1958. *De Magnete*. Trans. P. Fleury Mottelay. New York: Dover.

Gingerich, Owen. 1993. *The Eye of Heaven: Ptolemy, Copernicus, Kepler*. New York: American Institute of Physics.

Gingerich, Owen. 2002. *An Annotated Census of Copernicus' De Revolutionibus (Nuremberg, 1543 and Basel, 1566)*. Leiden: Brill.

Glarean, Heinrich (Glareanus, Henricus). 1965. *Dodecachordon*. Trans. Clement A. Miller. Rome: American Institute of Musicology.

Gordon, Cyrus Herzl. 1982. *Forgotten Scripts: Their Ongoing Discovery and Decipherment*. New York: Basic Books.

Gorham, Geoffrey. 1991. Planck's principle and Jeans's conversion. *Studies in History and Philosophy of Science* 22:471–497.

Gosselin, Guillaume. 1577. *De arte magna, seu de occulta parte mumerorum, quae & Algebra, & Almucabala vulgo dicitur*. Paris: Aegidium Beys.

Gouk, Penelope. 1982. *Music in the Natural Philosophy of the Early Royal Society*. London: University of London.

Gouk, Penelope. 1999. *Music, Science, and Natural Magic in Seventeenth-Century England*. New Haven: Yale University Press.

Gowers, Timothy, ed. 2008. *The Princeton Companion to Mathematics*. Princeton: Princeton University Press.

Gozza, Paolo. 2000. A Renaissance mathematics: The music of Descartes. In *Number to Sound: The Musical Way to the Scientific Revolution*, ed. Paolo Gozza. Dordrecht: Kluwer Academic.

Grafton, Anthony. 1991. *Defenders of the Text: The Traditions of Scholarship in an Age of Science, 1450–1800*. Cambridge, MA: Harvard University Press.

Grant, Edward. 1965. Part I of Nicole Oresme's Algorismus Proportionum. *Isis* 56:327–341.

Grassmann, Hermann. 1854. On the theory of compound colours. *Philosophical Magazine* 7(4):254–264.

Grattan-Guinness, I., and Roger Cooke, eds. 2005. *Landmark Writings in Western Mathematics 1640–1940*. Amsterdam: Elsevier.

Gray, Jeremy. 1989. *Ideas of Space: Euclidean, Non-Euclidean, and Relativistic*, 2nd ed. Oxford: Clarendon Press.

Green, Burdette Lamar. 1969. The Harmonic Series from Mersenne to Rameau: An historical study of circumstances leading to its recognition and application to music. PhD diss., Ohio State University.

Greene, Brian. 2000. *The Elegant Universe: Superstrings, Hidden Dimensions, and the Quest for the Ultimate Reality*. New York: Vintage Books.

Grosslight, Justin. 2013. Small skills, big networks: Marin Mersenne as mathematical intelligencer. *History of Science* 51:337–374.

Guilford, Francis North, and Jamie Croy Kassler. 2004. *The Beginnings of the Modern Philosophy of Music in England: Francis North's A Philosophical Essay of Musick (1677) with Comments of Isaac Newton, Roger North, and in the Philosophical Transactions*. Aldershot: Ashgate.

Gurney, Hudson. 1831. *Memoir of the Life of Thomas Young*. London: John & Arthur Arch.

Hagel, Stefan. 2010. *Ancient Greek Music: A New Technical History*. Cambridge: Cambridge University Press.

Hakfoort, Casper. 1995. *Optics in the Age of Euler: Conceptions of the Nature of Light, 1700–1795*. Cambridge: Cambridge University Press.

Hall, A. Rupert. 1993. *All Was Light: An Introduction to Newton's* Opticks. Oxford: Clarendon Press.

Harburger, W. 1980. *Johannes Keplers kosmische Harmonie*. Frankfurt: Insel.

Harrison, Edward. 1987. *Darkness at Night: A Riddle of the Universe*. Cambridge, MA: Harvard University Press.

Hashimoto, T. 1983. Ampère vs. Biot: Two mathematizing routes to electromagnetic theory. *Historia Scientiarum* 24:29–51.

Hatfield, Gary. 1990. *The Natural and the Normative: Theories of Spatial Perception from Kant to Helmholtz*. Cambridge, MA: MIT Press.

Hatfield, Gary. 1993. Helmholtz and classicism: The science of aesthetics and the aesthetics of science. In *Helmholtz and the Foundations of Nineteenth-Century Science*, ed. David Cahan, 522–558. Berkeley, CA: University of California Press.

Hawkins, Thomas. 1994. The birth of Lie's theory of groups. *Mathematical Intelligencer* 16:6–17.

Hawkins, Thomas. 2000. *The Emergence of the Theory of Lie Groups: An Essay in the History of Mathematics, 1869–1926*. New York: Springer.

Heath, Thomas Little. 1966. *Aristarchus of Samos, the Ancient Copernicus*. Oxford: Clarendon Press.

Heilbron, J. L. 1986. *The Dilemmas of an Upright Man: Max Planck as Spokesman for German Science*. Berkeley, CA: University of California Press.

Heilbron, J. L. 2010. *Galileo*. Oxford: Oxford University Press.

Heinzmann, Gerhard. 1992. Helmholtz and Poincaré's considerations on the genesis of geometry. In *1830–1930: A Century of Geometry*, ed. L. Boi, D. Flament, and J.-M. Salanskis, 245–249. Berlin: Springer-Verlag.

Heinzmann, Gerhard. 2001. The foundations of geometry and the concept of motion: Helmholtz and Poincaré. *Science in Context* 14:457–470.

Heisenberg, Werner. 1971. *Physics and Beyond: Encounters and Conversations*. New York: Harper & Row.

Hell, Helmut, and Horst Leuchtmann, eds. 1982. *Orlando di Lasso: Musik der Renaissance am Münchner Fürstenhof*. Wiesbaden: Reichert.

Heller-Roazen, Daniel. 2011. *The Fifth Hammer: Pythagoras and the Disharmony of the World*. New York: Zone Books.

Helmholtz, Hermann von. 1865. *Die Lehre von den Tonempfindungen als physiologische Grundlage für die Theorie der Musik*, 2nd ed. Braunschweig: Vieweg.

Helmholtz, Hermann von. 1867. *Handbuch der physiologischen Optik*. Leipzig: L. Voss.

Helmholtz, Hermann von. 1868. Ueber die thatsächlichen Grundlagen der Geometrie. *Verhandlungen des Naturhistorisch-medicinischen Vereins zu Heidelberg* 4:197–202.

Helmholtz, Hermann von. 1870. *Die Lehre von den Tonempfindungen als physiologische Grundlage für die Theorie der Musik*, 3rd ed. Braunschweig: Vieweg.

Helmholtz, Hermann von. 1873. *The Mechanism of the Ossicles of the Ear and Membrana Tympani*. Trans. Albert H. Buck and Normand Smith. New York: William Wood.

Helmholtz, Hermann von. 1882. *Wissenschaftliche Abhandlungen*. Leipzig: J. A. Barth.

Helmholtz, Hermann von. 1896. *Handbuch der Physiologischen Optik*, 2nd ed. Ed. Arthur König. Hamburg: L. Voss.

Helmholtz, Hermann von. 1903. *Vorträge und Reden*. Vieweg.

Helmholtz, Hermann von. 1954. *On the Sensations of Tone as a Physiological Basis for the Theory of Music*, 2nd. English ed. Ed. Alexander John Ellis. New York: Dover.

Helmholtz, Hermann von. 1962a. *Helmholtz's Treatise on Physiological Optics*. Ed. James P. C. Southall. New York: Dover.

Helmholtz, Hermann von. 1962b. *Popular Scientific Lectures*. New York: Dover.

Helmholtz, Hermann von. 1971. *Selected Writings of Hermann von Helmholtz*. Ed. Russell Kahl. Middletown, CT: Wesleyan University Press.

Helmholtz, Hermann von. 1977. *Epistemological Writings*. Ed. Paul Hertz, Moritz Schlick, R. S. Cohen, and Yehuda Elkana, trans. Malcolm F. Lowe. Dordrecht: D. Reidel.

Helmholtz, Hermann von. 1993. *Letters of Hermann von Helmholtz to His Parents: The Medical Education of a German Scientist, 1837–1846*. Ed. David Cahan. Stuttgart: Franz Steiner Verlag.

Helmholtz, Hermann von. 1995. *Science and Culture: Popular and Philosophical Essays*. Ed. David Cahan. Chicago: University of Chicago Press.

Hentschel, Frank, ed. 1998. *Musik und die Geschichte der Philosophie und Naturwissenschaften in Mittelalter: Fragen zur Wechselwirkung von "Musica" und "Philosophia" in Mittelalter*. Leiden: Brill.

Hermann, Armin. 1971. *The Genesis of Quantum Theory (1899–1913)*. Cambridge, MA: MIT Press.

Hermann, Armin. 1973. *Max Planck in Selbstzeugnissen und Bilddokumenten*. Reinbek bei Hamburg: Rowohlt.

Hermann, Thomas, Andy Hunt, and John G. Neuhoff, eds. 2011. *The Sonification Handbook*. Berlin: Logos Verlag.

Hertz, Heinrich. 1962. *Electric Waves: Being Researches on the Propagation of Electric Action with Finite Velocity through Space*. New York: Dover.

Hiebert, Elfrieda, and Erwin Hiebert. 1994. Musical thought and practice: Links to Helmholtz's Tonempfindungen. In *Universalgenie Helmholtz: Rückblick nach 100 Jahren*, ed. Lorenz Krüger, 295–311. Berlin: Akademie Verlag.

Hiebert, Erwin. 2003. Science and music in the culture of late 19th century physicists: The role and limits of the scientific analysis of music. In *Science and Cultural Diversity: Proceedings of the XXIst International Congress of History of Science, Mexico City, 7–14 July 2001*, ed. Juan José Saldaña, 97–109. Mexico City: Sociedad Mexicana de Historia de la Ciencia y la Tecnología.

Hilts, Victor L. 1978. Thomas Young's "Autobiographical Sketch." *Proceedings of the American Philosophical Society* 122:248–260.

Hine, William L. 1973. Mersenne and Copernicanism. *Isis* 64:18–32.

Hirshfeld, Alan. 2006. *The Electric Life of Michael Faraday*. New York: Walker.

Holton, Gerald. 1988. *Thematic Origins of Scientific Thought: Kepler to Einstein*. Rev. ed. Cambridge, MA: Harvard University Press.

Holton, Gerald. 1996. *Einstein, History, and Other Passions: The Rebellion Against Science at the End of the Twentieth Century*. Reading, MA: Addison-Wesley.

Holton, Gerald. 2009. George Sarton, his Isis, and the aftermath. *Isis* 100:79–88.

Home, R. W. 1988. Leonhard Euler's "anti-Newtonian" theory of light. *Annals of Science* 45:521–533.

Hoppin, Richard. 1978. *Medieval Music*. New York: W. W. Norton.

Horden, Penelope, ed. 2000. *Music as Medicine: The History of Music Therapy since Antiquity*. Aldershot: Ashgate.

Howard, Don. 2005. Albert Einstein as a philosopher of science. *Physics Today* 58(12):34–40.

Hubbard, Geoffrey. 1968. *Cooke and Wheatstone and the Invention of the Electric Telegraph*. New York: Augustus M. Kelley.

Hübner, Jürgen. 1975. *Die Theologie Johannes Keplers zwischen Orthodoxie und Naturwissenschaft*. Tübingen: Mohr.

Hucbald of Saint Amand, Guido d'Arezzo, and Johannes Afflighemensis. 1978. *Hucbald, Guido, and John on Music: Three Medieval Treatises*. Ed. Claude V. Palisca, trans. Warren Babb. New Haven: Yale University Press.

Hui, Alexandra. 2008. Hearing sound as music: Psychophysical studies of sound sensation and the music culture of Germany, 1860–1910. PhD diss., UCLA.

Hui, Alexandra. 2013a. *The Psychophysical Ear: Musical Experiments, Experimental Sounds, 1840–1910*. Cambridge, MA: MIT Press.

Hui, Alexandra. 2013b. Changeable ears: Ernst Mach's and Max Planck's studies of accommodation in hearing. *Osiris* 28:119–145.

Hui, Alexandra, Julia Kursell, and Myles W. Jackson, eds. 2013. *Music, Sound, and the Laboratory from 1750–1980*. Osiris 28.

Husserl, Edmund. 1939. Der Ursprung der Geometrie als intentional-historisches Problem. *Revue internationale de philosophie* 1:203–225.

Husserl, Edmund. 1970. *The Crisis of European Sciences and Transcendental Phenomenology: An Introduction to Phenomenological Philosophy*. Evanston: Northwestern University Press.

Hyder, David Jalal. 1999. Helmholtz's naturalized conception of geometry and his spatial theory of signs. *Philosophy of Science* 66:273–286.

Ionescu-Pallas, Nicholas, and Liviu Sofonea. 1986. Bernhard Riemann: A forerunner of classical electrodynamics. *Organon* 22:219–272.

Isichei, Elizabeth Allo. 1970. *Victorian Quakers*. London: Oxford University Press.

Jackson, Myles W. 2000. *Spectrum of Belief: Joseph von Fraunhofer and the Craft of Precision Optics*. Cambridge, MA: MIT Press.

Jackson, Myles W. 2006. *Harmonious Triads: Physicists, Musicians, and Instrument Makers in Nineteenth-Century Germany*. Cambridge, MA: MIT Press.

Jaeger, Werner Wilhelm. 1969. *Paideia: The Ideals of Greek Culture*, 2nd ed. New York: Oxford University Press.

James, Frank A. J. L. 1984. The physical interpretation of the wave theory of light. *British Journal for the History of Science* 17:47–60.

Jay, Martin. 1994. *Downcast Eyes: The Denigration of Vision in Twentieth-Century French Thought*. Berkeley, CA: University of California Press.

Jeans, James. 1968. *Science and Music*. New York: Dover.

Jonas, Oswald. 1982. *Introduction to the Theory of Heinrich Schenker: The Nature of the Musical Work of Art*. Trans. John Rothgeb. New York: Longman.

Jones, Arthur Taber. 1945. Singing flames. *Journal of the Acoustical Society of America* 16:254–266.

Jones, Bence. 1975. *The Royal Institution: Its Founder and Its First Professors*. New York: Arno Press.

Joost-Gaugier, Christiane L. 2006. *Measuring Heaven: Pythagoras and His Influence on Thought and Art in Antiquity and the Middle Ages*. Ithaca: Cornell University Press.

Joost-Gaugier, Christiane L. 2009. *Pythagoras and Renaissance Europe: Finding Heaven*. Cambridge: Cambridge University Press.

Kangro, Hans. 1976. *Early History of Planck's Radiation Law*. London: Taylor and Francis.

Kargon, Robert Hugh. 1966. *Atomism in England from Hariot to Newton*. Oxford: Clarendon Press.

Kassler, Jamie C. 1995. *Inner Music: Hobbes, Hooke, and North on Internal Character*. Madison: Fairleigh Dickinson University Press.

Kassler, Jamie C. 2001. *Music, Science, Philosophy: Models in the Universe of Thought*. Aldershot: Ashgate.

Kaufmann, Henry W. 1966. *The Life and Works of Nicola Vicentino, 1511–c. 1576*. Washington, DC: American Institute of Musicology.

Kepler, Johannes. 1937. *Gesammelte Werke*. Ed. Walther von Dyck, Max Caspar, and Franz Hammer. Munich: C. H. Beck.

Kepler, Johannes. 1981. *The Secret of the Universe: Mysterium Cosmographicum*. Ed. E. J. Aiton, trans. A. M. Duncan. New York: Abaris Books.

Kepler, Johannes. 1995. *Epitome of Copernican Astronomy; & Harmonies of the World*. Trans. Charles Glenn Wallis. Amherst, NY: Promethus Books.

Kepler, Johannes. 1997. *The Harmony of the World*. Trans. E. J. Aiton, A. M. Duncan, and J. V. Field. Philadelphia: American Philosophical Society.

Kipnis, Naum S. 1991. *History of the Principle of Interference of Light*. Basel: Birkhäuser.

Kircher, Athanasius. 2011. *A Study of the Life and Works of Athanasius Kircher, "Germanus Incredibilis."* Ed. Elizabeth Fletcher, trans. G. W Trompf and John Edward Fletcher. Leiden: Brill.

Kittler, Friedrich A. 1999. *Gramophone, Film, Typewriter*. Stanford, CA: Stanford University Press.

Kittler, Friedrich A. 2006. *Musik und Mathematik*. Munich: Wilhelm Fink.

Kleban, Matthew. 2011. Cosmic bubble collisions. http://arxiv.org/abs/1107.2593.

Klein, Jacob. 1985. *Lectures and Essays*. Ed. Robert B. Williamson and Elliott Zuckerman. Annapolis, MD: St. John's College Press.

Klein, Jacob. 1992. *Greek Mathematical Thought and the Origin of Algebra*. Trans. Eva Brann. New York: Dover.

Kmetz, John. 1994. *Stefano Rosetti at the Imperial Court*. Florence: Olschki.

Knobloch, Eberhard. 2008. Euler transgressing limits: The infinite and music theory. *Quaderns d'Història de l'Enginyeria* 9:9–24.

Knoll, Paul W. 1975. The arts faculty of the University of Cracow at the end of the fifteenth century. In *The Copernican Achievement*, ed. Robert S. Westman, 137–156. Berkeley, CA: University of California Press.

Knorr, Wilbur Richard. 1975. *The Evolution of the Euclidean Elements: a Study of the Theory of Incommensurable Magnitudes and Its Significance for Early Greek Geometry*. Dordrecht: D. Reidel.

Knuth, Donald Ervin. 1974. *Surreal Numbers: How Two Ex-students Turned On to Pure Mathematics and Found Total Happiness*. Reading, MA: Addison-Wesley.

Koenigsberger, Leo. 1965. *Hermann von Helmholtz*. New York: Dover.

Koertge, Noretta. 2008. *New Dictionary of Scientific Biography*. Detroit: Charles Scribner's Sons.

Koyré, Alexandre. 1957. *From the Closed World to the Infinite Universe*. Baltimore, MD: Johns Hopkins University Press.

Koyré, Alexandre. 1978. *Galileo Studies*. Atlantic Highlands, NJ: Humanities Press.

Koyré, Alexandre. 1992. *The Astronomical Revolution: Copernicus, Kepler, Borelli*. New York: Dover.

Kramer, Gregory, ed. 1994. *Auditory Display: Sonification, Audification, and Auditory Interfaces*. Reading, MA: Addison-Wesley.

Kramer, Gregory, Bruce Walker, Perry Cook, and Nadine Miner, eds. 2010. *Sonification Report: Status of the Field and Research Agenda*. DigitalCommons@University of Nebraska–Lincoln. http://digitalcommons.unl.edu/psychfacpub/444.

Krüger, Lorenz, ed. 1994. *Universalgenie Helmholtz: Rückblick nach 100 Jahren*. Berlin: Akademie Verlag.

Kursell, Julia. 2013. Experiments on sound color in music and acoustics: Helmholtz, Schoenberg, and Klangfarbenmelodie. *Osiris* 28:191–211.

Laertius, Diogenes. 1972. *Lives of Eminent Philosophers*. Trans. Robert Drew Hicks. Cambridge, MA: Harvard University Press.

References

Lasso, Orlando di. 1894. *Sämtliche Werke*. Ed. Fr. X. Haberl and Adolf Sandberger. Leipzig: Breitkopf & Härtel.

Laugwitz, Detlef. 1999. *Bernhard Riemann, 1826–1866: Turning Points in the Conception of Mathematics*. Basel: Birkhäuser.

Lawrence, Christopher. 1990. The power and the glory: Humphrey Davy and Romanticism. In *Romanticism and the Sciences*, ed. Andrew Cunningham and Nicholas Jardine, 213–227. Cambridge: Cambridge University Press.

Lefèvre d'Étaples, Jacques, and Nemorarius Jordanus. 1496. *Elementa Musicalia*. Paris.

Leibniz, Gottfried Wilhelm. 1989. *Philosophical Essays*. Trans. Roger Ariew and Daniel Garber. Indianapolis: Hackett.

Lenoir, Timothy. 1993. The eye as mathematician: Clinical practice, instrumentation, and Helmholtz's construction of an empiricist theory of vision. In *Helmholtz and the Foundations of Nineteenth-Century Science*, ed. David Cahan, 109–153. Berkeley, CA: University of California Press.

Lenoir, Timothy. 2006. Operationalizing Kant: Manifolds, models, and mathematics in Helmholtz's theories of perception. In *The Kantian Legacy in Nineteenth-Century Science*, ed. Michael Friedman and Alfred Nordmann, 141–210. Cambridge, MA: MIT Press.

Léry, Jean de. 1990. *History of a Voyage to the Land of Brazil, Otherwise Called America*. Berkeley, CA: University of California Press.

Levin, Flora R. 2009. *Greek Reflections on the Nature of Music*. Cambridge: Cambridge University Press.

Lewis, John. 2006. *Galileo in France: French Reactions to the Theories and Trial of Galileo*. New York: Peter Lang.

Libre, Pearl. 1969. The quadrivium in the thirteenth century universities (with special reference to Paris). In *Arts libéraux et philosophie au Moyen Age*, 175–191. Montreal: Institut d'études médiévales.

Lichtenberg, Georg Christoph. 1997. *Observationes: die lateinischen Schriften*. Göttingen: Wallstein.

Lichtenberg, Georg Christoph. 2000. *The Waste Books*. Trans. R. J. Hollingdale. New York: New York Review of Books.

Liessem, Franz. 1969. *Musik und Alchemie*. Tutzing: H. Schneider.

Lindell, Robert. 1994. Music and patronage at the Court of Rudolf II. In *Music in the German Renaissance: Sources, Styles, and Contexts*, ed. John Kmetz, 254–271. Cambridge: Cambridge University Press.

Lindley, Mark. 1982. Chromatic systems (or non-systems) from Vicentino to Monteverdi. *Early Music History* 2:377–404.

Lloyd, G. E. R. 1987. *The Revolutions of Wisdom: Studies in the Claims and Practice of Ancient Greek Science*. Berkeley, CA: University of California Press.

Lossius, Lucas. 1570. *Erotema musicae practicae*. Nuremberg.

Lundberg, Mattias. 2011. *Tonus Peregrinus: The History of a Psalm-Tone and Its Use in Polyphonic Music*. Farnham, Surrey: Ashgate.

Lundberg, Robert. 1992. In tune with the universe: The physics and metaphysics of Galileo's lute. In *Music and Science in the Age of Galileo*, ed. Victor Coelho, 211–239. Dordrecht: Kluwer Academic.

MacAdam, David L., ed. 1970. *Sources of Color Science*. Cambridge, MA: MIT Press.

Macey, Patrick. 2009. Josquin and Champion: Conflicting attributions for the Psalm Motet *De profundis clamavi*. In Uno gentile et subtile ingenio: *Studies in Renaissance Music in Honour of Bonnie J. Blackburn*, ed. Gioia Filocamo and M. Jennifer Bloxam, 453–468. Turnhout: Brepols.

Mahr, Bernd, and Wladimir Velminski. 2010. Denken in Modellen: Zur Lösung des Königsberger Brückenproblems. In *Mathesis & Graphé: Leonhard Euler und die Entfaltung der Wissensysteme*, ed. Horst Bredekamp and Wladimir Velminski, 85–100. Berlin: Akademie Verlag.

Maier, Michael. 1989. *Atalanta Fugiens: An Edition of the Fugues, Emblems, and Epigrams*. Ed. Hildemarie Streich, trans. Joscelyn Godwin. Grand Rapids, MI: Phanes Press.

Malebranche, Nicolas. 1997. *The Search after Truth*. Trans. Thomas M. Lennon and Paul J. Olscamp. Cambridge: Cambridge University Press.

Mambella, Guido. 2008. Corpo sonoro, geometria e temperamenti: Zarlino e la crisi del fondamento numerico della musica. In *Music and Mathematics: In Late Medieval and Early Modern Europe*, ed. Philippe Vendrix, 185–234. Turnhout: Brepols.

Martens, Rhonda. 2000. *Kepler's Philosophy and the New Astronomy*. Princeton, NJ: Princeton University Press.

Martínez, Alberto A. 2012. *The Cult of Pythagoras: Math and Myths*. Pittsburgh, PA: University of Pittsburgh Press.

Mathiesen, Thomas J. 1999. *Apollo's Lyre: Greek Music and Music Theory in Antiquity and the Middle Ages*. Lincoln: University of Nebraska Press.

Mathieson, Genevieve. 2007. Thomas Young, Quaker scientist. MA thesis, Case Western Reserve University.

Maury, Jean-Pierre. 2003. *A l'origine de la recherche scientifique: Mersenne*. Ed. Sylvie Taussig. Paris: Vuibert.

Mautner, Franz H., and Franklin Miller. 1952. Remarks on G. C. Lichtenberg, humanist-scientist. *Isis* 43:223–231.

Maxham, Robert Eugene. 1976. The contributions of Joseph Sauveur (1653–1716) to acoustics. PhD diss., University of Rochester.

Maxwell, James Clerk. 1868. On a method of making a direct comparison of electrostatic with electromagnetic force; with a note on the electromagnetic theory of light. *Philosophical Transactions of the Royal Society of London* 158:643–657.

Maxwell, James Clerk. 1890. *The Scientific Papers of James Clerk Maxwell*. Cambridge: Cambridge University Press.

McGucken, William. 1969. *Nineteenth-Century Spectroscopy: Development of the Understanding of Spectra, 1802–1897*. Baltimore: Johns Hopkins Press.

McGuire, J. E., and P. M. Rattansi. 1966. Newton and the "Pipes of Pan." *Notes and Records of the Royal Society of London* 21:108–143.

McKay, John Zachary. 2012. Universal music-making: Athanasius Kircher and musical thought in the seventeenth century. PhD diss., Harvard University.

McKinney, Timothy R. 2005. Point/counterpoint: Vicentino's musical rebuttal to Lusitano. *Early Music* 33:393–411.

McKinzie, Mark. 2007. Euler's observations on harmonic progressions. In *Euler at 300: An Appreciation*, ed. Robert E. Bradley, Lawrence A. D'Antonio, and C. Edward Sandifer, 131–141. Washington, DC: Mathematical Association of America.

McLaren, K. 1985. Newton's indigo. *Color Research and Application* 10:225–229.

Meinel, Christoph, ed. 1986. Alchemie und Musik. In *Die Alchemie in der Europäischen Kultur- und Wissenschaftsgeschichte*, 201–227. Wiesbaden: O. Harrassowitz.

Mellon, Elizabeth A. 2011. Inscribing sound: Medieval remakings of Boethius's "De Institutione Musica." PhD diss., University of Pennsylvania.

Mersenne, Marin. 1623. *Quaestiones Celeberrimae in Genesim*. Paris: Sebastian Cramoisy.

Mersenne, Marin. 1957. *Harmonie Universelle: The Books on Instruments*. Trans. Roger E. Chapman. The Hague: M. Nijhoff.

Mersenne, Marin. 1963. *Harmonie Universelle: Contenant la théorie et la pratique de la musique*. Paris: Centre national de la recherche scientifique.

Mersenne, Marin. 1972. *Questions Harmoniques*. Stuttgart: F. Frommann.

Methuen, Charlotte. 1998. *Kepler's Tübingen: Stimulus to a Theological Mathematics*. Aldershot: Ashgate.

Meulders, Michel. 2010. *Helmholtz: From Enlightenment to Neuroscience*. Cambridge, MA: MIT Press.

Mews, Constant J. 2009. Liturgists and dance in the twelfth century: the witness of John Beleth and Sicard of Cremona. *Church History* 78:512–548.

Mollon, J. D. 2002. The origins of the concept of interference. *Philosophical Transactions: Mathematical, Physical, and Engineering Sciences* 360:807–819.

Moore, Walter John. 1989. *Schrödinger: Life and Thought*. Cambridge: Cambridge University Press.

Moreno, Jairo. 2004. *Musical Representations, Subjects, and Objects the Construction of Musical Thought in Zarlino, Descartes, Rameau, and Weber*. Bloomington: Indiana University Press.

Moyer, Ann E. 1992. *Musica Scientia: Musical Scholarship in the Italian Renaissance*. Ithaca: Cornell University Press.

Moyer, Ann E. 2008. Music, mathematics, and aesthetics: The case of the visual arts in the Renaissance. In *Music and Mathematics: In Late Medieval and Early Modern Europe*, ed. Philippe Vendrix, 111–146. Turnhout: Brepols.

Moyer, Ann E. 2011. Reading Boethius on proportion: Renaissance editions, epitomes, and versions of the arithmetic and music. In *Proportions: Science, musique, peinture & architecture*, ed. Sabine Rommevaux, Philippe Vendrix, and Vasco Zara, 51–68. Turnhout: Brepols.

Mullaly, Robert. 2011. *The Carole: A Study of a Medieval Dance*. Farnham, Surrey: Ashgate.

Mulligan, Joseph F. 1994. Max Planck and the "Black Year" of German physics. *American Journal of Physics* 62:1089–1097.

Muzzulini, Daniel. 1994. Leonhard Eulers Konsonanztheorie. *Musiktheorie* 9:135–146.

Nambu, Yoichiro. 2012. From the S-matrix to string theory. In *The Birth of String Theory*, ed. Andrea Cappelli, et al., 275–282. Cambridge: Cambridge University Press.

Neal, Katherine. 2002. *From Discrete to Continuous: The Broadening of Number Concepts in Early Modern England*. Dordrecht: Kluwer Academic.

Newman, James Roy, ed. 1956. *The World of Mathematics*. New York: Simon and Schuster.

Newton, Isaac. 1665. Cambridge University Library Add. Ms. 4000.

Newton, Isaac. 1959. *The Correspondence of Isaac Newton*. Ed. H. W. Turnbull and J. F. Scott. Cambridge: Published for the Royal Society at the University Press.

Newton, Isaac. 1978. *Isaac Newton's Papers & Letters on Natural Philosophy and Related Documents*, 2nd ed. Ed. I. Bernard Cohen, Robert E. Schofield, and Marie Boas Hall. Cambridge, MA: Harvard University Press.

Newton, Isaac. 1979. *Opticks: Or A Treatise of the Reflections, Refractions, Inflections, and Colours of Light*. New York: Dover.

Newton, Isaac. 1983. *Certain Philosophical Questions: Newton's Trinity Notebook*. Ed. Martin Tamny and J. E. McGuire. Cambridge: Cambridge University Press.

Newton, Isaac. Shapiro, Alan E., ed. 1984. *The Optical Papers of Isaac Newton*. Cambridge: Cambridge University Press.

Nicomachus of Gerasa. 1994. *The Manual of Harmonics of Nicomachus the Pythagorean*. Trans. Flora R. Levin. Grand Rapids, MI: Phanes Press.

Nielsen, Holger G. 2012. The string picture of the Veneziano model. In *The Birth of String Theory*, ed. Andrea Cappelli, et al., 266–274. Cambridge: Cambridge University Press.

Nowak, Gregory. 1989. Riemann's Habilitationsvortrag and the synthetic a priori status of geometry. In *The History of Modern Mathematics*, ed. David E. Rowe and John McCleary. Boston: Academic Press.

O'Raifeartaigh, Lochlainn, ed. 1997. *The Dawning of Gauge Theory*. Princeton: Princeton University Press.

Oettinger, Rebecca Wagner. 2003. Thomas Murner, Michael Stifel, and songs as polemic in the early Reformation. *Journal of Musicological Research* 22:45–100.

Oreskes, Naomi. 2013. Why I am a presentist. *Science in Context* 26:595–609.

Oresme, Nicole. 1966. *De Proportionibus Proportionum and Ad Pauca Respicientes*. Trans. Edward Grant. Madison: University of Wisconsin Press.

Oresme, Nicole. 1968a. *Le livre du ciel et du monde*. Madison: University of Wisconsin Press.

Oresme, Nicole. 1968b. *Nicole Oresme and the Medieval Geometry of Qualities and Motions: A Treatise on the Uniformity and Difformity of Intensities Known as Tractatus de configurationibus qualitatum et motuum*. Madison: University of Wisconsin Press.

Oresme, Nicole. 1971. *Nicole Oresme and the Kinematics of Circular Motion: Tractatus de commensurabilitate vel incommensurabilitate motuum celi*. Madison: University of Wisconsin Press.

Ørsted, Hans Christian. 1998. *Selected Scientific Works of Hans Christian Ørsted*. Princeton, NJ: Princeton University Press.

Ørsted, Hans Christian. 2011. *The Travel Letters of H. C. Ørsted*. Trans. Karen Jelved and Andrew D. Jackson. Copenhagen: Det Kongelige Danske Videnskabernes Selskab.

Pais, Abraham. 1986. *Inward Bound: Of Matter and Forces in the Physical World*. Oxford: Clarendon Press.

Palisca, Claude V. 1961. Scientific empiricism in musical thought. In *Seventeenth Century Science and the Arts*, ed. Stephen Edelston Toulmin and Hedley Howell Rhys, 91–137. Princeton: Princeton University Press.

Palisca, Claude V. 1972. *Ut Oratoria Musica*: The rhetorical basis of musical mannerism. In *The Meaning of Mannerism*, ed. Stephen G. Nichols and Franklin Westcott Robinson, 37–65. Hanover, NH: University Press of New England.

Palisca, Claude V. 1985. *Humanism in Italian Renaissance Musical Thought*. New Haven: Yale University Press.

Palisca, Claude V. 1992. Was Galileo's father an experimental scientist? In *Music and Science in the Age of Galileo*, ed. Victor Coelho, 143–151. Dordrecht: Kluwer Academic.

Pangrazi, Tiziana. 2009. *La Musurgia Universalis di Athanasius Kircher: Contenuti, fonti, terminologia*. Florence: Olschki.

Panofsky, Erwin. 1954. *Galileo as a Critic of the Arts*. The Hague: M. Nijhoff.

Panofsky, Erwin. 1956a. Galileo as a critic of the arts: Aesthetic attitude and scientific thought. *Isis* 47:3–15.

Panofsky, Erwin. 1956b. More on Galileo and the arts. *Isis* 47:182–185.

Pantalony, David. 2009. *Altered Sensations: Rudolph Koenig's Acoustical Workshop in Nineteenth-Century Paris*. Dordrecht: Springer.

Panza, Marco. 2006. François Viète: Between analysis and cryptanalysis. *Studies in History and Philosophy of Science* 37:267–289.

Park, David Allen. 1997. *The Fire within the Eye: A Historical Essay on the Nature and Meaning of Light*. Princeton, NJ: Princeton University Press.

Partington, J. R. 2004. Albertus Magnus on alchemy. In *Alchemy and Early Modern Chemistry: Papers from Ambix*, ed. Allen G. Debus, 45–62. London: Jeremy Mills.

Peacock, George. 1855. *Life of Thomas Young, M.D., F.R.S., &c.; and One of the Eight Foreign Associates of the National Institute of France*. London: J. Murray.

Pedersen, Kurt Møller. 2008. Leonhard Euler's wave theory of light. *Perspectives on Science* 16:392–416.

Pelseneer, Jean. 1951. Une lettre inédite d'Euler à Rameau. *Bulletin de la Classe des sciences: Académie royale de Belgique* 5:480–482.

Pesic, Peter. 1997a. François Viète, father of modern cryptanalysis—two new manuscripts. *Cryptologia* 21:1–29.

Pesic, Peter. 1997b. Secrets, symbols, and systems: Parallels between cryptanalysis and algebra, 1580–1700. *Isis* 88:674–692.

Pesic, Peter. 2000a. *Labyrinth: A Search for the Hidden Meaning of Science*. Cambridge, MA: MIT Press.

Pesic, Peter. 2000b. Kepler's critique of algebra. *Mathematical Intelligencer* 22(4):54–59.

Pesic, Peter. 2002. *Seeing Double: Shared Identities in Physics, Philosophy, and Literature*. Cambridge, MA: MIT Press.

Pesic, Peter. 2003. *Abel's Proof: An Essay on the Sources and Meaning of Mathematical Unsolvability*. Cambridge, MA: MIT Press.

Pesic, Peter. 2004. Plato and zero. *Graduate Faculty Philosophy Journal* 25(2):1–18.

Pesic, Peter. 2005. *Sky in a Bottle*. Cambridge, MA: MIT Press.

Pesic, Peter. 2006. Isaac Newton and the mystery of the major sixth: A transcription of his manuscript "Of Musick" with commentary. *Interdisciplinary Science Reviews* 31:291–306.

Pesic, Peter, ed. 2007. *Beyond Geometry: Classic Papers from Riemann to Einstein*. Mineola, NY: Dover.

Pesic, Peter. 2010. Hearing the irrational: Music and the development of the modern concept of number. *Isis* 101:501–530.

Pesic, Peter. 2013a. Euler's musical mathematics. *Mathematical Intelligencer* 35(2):35–43.

References

Pesic, Peter. 2013b. Helmholtz, Riemann, and the sirens: Sound, color, and the "problem of space." *Physics in Perspective* 15:256–294.

Pesic, Peter. 2013c. Thomas Young's musical optics: Translating between hearing and seeing. *Osiris* 28:15–39.

Pesic, Peter. 2014a. Bacon, violence, and the motion of liberty: The Aristotelian background. *Journal of the History of Ideas* 75:69–90.

Pesic, Peter. 2014b. Thomas Young and eighteenth century tempi. *Performance Practice Review* 18(1) http://scholarship.claremont.edu/ppr/vol28/iss1/2.

Pesic, Peter. 2014c. Max Planck's writings on music: A translation and commentary. *Theoria* 21.

Pesic, Peter. In preparation-a. *The Polyphonic Mind*.

Pesic, Peter. In preparation-b. *Music, Science, Passion*.

Peterson, Mark A. 2011. *Galileo's Muse: Renaissance Mathematics and the Arts*. Cambridge, MA: Harvard University Press.

Pirro, André. 1973. *Descartes et la musique*. Geneva: Minkoff Reprint.

Pirrotta, Nino. 1984. *Music and Culture in Italy from the Middle Ages to the Baroque: a Collection of Essays*. Cambridge, MA: Harvard University Press.

Planck, Max. 1893. Die natürliche Stimmung in der modernen Vokalmusik. *Vierteljahrsschrift für Musikwissenschaft* 9:418–444.

Planck, Max. 1899. Über irreversible Strahlungsvorgänge. *Sitzungsberichte der Königlich Preußischen Akademie der Wissenschaften zu Berlin* 5:440–480.

Planck, Max. 1949. *Scientific Autobiography and Other Papers*. New York: Philosophical Library.

Planck, Max. 1958. *Physikalische Abhandlungen und Vorträge*. Braunschweig: Vieweg.

Planck, Max. 1981. *Where Is Science Going?* Woodbridge, CT: Ox Bow Press.

Planck, Max. 1998. *Eight Lectures on Theoretical Physics*. Ed. Peter Pesic, trans. A. P. Wills. Mineola, NY: Dover.

Plato. 2006. *The Republic*. Trans. R. E. Allen. New Haven, CT: Yale University Press.

Pollitt, J. J. 1972. *Art and Experience in Classical Greece*. Cambridge: Cambridge University Press.

Polzonetti, Pierpaolo. 2001. *Tartini e la musica secondo natura*. Lucca: LIM.

Principe, Lawrence. 2013. *The Secrets of Alchemy*. Chicago: University of Chicago Press.

Proceedings of the Royal Institution. 1829. *Quarterly Journal of Science* 27:379–380.

Ptolemy. 1998. *Ptolemy's Almagest*. Trans. G. J. Toomer. Princeton, NJ: Princeton University Press.

Quintilianus, Aristides. 1983. *On Music, in Three Books*. Trans. Thomas J. Mathiesen. New Haven: Yale University Press.

Raasted, Jørgen. 1979. A neglected version of the anecdote about Pythagoras's hammer experiments. *Cahiers de l'Institut du moyen-âge grec et latin* 31a–31b:1–9.

Rasch, Rudolf. 2002. Why were enharmonic keyboards built? From Nicola Vicentino (1555) to Michael Bulyowsky (1699). *Schweizer Jahrbuch für Musikwissenschaft* 22:35–93.

Rasch, Rudolf. 2008. Simon Stevin and the calculation of equal temperament. In *Music and Mathematics: In Late Medieval and Early Modern Europe*, ed. Philippe Vendrix, 253–320. Turnhout: Brepols.

Rayleigh, John William Strutt. 1945. *The Theory of Sound*, 2nd ed. New York: Dover.

Recorde, Robert. 1557. *The Whetstone of Witte*. London.

Rehding, Alexander. 2003. *Hugo Riemann and the Birth of Modern Musical Thought*. Cambridge: Cambridge University Press.

Rehding, Alexander. 2014. Of sirens old and new. In *The Oxford Handbook of Mobile Music*, ed. Sumanth S. Gopinath and Jason Stanyek. Oxford: Oxford University Press.

Reichenbach, Hans. 1970. The philosophical significance of the theory of relativity. In *Albert Einstein: Philosopher-Scientist*, ed. Paul Arthur Schilpp, 3rd ed., 289–311. La Salle, IL: Open Court.

Reichenbach, Hans. 2006. *Experience and Prediction: An Analysis of the Foundations and the Structure of Knowledge*. Notre Dame, IN: University of Notre Dame Press.

Richards, Joan L. 1977. The evolution of empiricism: Hermann von Helmholtz and the foundations of geometry. *British Journal for the Philosophy of Science* 28:235–253.

Richeson, David S. 2008. *Euler's Gem: The Polyhedron Formula and the Birth of Topology*. Princeton, NJ: Princeton University Press.

Riemann, Bernhard. 1867. Ueber die Hypothesen, welche die Geometrie zu Grunde Liegen. *Abhandlungen der Königlichen Gesellschaft der Wissenschaften in Göttingen* 13:133–152.

Riemann, Bernhard. 1984. The mechanism of the ear. Trans. David Cherry, Robert Gallagher, and John Siegerson. *Fusion* 6(3):31–38.

Riemann, Bernhard. 1990. *Gesammelte mathematische Werke, wissenschaftlicher Nachlass und Nachträge: Collected Papers*. Ed. Heinrich Weber, Richard Dedekind, and Raghavan Narasimhan. Berlin: Springer-Verlag.

Ritchey, Tom. 1991. On scientific method—based on a study by Bernhard Riemann. *Systems Research* 8:21–41.

Ritter, Johann Wilhelm. 1806. *Physisch-chemische Abhandlungen in chronologischer Folge*. Leipzig: Reclam.

Ritter, Johann Wilhelm. 2010. *Key Texts of Johann Wilhelm Ritter (1776–1810) on the Science and Art of Nature*. Trans. Jocelyn Holland. Leiden: Brill.

Robinson, Andrew. 2006. *The Last Man Who Knew Everything: Thomas Young, the Anonymous Polymath Who Proved Newton Wrong, Explained How We See, Cured the Sick, and Deciphered the Rosetta Stone, Among Other Feats of Genius*. New York: Pi Press.

Ronald, Calinger. 1996. Leonhard Euler: The first St. Petersburg years (1727–1741). *Historia Mathematica* 23:121–166.

Rosen, Charles. 1995. *The Romantic Generation*. Cambridge, MA: Harvard University Press.

Rosen, Edward. 1956. Review of Panofsky, *Galileo as a Critic of the Arts*. *Isis* 47:78–80.

Rosen, Edward. 1984. Kepler's attitude toward astrology and mysticism. In *Occult and Scientific Mentalities in the Renaissance*, ed. Brian Vickers, 253–272. Cambridge: Cambridge University Press.

Rosenfeld, B. A. 1987. *The History of Non-Euclidean Geometry: Evolution of the Concept of a Geometric Space*. Trans. Abe Shenitzer. New York: Springer-Verlag.

Rowe, David E. 1992. Klein, Lie, and the "Erlanger Programm." In *1830–1930: A Century of Geometry*, ed. L. Boi, D. Flament, and J.-M. Salanskis, 45–54. Berlin: Springer-Verlag.

Ruhnke, Martin. 1955. *Joachim Burmeister: Ein Beitrag zur Musiklehre um 1600*. Kassel: Bärenreiter.

Rydberg, J. R. 1890. Recherches sur la constitution des spectres d'émission des éléments chimiques. *Kongliche Svenska vetenskaps-akademiens handlingar* 23(11):1–155.

Sabra, A. I. 1981. *Theories of Light, from Descartes to Newton*. Cambridge: Cambridge University Press.

Sabra, A. I. 1963. Newton and the "bigness" of vibrations. *Isis* 54:267–268.

Sachs, Horst, Michael Stiebitz, and Robin J. Wilson. 1988. An historical note: Euler's Königsberg letter. *Journal of Graph Theory* 12:133–139.

Salem, Michael P. 2011. Bubble collisions and measures of the multiverse. http://arxiv.org/abs/1108.0040.

Salinas, Francisco. 1958. *De musica libri septem*. Kassel: Bärenreiter-Verlag.

Sandifer, C. Edward. 2007a. *The Early Mathematics of Leonhard Euler*. Washington, DC: Mathematical Association of America.

Sandifer, C. Edward. 2007b. Euler rows the boat. In *Euler at 300: An Appreciation*, ed. Robert E. Bradley, Lawrence A. D'Antonio, and C. Edward Sandifer, 273–279. Washington, DC: Mathematical Association of America.

Sandifer, C. Edward. 2007c. *How Euler Did It*. Washington, DC: Mathematical Association of America.

Saunders, Steven. 1995. *Cross, Sword, and Lyre: Sacred Music at the Imperial Court of Ferdinand II of Habsburg (1619–1637)*. Oxford: Clarendon Press.

Savart, Félix. 1819. *Mémoire sur la construction des instruments à cordes et à archet*. Paris: Librairie Encyclopédique de Roret.

Schiffer, Michael B. 2003. *Draw the Lightning Down: Benjamin Franklin and Electrical Technology in the Age of Enlightenment*. Berkeley, CA: University of California Press.

Scholz, Erhard. 1980. *Geschichte des Mannigfaltigkeitsbegriffs von Riemann bis Poincaré*. Basel: Birkhäuser.

Scholz, Erhard. 1982. Riemanns frühe Notizen zum Mannigfaltigkeitsbegriff und zu den Grundlagen der Geometrie. *Archive for History of Exact Sciences* 27:213–282.

Schrödinger, Erwin. 1928. *Four Lectures on Wave Mechanics*. London: Blackie.

Schrödinger, Erwin. 1982. *Collected Papers on Wave Mechanics: Together with His Four Lectures on Wave Mechanics*. New York: Chelsea.

Schüller, Volkmar. 1994. Das Helmholtz-Liesche Raumproblem und seine ersten Lösung. In *Universalgenie Helmholtz: Rückblick nach 100 Jahren*, ed. Lorenz Krüger, 260–275. Berlin: Akademie Verlag.

Schuster, John A., and Judit Brody. 2013. Descartes and sunspots: Matters of fact and systematizing strategies in the *Principia Philosophiae*. *Annals of Science* 70:1–45.

Schwartz, Hillel. 2011. *Making Noise: From Babel to the Big Bang & Beyond*. Brooklyn, NY: Zone Books.

Sedlar, Jean W. 1980. *India and the Greek World: A Study in the Transmission of Culture*. Totowa, NJ: Rowman & Littlefield.

Semmens, Richard. 1987. Joseph Sauveur's "Treatise on the Theory of Music": A study, diplomatic transcription, and annotated translation. PhD diss., University of Western Ontario.

Sepper, Dennis L. 1994. *Newton's Optical Writings: A Guided Study*. New Brunswick, NJ: Rutgers University Press.

Shankman, Steven, and Stephen W. Durrant, eds. 2002. *Early China/Ancient Greece: Thinking through Comparisons*. Albany: SUNY Press.

Shapin, Steven. 1994. *A Social History of Truth: Civility and Science in Seventeenth-Century England*. Chicago: University of Chicago Press.

Shapin, Steven. 2010. *Never Pure: Historical Studies of Science as If It Was Produced by People with Bodies, Situated in Time, Space, Culture, and Society, and Struggling for Credibility and Authority*. Baltimore: Johns Hopkins University Press.

Shapiro, Alan E. 1980. The evolving structure of Newton's theory of white light and color. *Isis* 71:211–235.

Shapiro, Alan E. 1993. *Fits, Passions, and Paroxysms: Physics, Method, and Chemistry and Newton's Theories of Colored Bodies and Fits of Easy Reflection*. Cambridge: Cambridge University Press.

Shapiro, Alan E. 1994. Experiment and mathematics in Newton's theory of color. *Physics Today* 37(9):34–42.

Shapiro, Alan E. 2001. Newton's experiments on diffraction and the delayed publication of the *Opticks*. In *Isaac Newton's Natural Philosophy*, ed. Jed Z. Buchwald and I. Bernard Cohen, 47–76. Cambridge, MA: MIT Press.

Shapiro, Alan E. 2002. Newton's optics and atomism. In *The Cambridge Companion to Newton*, ed. I. Bernard Cohen and George E. Smith, 227–255. Cambridge: Cambridge University Press.

Smith, Anne. 2011. *The Performance of 16th-Century Music: Learning from the Theorists*. New York: Oxford University Press.

Smith, Bruce R. 1999. *The Acoustic World of Early Modern England: Attending to the O-factor*. Chicago: University of Chicago Press.

Smith, Charles Samuel. 1960. Leonhard Euler's *Tentamen novae theoriae musicae*: A translation and commentary. PhD diss., Indiana University.

Smith, Robert. 1749. *Harmonics, or The Philosophy of Musical Sounds*. Cambridge: J. Bentham.

Smolin, Lee. 2006. *The Trouble with Physics: The Rise of String Theory, the Fall of a Science, and What Comes Next*. Boston: Houghton Mifflin.

Snelders, H. A. M. 1990. Oersted's discovery of electromagnetism. In *Romanticism and the Sciences*, ed. Andrew Cunningham and Nicholas Jardine, 228–240. Cambridge: Cambridge University Press.

Solovine, Maurice, and Albert Einstein. 1956. *Lettres à Maurice Solovine*. Paris: Gauthier-Villars.

Sommerfeld, Arnold. 1934. *Atomic Structure and Spectral Lines*, 3rd ed. Trans. Henry L. Brose. London: Methuen.

Sonnert, Gerhard. 2005. *Einstein and Culture*. Amherst, NY: Humanity Books.

Spitzer, Leo. 1963. *Classical and Christian Ideas of World Harmony: Prolegomena to an Interpretation of the Word "Stimmung."*. Baltimore, MD: Johns Hopkins University Press.

Steege, Benjamin. 2012. *Helmholtz and the Modern Listener*. Cambridge: Cambridge University Press.

Stein, Howard. 1977. Some philosophical prehistory of general relativity. In *Foundations of Space-Time Theories*, ed. John S. Earman, Clark N. Glymour, and John J. Stachel, 3–49. Minneapolis, MN: University of Minnesota Press.

Stephenson, Bruce. 1994a. *The Music of the Heavens: Kepler's Harmonic Astronomy*. Princeton, NJ: Princeton University Press.

Stephenson, Bruce. 1994b. *Kepler's Physical Astronomy*. Princeton, NJ: Princeton University Press.

Sterne, Jonathan. 2003. *The Audible Past: Cultural Origins of Sound Reproduction*. Durham: Duke University Press.

Stifel, Michael. 1544. *Arithmetica Integra*. Nuremburg: J. Petreium.

Stoney, G. Johnstone. 1871. On the cause of the interrupted spectra of gases. *Philosophical Magazine* 41:291–296.

Stoney, G. Johnstone. 1880. On a new harmonic relation between the lines of hydrogen. *Nature* 21:508.

Stoney, G. Johnstone, and J. Emerson Reynolds. 1871. An inquiry into the cause of the interrupted spectra gases—part II. On the absorption-spectrum of chlorochromic anhydride. *Philosophical Magazine* 42:41–52.

Strässle, Thomas. 2004. "Das Hören ist ein Sehen von und durch Innen": Johann Wilhelm Ritter and the aesthetics of music. In *Music and Literature in German Romanticism*, ed. Siobhán Donovan and Robin Elliott, 27–42. Rochester, NY: Camden House.

Strickland, Stuart Walker. 1998. The ideology of self-knowledge and the practice of self-experimentation. *Eighteenth-Century Studies* 31:453–471.

Strunk, W. Oliver, and Leo Treitler, eds. 1998. *Source Readings in Music History*. New York: Norton.

Supper, Alexandra. 2012. Lobbying for the ear: The public fascination with and academic legitimacy of the sonification of scientific data. PhD diss., Maastricht University.

Susskind, Leonard. 2005. *Cosmic Landscape: String Theory and the Illusion of Intelligent Design*. New York: Little, Brown.

Susskind, Leonard. 2012. The first string theory: Personal recollections. In *The Birth of String Theory*, ed. Andrea Cappelli, et al., 262–265. Cambridge: Cambridge University Press.

Swafford, Jan. 1997. *Johannes Brahms: A Biography*. New York: Alfred A. Knopf.

Szabó, Árpád. 1978. *The Beginnings of Greek Mathematics*. Dordrecht: D. Reidel.

Takahashi, Yuzo. 1979. Two hundred years of Lichtenberg figures. *Journal of Electrostatics* 6:1–13.

Tannery, Paul. 1902. Du rôle de la musique greque dans le développement de la mathématique pure. *Mémoire scientifique* 3:161–175.

Tartaglia, Niccolò. 1578. *L'arithmetiqve*. Trans. Guillaume Gosselin. Paris: Gilles Beys.

Taschow, Ulrich. 1999. Die Bedeutung der Musik als Modell für Nicoles Oresme Theorie. *Early Science and Medicine* 4:37–90.

Taschow, Ulrich. 2003. *Nicole Oresme und der Frühling der Moderne: Die Ursprünge unserer modernen quantitativ-matrischen Weltaneignungsstrategien und neuzeitlichen Bewusstseins- und Wissenschaftskultur*. Halle: Avox Medien-Verlag.

Thomas, Jennifer. 2009. *Absalon fili mi*, Josquin, and the French royal court: Attribution, authenticity, context, and conjecture. In *Uno gentile et subtile ingenio: Studies in Renaissance Music in Honour of Bonnie J. Blackburn*, ed. Gioia Filocamo and M. Jennifer Bloxam, 477–489. Turnhout: Brepols.

Thompson, Emily Ann. 2002. *The Soundscape of Modernity: Architectural Acoustics and the Culture of Listening in America, 1900–1933*. Cambridge, MA: MIT Press.

Tiella, Marco. 1975. The Archicembalo of Nicola Vicentino. *English Harpsichord Magazine* 1:134–144.

Tinctoris, Johannes. 1961. *The Art of Counterpoint: Liber de Arte Contrapuncti*. Trans. Albert Seay. Rome: American Institute of Musicology.

Tomlinson, Gary. 1993. *Music in Renaissance Magic: Toward a Historiography of Others*. Chicago: University of Chicago Press.

Tonietti, Tito M. 2000. Does Newton's musical model of gravitation work? A mistake and its meaning. *Centaurus* 42:135–149.

Tonietti, Tito M. 2011. Music between hearing and counting (a historical case chosen within continuous long-lasting conflicts). In *Mathematics and Computation in Music*, ed. Carlos Agon, Moreno Andreatta, Gérard Assayag, Emmanuel Amiot, Jean Bresson, and John Mandereau, 285–296. Berlin: Springer. http://www.springerlink.com/content/a576762u0w2j4608/.

Torretti, Roberto. 1978. *Philosophy of Geometry from Riemann to Poincaré*. Dordrecht, Holland: D. Reidel.

Tserlyuk-Askadskaya, S. S. 2007. Euler's music-theoretical manuscripts and the formation of his conception of the theory of music. In *Euler and Modern Science*, ed. A. P. Yushkevich, N. N. Bogolyubov, and G. K. Mikhaĭlov, trans. Robert Burns, 349–360. Washington, DC: Mathematical Association of America.

Tweney, Ryan D. 1992a. Stopping time: Faraday and the scientific creation of perceptual order. *Physis* 29:149–164.

Tweney, Ryan D. 1992b. Inventing the field: Michael Faraday and the creative engineering of electromagnetic field theory. In *Inventive Minds: Creativity in Technology*, ed. D. Perkins and R. Weber, 31–47. Oxford: Oxford University Press.

Tyndall, John. 1898. *Sound*, 3rd ed. New York: D. Appleton.

Tyndall, John. 1961. *Faraday as a Discoverer*. New York: Crowell.

Van Berkel, Klaas. 2000. Descartes' debt to Beeckman: Inspiration, cooperation, conflict. In *Descartes' Natural Philosophy*, ed. Stephen Gaukroger, John Schuster, and John Sutton, 46–59. London: Routledge.

Van Wymeersch, Brigitte. 1999. *Descartes et l'évolution de l'esthétique musicale*. Sprimont: Mardaga.

Van Wymeersch, Brigitte. 2008. Qu'entend-on par "nombre sourd"? In *Music and Mathematics in Late Medieval and Early Modern Europe*, ed. Philippe Vendrix, 97–110. Turnhout: Brepols.

Van Wymeersch, Brigitte. 2011. Proportion, harmonie et beauté chez Mersenne: entre lecture analogique et lecture physico-mathématique de la musique. In *Proportions: Science, musique, peinture & architecture*, ed. Sabine Rommevaux, Philippe Vendrix, and Vasco Zara, 261–274. Turnhout: Brepols.

Varadarajan, V. S. 2006. *Euler through Time: A New Look at Old Themes*. Providence, RI: American Mathematical Society.

Vecchia, P. Di. 2008. The birth of string theory. In *String Theory and Fundamental Interactions*, vol. 737. ed. Maurizio Gasperini and Jnan Maharana, 59–118. Lecture Notes in Physics. Berlin: Springer.

Velminski, Wladimir, ed. 2009a. *Leonhard Euler: Die Geburt der Graphentheorie*. Berlin: Kulturverlag Kadmos.

Velminski, Wladimir. 2009b. *Form, Zahl, Symbol: Leonhard Eulers Strategien der Anschaulichkeit*. Berlin: Akademie-Verlag.

Vendrix, Philippe, ed. 2008. *Music and Mathematics: In Late Medieval and Early Modern Europe*. Turnhout: Brepols.

Vicentino, Nicola. 1996. *Ancient Music Adapted to Modern Practice*, ed. Claude V. Palisca, trans. Maria Rika Maniates. New Haven: Yale University Press.

Vickers, Brian. 1984. Analogy versus identity: The rejection of occult symbolism, 1580–1680. In *Occult and Scientific Mentalities in the Renaissance*, ed. Brian Vickers, 273–296. Cambridge: Cambridge University Press.

Vogel, Stephan. 1993. Sensations of tone, perception of sound, and empiricism. In *Helmholtz and the Foundations of Nineteenth-Century Science*, ed. David Cahan, 259–287. Berkeley, CA: University of California Press.

Volkert, Klaus. 1993. On Helmholtz' paper "Ueber die thatsächlichen Grundlagen der Geometrie." *Historia Mathematica* 20:307–309.

Volkert, Klaus. 1996. Hermann von Helmholtz und die Grundlagen der Geometrie. In *Hermann von Helmholtz: Vorträge eines Heidelberger Symposium anlässlich des einhundersten Todestages*, ed. Wolfgang U. Eckart and Klaus Volkert, 177–207. Pfaffenweiler: Centaurus-Verlagsgesellschaft.

Volmar, Axel. 2012. Klang-Experimente: Eine Geschichte der auditiven Kultur der Naturwissenschaften seit 1800. PhD diss., University of Siegen.

Volmar, Axel. 2013a. Listening to the cold war: The nuclear test ban negotiations, seismology, and psychoacoustics, 1958–1963. *Osiris* 28:80–102.

Volmar, Axel. 2013b. Sonic facts for sound arguments: Medicine, experimental physiology, and the auditory construction of knowledge in the 19th century. *Journal of Sonic Studies* 4(1). http://journal.sonicstudies.org/vol04/nr01/a13.

Voltaire. 1967. *The Elements of Sir Isaac Newton's Philosophy*. London: Cass.

Wahsner, Renate. 1994. Apriorische Funktion und aposteriorische Herkunft: Hermann von Helmholtz' Untersuchungen zum Erfahrungsstatus der Geometrie. In *Universalgenie Helmholtz*, ed. Lorenz Krüger, 245–259. Berlin: Akademie Verlag.

Walker, D. P. 1978. *Studies in Musical Science in the Late Renaissance*. London: Warburg Institute, University of London.

Ward, Tom R. 2013. Music and music theory in the universities of central Europe during the fifteenth century. In *Musical Theory in the Renaissance*, ed. Cristle Collins Judd, 563–571. Farnham, Surrey: Ashgate.

Wardhaugh, Benjamin. 2008. *Music, Experiment, and Mathematics in England, 1653–1705*. Farnham: Ashgate.

Warusfel, André. 2009. *Euler: Les mathématiques et la vie*. Paris: Vuibert.

Watson, Gerard. 1973. *Plato's Unwritten Teaching*. Dublin: Talbot Press.

Weil, André. 1984. *Number Theory: An Approach through History from Hammurapi to Legendre*. Basel: Birkhäuser.

Welsh, Caroline. 2010. Ätherschwingungen und Nervenvibrationen: Leonhard Eulers Seele in der camera obscura des Körpers und die Probleme der Wahrnehmung im 18. Jahrhundert. In *Mathesis & Graphé: Leonhard Euler und die Entfaltung der Wissensystems*, ed. Horst Bredekamp and Wladimir Velminski, 224–237. Berlin: Akademie Verlag.

Werner, Eric. 1956. The mathematical foundation of Philippe de Vitri's "Ars Nova." *Journal of the American Musicological Society* 9:128–132.

Werner, Eric. 1978. The last Pythagorean musician: Johannes Kepler. In *Aspects of Medieval and Renaissance Music: A Birthday Offering to Gustave Reese*, ed. Jan LaRue, 867–882. New York: Pendragon Press.

West, M. L. 1992. *Ancient Greek Music*. Oxford: Clarendon Press.

Westfall, Richard S. 1967. Uneasily fitful reflections on fits of easy transmission. *Texas Quarterly* 10:86–107.

Westfall, Richard S. 1980. *Never at Rest: A Biography of Isaac Newton*. Cambridge: Cambridge University Press.

Westman, Robert S. 1975. Three responses to the Copernican Theory: Johannes Praetorius, Tycho Brahe, and Michael Maestlin. In *The Copernican Achievement*, ed. Robert S. Westman, 285–345. Berkeley, CA: University of California Press.

Westman, Robert S. 2011. *The Copernican Question: Prognostication, Skepticism, and Celestial Order*. Berkeley, CA: University of California Press.

Wetzels, Walter D. 1990. Johann Wilhelm Ritter: Romantic physics in Germany. In *Romanticism and the Sciences*, ed. Andrew Cunningham and Nicholas Jardine, 199–212. Cambridge: Cambridge University Press.

Weyl, Hermann. 2009. *Mind and Nature: Selected Writings on Philosophy, Mathematics, and Physics*. Ed. Peter Pesic. Princeton, NJ: Princeton University Press.

Wheatstone, Charles. 1879. *The Scientific Papers of Sir Charles Wheatstone*. London: Taylor & Francis.

Whitehead, Alfred North. 1967. *Science and the Modern World*. New York: Free Press.

Wilczek, Frank, and Betsy Devine. 1988. *Longing for the Harmonies: Themes and Variations from Modern Physics*. New York: Norton.

Williams, L. Pearce. 1971. *Michael Faraday: A Biography*. New York: Simon & Schuster.

Wilson, Curtis. 1989. *Astronomy from Kepler to Newton: Historical Studies*. London: Variorum Reprints.

References

Woit, Peter. 2006. *Not Even Wrong: The Failure of String Theory and the Search for Unity in Physical Law.* New York: Basic Books.

Wood, Alexander, and Frank Oldham. 1954. *Thomas Young, Natural Philosopher, 1773–1829.* Cambridge: Cambridge University Press.

Worrall, John. 1976. Thomas Young and the "refutation" of Newtonian optics: A case-study in the interaction of philosophy of science and history of science. In *Method and Appraisal in the Physical Sciences*, ed. Colin Howson. Cambridge: Cambridge University Press.

Wright, Craig. 2001. *The Maze and the Warrior: Symbols in Architecture, Theology, and Music.* Cambridge, MA: Harvard University Press.

Wright, Denzil. 2002. The cimbalo cromatico and other Italian string keyboard instruments with divided accidentals. *Schweizer Jahrbuch für Musikwissenschaft* 22:105–136.

Yates, Frances Amelia. 1991. *Giordano Bruno and the Hermetic Tradition.* Chicago: University of Chicago Press.

Young, Thomas. 1855. *Miscellaneous Works of the Late Thomas Young.* London: J. Murray.

Young, Thomas. 2002. *Thomas Young's Lectures on Natural Philosophy and the Mechanical Arts.* Bristol: Thoemmes.

Yushkevich, A. P., N. N. Bogolyubov, and G. K. Mikhaïlov, eds. 2007. *Euler and Modern Science.* Trans. Robert Burns. Washington, DC: Mathematical Association of America.

Zarlino, Gioseffo. 1579. *Discorso del reverendo M. Gioseffo Zarlino ... intorno il vero anno, & il vero giorno, nel quale fu crucifisso Il N.S. Giesv Christo redentor del mondo.* Venice: Domenico Nicolini.

Zarlino, Gioseffo. 1588. *De tutte l'opere del r.m. Gioseffo Zarlino.* Venice: Francesco de' Franceschi Senese.

Zarlino, Gioseffo. 1968. *The Art of Counterpoint. Part Three of* Le istitutioni harmoniche, *1558.* New Haven: Yale University Press.

Zarlino, Gioseffo. 1983. *On the Modes: Part Four of* Le istitutioni harmoniche, *1558.* Ed. Claude V. Palisca, trans. Vered Cohen. New Haven: Yale University Press.

Zarlino, Gioseffo. 2011. *L'istituzioni armoniche.* Treviso: Diasteme.

Ziemer, Hansjakob. 2008. *Die moderne Hören: Das Konzert als urbanes Forum, 1890–1940.* Frankfurt: Campus.

Zoubov, V. P. 1961. Nicole Oresme et la musique. *Mediaeval and Renaissance Studies* 5:96–107.

Sources and Illustration Credits

Portions of this book appeared originally in the following journals, which have kindly given permission for the appearance of the material here: *Interdisciplinary Science Reviews* (© 2006 Institute of Materials, Minerals, and Mining); *Isis* and *Osiris* (© The History of Science Society, University of Chicago Press, 2010, 2013); *Journal of Seventeenth-Century Music* (http://www.sscm-jscm.org/) [11, 1 (2005)] (© 2005 University of Illinois Press); *Mathematical Intelligencer* (© 2000, 2013 Springer Science + Business Media New York); and *Physics in Perspective* (© 2013 Springer Basel AG).

Permission for the use of the figures has kindly been given by the following: The American Institute of Musicology, Verlag Corpusmusicae, GmbH (boxes 3.1, 3.2); ATLAS experiment © 2013 CERN (fig. 18.4b); Bibliothèque nationale de France (figs. 2.1, 2.2); British Museum (fig. 4.6); John Carter Brown Library at Brown University (fig. 3.2b); Syndics of the Cambridge University Library (fig. 8.1); Collection of Historical Scientific Instruments, Harvard University (figs. 6.1, 14.9, 14.10b); Deutsches Museum, Munich (figs. 12.3, 17.1); Keith B. MacAdam (fig. 18.1); George Peabody Library, The Sheridan Libraries, The Johns Hopkins University (figs. 9.4, 9.5, 10.2, 12.5a–b, 13.8a–c, 13.9a, 13.10a–c, 13.11, 16.1b); Huntington Library, San Marino, California (figs. 4.1 [Huntington, RB67813], 4.2 [Huntington, RB 707254]); Royal Institution (fig. 13.12); Anne Smith, *The Performance of Sixteenth-Century Music* (2011), by permission of Oxford University Press, USA (figs. 5.2b, 5.5); Springer-Verlag (fig. 8.3a); Marco Tiella (fig. 4.5a); and Warburg Institute (fig. 7.2a). Special thanks to Alexei Pesic for his expert help in preparing the figures.

Acknowledgments

I thank Marguerite Avery and her colleagues at the MIT Press for their wonderful support and collaboration, including Judy Feldmann, Gita Manaktala, Cristina Sanmartin, and many others. I feel very fortunate to have worked with them now over five books and many years, beginning with Larry Cohen, the editor who gave me my first chance and never stopped supporting and helping me. My admiration and thanks know no bounds.

St. John's College in Santa Fe has been an ideal home in which the questions and thoughts that led to this book could grow in an open, sympathetic environment. The emphasis on questioning and discussion, the absence of the usual disciplinary barriers, the great books we read together all encouraged me in my quest. I am profoundly grateful to my colleagues and fellow students; I hope this book can show something of what this remarkable college can foster.

I thank the John Simon Guggenheim Memorial Foundation for their support at a formative stage in this project. Melissa Franklin, Alexander Rehding, and Anne Shreffler graciously welcomed me as a visitor to Harvard, enabling me to use their libraries, swim in their pools, and discuss my ideas with new interlocutors.

I sincerely thank those who kindly read and commented on parts of the book: Jed Buchwald, H. Floris Cohen, Sean Gallagher, Jean-François Gauvin, Owen Gingerich, Marie Louise Göllner, Justin Grosslight, James Haar, Myles Jackson, Thomas Mathiesen, Andrei Pesic, Benjamin Wardhaugh, and especially Curtis Wilson, whose memory I here honor with deep gratitude. Special thanks to Veit Erlmann, Alexandra Hui, Julia Kursell, and Victoria Tkaczyk, who generously read the whole work and gave me good advice. I especially thank Gerald Holton for his continued encouragement over many years.

This book owes a special debt to Alexei Pesic, whose creativity and skill has opened new doors in allowing readers to have direct access to the sound examples through a single touch in the iBook version he designed. These new possibilities owe everything to his hard and devoted work.

Andrei, Ssu, and Alexei are the inspirations for all I do and the ones whose love and support make it possible and worthwhile.

Index

Abbott, Edwin (1838–1926), 242
Acoucryptophone. *See* Enchanted Lyre
Acoustics, 4, 112, 158, 184, 250–252, 257, 275–277
Adrastus of Aphrodisias (second century C.E.), 17, 287n30
Alchemy, 45, 74
Algebra, 3, 55–56, 59–60, 70, 72, 77, 93, 148–149, 179
Ampère, André-Marie (1775–1836), 178–179, 194, 196, 198, 213
 Ampère's law, 196
Ångström, Anders Jonas (1814–1874), 245, 247–248, 250–251
Anthropic principle, 282
Apollo, 28–34
Arago, François Jean Dominique (1786–1853), 177–179
Archicembalo, 62–63, 65
Archytas (428–347 B.C.E.), 9, 14
Aristarchus, 53, 104
Aristides Quintilianus (late third to early fourth century C.E.), 65
Aristotle (384–322 B.C.E.), 21, 35, 40, 58, 95, 123
 on cosmology, 14, 22, 27, 45, 90
 on music, 18, 25, 38, 66, 114, 116, 285n8
 on physics, 31, 44–46, 58, 98
 on tragedy, 140
Aristoxenus (fl. 335 B.C.E.), 18, 66–67, 135
Arithmetic, 1–2, 9, 14–22, 28–35, 47, 56–60, 68–69, 72, 90–92, 144–145
Arithmos. *See* Number
Ars antiqua, 31–32
Ars nova, 2, 31–33, 288n21
Arts
 fine, 135
 liberal, 19, 135
Artusi, Giovanni Maria (ca. 1540–1613), 70, 72
Astrolabe, 90–91
Astronomy, 55, 109, 181–182, 280. *See also* Cosmology

Copernican controversy, 19, 22, 24, 27, 30, 33, 40, 46–52
 and Kepler, 73, 86–89
 in Oresme, 18–29, 33
 in quadrivium, 2–3, 9, 14, 18, 54
Atomic theory, 3, 5, 98, 119–120, 245, 248–249, 256, 265, 268, 273, 274
 Bohr–Sommerfeld theory, 275–276
Augustine (354–430 C.E.), 24
Aulos, 66–67

Bach, Johann Sebastian (1685–1750), 263, 272, 299n26
Bacon, Francis (1561–1626), 6, 45, 90, 203, 211, 285n6
Balmer, Johann (1825–1898), 5, 249–255, 266, 275
 Balmer formula, 250–251
Barker, Andrew, 19
Battery (Voltaic pile), 189, 211
Beat, 92, 111–114, 144, 164, 166, 170, 227, 277
Beauty, 29–30, 33, 47, 57, 59, 61, 85, 88, 136, 144, 191, 196, 200, 203, 207, 209–210, 271, 278–280
Beeckman, Isaac (1588–1637), 89–90, 103, 111
Bells, 12, 118–119, 182
Beltrami, Eugenio (1835–1899), 237–238, 240
Benedetti, G. B. (1530–1590), 111
Benjamin, Walter (1892–1940), 190
Bernoulli, Daniel (1700–1782), 134–135
Bernoulli, Johann I (1667–1748), 133
Bernoulli, Johann II (1710–1790), 133, 152–153, 182
Bessel, Friedrich (1784–1846), 251
Billroth, Theodor (1829–1894), 257
Biot, Jean-Baptiste (1774–1862), 181, 185, 194
Biot-Savart law, 181, 194
Black body radiation, 5, 263–269
Blacksmith shop, myth of, 9–13, 60, 117, 282, 286n9, 286n11
Bloch, Felix (1905–1983), 273, 275
Bodies, rigid, 237–238, 240, 243

Boethius, Anicius Manlius Severinus (ca. 480–526 C.E.), 1–2, 10–13, 18–21, 27, 32, 35, 50, 56, 60–63, 66, 77, 117, 123, 135
Bohr, Niels (1885–1962), 275–276
Boltzmann, Ludwig (1844–1906), 256, 263, 265
Bólyai, János (1802–1860), 232
Brahe, Tycho (1546–1601), 104
Brahms, Johannes (1833–1897), 229, 257–258, 260
Brentano, Clemens (1778–1842), 188
Broglie, Louis de (1892–1987), 273, 275
Bruno, Giordano (1548–1600), 63
Bubbles, 156, 282–283
Bunsen, Robert (1811–1899), 248
Buridan, Jean (ca. 1300–after 1358), 46, 293n44
Burmeister, Joachim (ca. 1566–1629), 78

Cadence, 3, 39–45, 80, 82, 84–88, 158
Cage, John (1912–1992), 282
Calendar, 49, 52
Camerata, 48, 70
Cardano, Girolamo (1501–1576), 3, 55–60, 69–70
Carole, 22, 46
Cassiodorus, Flavius Magnus Aurelius Senator (ca. 485–ca. 585 C.E.), 27
Champollion, Jean-François (1790–1832), 179
Chant
 Gregorian, 31, 38, 76–78
 Turkish, 75–78, 80–81, 84
Charles V (king of France; 1338–1380), 21–22
Chemistry, 188–191, 195, 245, 249
Chladni, Ernst (1756–1827), 4, 178, 181–193, 198, 200, 205, 207, 251, 279
Chords, 4, 84, 229, 272
 hexachord, 36
 seventh, 70, 72, 136–137, 158–159
Christianity, 25, 32, 63, 77–78, 105
Circle, 14, 22, 33, 49, 97–98, 200, 211, 219, 221, 288n31
Clark, Suzannah, 90
Clavius, Christopher (1538–1612), 53
Clifford, William Kingdon (1845–1879), 242
Clio, 81
Clock, 114, 214
Cohen, H. Floris, 6, 90, 92, 97
Color
 analogy with sound, 3, 121–130, 153–160, 164, 173–176, 189, 203, 222, 247
 parameters of, 219–220
 vision, 172–174, 217–222, 228, 231, 235, 238–241, 272–277
Combination tones, 166, 226–227, 234
Comedy, 139–140
Commensurability, 15, 21–22, 28–33, 66–67
Concertina, 205
Consonance, 10, 17, 29, 35, 47, 53, 60, 71, 73, 78, 80, 93, 96–97, 105, 109, 114, 135–137
Cooke, William (1806–1879), 213

Copernicus, Nicolaus (1473–1543), 2, 22–25, 46–53, 100, 104–107
Cosmology
 Copernican, 19, 22, 24, 27, 30, 33, 40, 46–52
 geocentric, 2, 19–28, 36, 49–52
 heliocentric, 2, 19, 22, 24, 28, 46–49, 52, 105, 109
Coulomb, Charles-Augustine de (1736–1806), 185, 188, 196
Counterpoint, 35, 255
Counting, 14–17, 144
Culture
 aural, 1, 6
 material, 6
Curvature, 215, 233, 273–274, 277
 Gaussian, 231

D'Alembert, Jean le Rond (1717–1783), 236
Dance, 4, 9, 15, 22, 24, 34, 46, 67, 74, 92, 133, 162
Dante, 38
Davy, Humphrey (1778–1829), 177, 195, 205
Deaf, music for the, 97, 112–113
Debye, Peter (1884–1966), 273, 275
Dedekind, Richard (1831–1916), 233
Degree of agreeableness, 4, 135–142
de la Rive, Charles Gaspard (1770–1834), 196
de Muris, Johannes (ca. 1290–ca. 1355), 22, 32, 46
Descartes, René (1596–1650), 3, 6, 89–102, 116, 121, 137, 151, 155, 196, 280
 and Mersenne, 90, 93–103, 116, 137
Desedimentation, 284
des Prez, Josquin (ca. 1450–1521), 2, 35, 38–45, 282, 288n6
de Vitry, Phillipe (1291–1361), 32
Dimensionality, 5, 191, 207, 231–242, 273, 274, 278, 279–280, 307
Dirac, P. A. M. (1902–1984), 278
Dissonance, 11, 47, 53–54, 105, 109, 114, 223, 272, 282
 and Euler, 135, 137, 139–140, 158
 and Kepler, 78, 80, 84, 86
Drake, Stillman (1910–1993), 6–7
Dyad, concept of, 16
Dynamics, 96, 188, 191, 193, 198, 200, 242, 255, 273, 276, 280

e, 143
Ear
 accommodation of, 264
 drum, 226, 236
 ossicles, 226, 234–236
 physiology of, 98, 156, 172, 189, 226, 233–236, 306n25
Earth, 2–3, 22, 24–25, 27–28, 31, 35–38, 46–49, 53–55, 73, 80–87, 99–101, 104–105, 116, 124, 188, 282
Education, 2, 9, 14, 19–20, 179, 195, 158, 265

Einstein, Albert (1879–1955), 5, 215, 232, 238, 241–243, 256, 271–273
 and Helmholtz, 242–243
 and history of science, 243
 and Riemann, 243
Eitz, Carl (1848–1924), 257–259, 265, 267–268
Electricity, 4, 181–196, 205–206, 210–214, 233, 266
 electric conflict, 193
 fluid theory of, 182, 196
 velocity of, 205–206, 210, 214
Electromagnetic chronoscope, 213–214
Electromagnetic clock, 214
Electromagnetism, 4, 181, 193–198, 207, 213–215, 233
 electromagnetic rotation, 197–198
Electron, 46, 273, 275–278, 281
Electrophorus, 182–183
Electro-tonic state, 196, 213
Enchanted Lyre, 200–201, 206, 280
Energy, 217, 242–243, 263, 265–268, 275–281
Engineering, 89, 107, 134, 144
Equation
 differential, 226, 232, 275
 Hamilton-Jacobi, 276
 Schrödinger, 272–278
Ergodic hypothesis, 87
Erlangen program, 242
Erlmann, Veit, 6
Étaples, Jacques Lefèvre de (ca. 1455–1536), 57
Ether, 93, 124, 151–152, 155, 172, 179, 214–215, 263
Euclid, 17, 22, 28, 32, 55, 57, 66, 77, 172, 231, 236–242
Euler, Leonhard (1707–1783), 4, 131–160, 162–164, 172–174, 182, 203, 223, 279, 282
 Euler characteristic, 148
 Euler's formula, 146–148
 and Newton, 133–134, 151–157
 number theory, 143–145
 theory of light, 151–158, 247, 276
 theory of music, 134–142, 158–160
 topology, 145–149
Evans, R. J. W., 74
Experiment, origin of, 9–13
Eye
 accommodation of, 162–163, 170, 172
 compared to ear, 127, 156, 160, 162–163, 172–175, 189, 227, 229
 physiology of, 162–163, 170–177, 217–222, 228–229, 234

Faraday, Michael (1791–1867), 4–5, 195–215, 279
 discovery of electromagnetic induction, 207, 212–214
 field theory of light, 214–215
 work on sound, 207–211
Farina, Carlo (ca. 1600–1639), 80

Faust (Goethe), 237
Fermat primes, 143
Feynman, Richard (1918–1988), 278
Fifth hammer, problem of, 17, 117
Figured bass, 141–142
Fizeau, Hippolyte (1819–1896), 206
Fludd, Robert (1574–1637), 3, 73, 104–108, 285n6
Flute, 13, 108, 134, 162, 200, 206, 298n5
Force, 4, 44, 89, 109, 120, 164, 185, 190–191, 193, 233–234, 239, 249
 lines of, 194, 196, 208–215, 277
Foucault, Léon (1819–1868), 206
Fraction, 17, 55, 58, 69, 93
 continued, 141–143
Fraunhofer, Joseph von (1787–1826), 244, 246
Frequency
 light, 172, 174, 178, 193, 221–222, 265–268, 277
 resonant, 203, 280–281
 sound, 3, 97, 111, 114, 131, 155–158, 166, 226, 251
Fresnel, Augustin-Jean (1788–1827), 177–179, 198
Frogs, 189, 217–218

Gabrieli, Andrea (1532–1585), 74
Gaffurius, Francinus (1451–1522), 12, 35–36
Galilei, Galileo (1564–1642), 2, 6–7, 24–25, 63, 89, 93, 96, 99, 107 111–114, 203
 and Copernicus, 2, 52–54
Galilei, Vincenzo (ca. 1520–1591), 2, 48–49, 52–53, 59, 70, 76, 111, 158
 and Copernicus, 2, 35, 53
 and Galileo, 6
Galvanism, 193
Gassendi, Pierre (1592–1655), 119–120
Gauss, Carl Friedrich (1777–1855), 231–232, 237
Génder, 203–204
Genus
 chromatic, 60–63, 68–69, 158
 diatonic, 36, 40, 60–61, 68–69, 96, 121, 123–124, 126, 224
 enharmonic, 61–63, 66–70, 291n48
Geometria situs. See Topology
Geometry, 1–2, 5, 9, 14–35, 56, 59, 68–70, 77, 84, 90, 93, 145–149, 210, 231–243, 273
 Euclidean, 77, 231, 236–242
 Lobachevskian, 242
 non-Euclidean, 231, 237–242, 273, 277–278
Gerson, Levi ben (Gersonides; 1288–1344), 32–33
Gilbert, William (1544–1603), 48, 53
Glarean, Heinrich (1488–1563), 2, 36–40, 49, 52, 74, 78, 289n7
Glass, vibrating, 11–12, 45, 182, 198, 200, 208, 211–212, 223 280–281
Glissando, 80–81
God, 10, 23–24, 29–33, 75, 77, 80, 82–83, 86–87, 105–107, 110, 120, 181, 233, 282

Goethe, Johann Wolfgang von (1749–1832), 181, 188, 237, 243
Goldbach, Christian (1690–1764), 143–144
Golden ratio, 83, 141
Gosselin, Guillaume (ca. 1536–ca. 1600), 70–71
Gouk, Penelope, 6
Gozza, Paolo, 6
Gradus suavitatis. *See* Degree of agreeableness
Grating, diffraction, 173–174
Gravity, 89, 131, 280, 282
Great Year, 22, 287n5
Greek music, 6, 18, 35, 38, 48–49, 52, 56, 158
Greek natural philosophy, 1–2
Grimaldi, Francesco Maria (1618–1663), 125, 128, 175
Grundton (fundamental tone), 250
Guimbarde, 203–204
Gymnastikē, 14

Hadrons, Hagedorn model of, 279
Hakfoort, Casper, 155
Hamlet, 1–2, 284
Handel, Georg Friedrich (1685–1759), 170
Harmonia, 9–10, 14, 46–47, 61
Harmonica, 205
Harmonic series, 145
Harmonium, 5, 205, 226, 256–259, 265–269, 279
Harmony
 cosmic, 2, 5, 14, 21, 29–35, 46–48, 53–54, 76, 80–86, 102, 104–111
 as criterion, 190, 193, 200, 229, 239, 245, 271, 278–284
 practical, 134–145, 156, 158–159, 173, 255
Hassler, Hans Leo (1564–1612), 74
Haydn, Joseph (1732–1809), 170
Heat, 123, 181, 188, 193, 247, 265–266
Heilbron, John, 256, 269
Heisenberg, Werner (1901–1976), 5, 271–273, 280
Helmholtz, Hermann von (1821–1894), 4–5, 137, 217–252, 255–267, 272–273, 276–277, 279
 on hearing, 223–229, 235
 on space, 235–242
 on vision, 219–222
Heptagon, 77
Heraclitus (ca. 535–ca. 475 B.C.E.), 280
Herschel, Sir John (1792–1871), 161
Herschel, William (1738–1822), 188
Hertz, Heinrich (1857–1894), 264, 266–267, 269
Hiebert, Erwin (1919–2012), 256
Hieroglyphics, 4, 177, 179
Hoffmann, E. T. A. (1776–1822), 190
Holton, Gerald, 278–279
Homer (seventh or eighth century B.C.E.), 9–10, 14–15, 19
Hooke, Robert (1635–1703), 6, 54, 123, 127, 285n6, 297n6
Horatio, 1–2, 284

Hui, Alexandra, 6, 256, 263–264
Humboldt, Alexander von (1769–1859), 188–189
Husserl, Edmund (1859–1938), 252–253, 255, 269, 271, 285n3
Huygens, Christiaan (1629–1695), 114, 151–155, 285n6
 Huygens's principle, 151–153
Hydrogen, 5, 190, 196, 247–252, 275–277

Incommensurability, 22, 28–34, 57, 140
Indexing, 4, 137, 146, 148–149
Induction
 electrical, 182–183
 electromagnetic, 4, 196, 198, 207, 212–214
Infinitesimal, 232, 237
Infinity, 3, 16, 22, 55, 58, 69, 77, 79, 81, 86–88, 95, 98, 143, 152, 236, 251
Instruments
 optical, 158, 168
 percussion, 112, 134
 string, 109, 112, 116–118, 135, 153, 166, 168, 187, 210
 wind, 112, 117, 134–135
Interference, 125, 152–154, 161, 168, 176
Internalism, 285n2
Intervals
 comma, 56, 167, 261–262
 diesis, 59, 61, 63–69, 84, 291n34
 fifth, 10–13, 17, 19, 33, 42–43, 60, 78, 93, 97, 116–17, 136, 141, 250
 fourth, 10–11, 13, 33, 60, 83–84, 93, 159, 167, 173–174
 octave, 3, 9–13, 17, 33, 36, 40, 49, 56, 60, 75, 77–78, 91, 93, 97, 116–117, 121–127, 131, 135, 141, 156–159, 164, 167–169, 173–174, 193, 222, 224, 229, 250, 257, 267–268
 overtone, 167, 221
 quarter tone, 3, 59, 61, 63, 66–69, 123 (*see also* Intervals, diesis)
 semitone, 36, 42, 49, 56–61, 63, 67–68, 71, 78, 81–84, 123, 125, 135, 221, 224, 267–268
 third, 42–43, 60, 76, 83, 93, 116–117, 159, 167, 173, 226, 257–258, 262, 264
 tone (whole step), 11, 17, 33, 36, 55–63, 123
 tritone, 17, 158
Invariance, 149, 229, 236–239, 243
Ippolito II d'Este (1509–1572), 61, 63, 69
Iridescence, 156
Irrational, 2–3, 9, 15–17, 28–32, 38, 55–72, 87, 92, 137, 141, 143, 282

Jackson, Myles, 6
Jesuits (Society of Jesus), 53, 63, 104
Jew's harp. *See Guimbarde*
Joshua, miracle in, 24
Jupiter, 80

Kabbalism, 75
Kaleidophone, 200, 202, 210
Kant, Immanuel (1724–1804), 170, 181, 185, 217, 231, 233, 241–242
Kassler, Jamie, 6
Keats, John (1795–1821), 279
Kepler, Johannes (1571–1630), 3, 6, 46, 48, 73–88, 99–100, 102–107, 116, 123, 127, 131, 279–280, 282
 and Descartes, 99–100
 and Galileo Galilei, 6, 52
 and Mersenne, 102–107, 116
Kircher, Athanasius (1601–1680), 285n6
Kirchhoff, Gustav (1824–1887), 248, 265
Kittler, Friedrich (1943–2011), 6
Klein, Felix (1849–1925), 242
Königsberg bridge problem, 4, 145–149
Kulturträger, 222, 233, 255, 271

Lagrange, Joseph Louis (1736–1813), 236, 252
Languages, 4, 161, 163, 170, 179, 190
Laplace, Pierre-Simon (1749–1827), 190
Lasso, Orlando di (1532–1594), 3, 63, 73–75, 78–82, 85–86, 102
Leibniz, Gottfried (1646–1716), 144–146, 151, 179
Lichtenberg, Georg Christoph (1742–1799), 162, 181–185, 188, 190
 Lichtenberg figures, 182–183, 185, 188, 190
Lie, Sophus (1842–1899), 237–238, 242
Light
 emission theory of, 151–154
 infrared, 157, 188
 medium theory, 152–155
 polarization of, 177–178, 215
 speed of, 151, 206, 214, 243, 268
 transverse nature of, 4, 153, 178–179, 193, 196, 198, 206–209, 215
 ultraviolet, 157, 188, 222, 249
 wave theory of, 3–4, 7, 125, 127, 131, 151–155, 161–179, 221
Lines of force. *See* Force, lines of
Lobachevsky, Nicolai Ivanovich (1792–1856), 232, 242
Logarithms, 141, 143
Logos, 15–16
Lusitano, Vincente (d. after 1561), 61, 63
Luther, Martin (1483–1546), 56, 69, 82

Magnetism
 Ampère's law, 4, 181, 185, 191–198, 207, 211–215, 233 196
 Biot-Savart law, 181, 194
Magnitude (*plēthos*), 15, 17–18, 30, 55, 66, 69
Maier, Michael (1568–1622), 74
Mairan, Dortous de (1678–1771), 153
Malebranche, Nicolas (1638–1715), 131
Malus, Étienne-Louise (1775–1812), 177–178

Manifold, 5, 148, 217, 219, 231–235, 238, 240–242, 273–274, 278
Marić, Mileva (1875–1948), 242
Mars, 33, 39, 84
Martianus Capella (fifth century C.E.), 46
Mathematics, 1–2, 5–9, 181, 271, 277–278, 280, 282, 284. *See also* Algebra; Arithmetic; Geometry; Topology
 ancient, 9, 16, 18, 27, 45
 Euler and Continental, 133–158, 163, 179
 and Helmholtz, 226, 229
 Renaissance, 49, 55, 59, 63, 69, 72, 74, 89–90, 93, 103, 120
 and Young, 161, 163, 176
Matrix, 25–27, 272–273
Maxwell, James Clerk (1831–1879), 22, 245, 248, 266, 269
Means (arithmetic, geometric, harmonic), 18, 56
Mechanics, 3, 94, 96, 114, 134, 136, 151, 256
 celestial, 73, 89
 classical, 226, 241, 255
 continuum, 3, 89, 98, 151, 223, 266
 matrix, 272
 quantum, 245, 263, 272–278
 statistical, 263
 wave, 5, 272–277
Mei, Girolamo (1519–1594), 48–49
Melody, 34, 40, 48, 60–61, 66, 69, 73, 77–78, 83, 86–87, 92, 134, 173
Mersenne, Marin (1588–1648), 3, 90, 93–120
 and Descartes, 90, 93–103, 116, 137
Metaphysics, 89, 134, 136
Metric, 273, 306n8
Miller, Clement, 59
Miracles, 13, 24, 39, 93
Mode, 14, 19, 35–36, 39–45, 83, 109, 123–126
 Aeolian, 40–45, 137, 139
 change of, 19, 35–40
 Dorian, 124
 Ionian, 40–45, 137
 Lydian, 38
 Mixolydian, 124
 Phrygian, 38–40, 43–45, 78, 81, 288n6
 tonus peregrinus, 38
 vibrational, 198, 202, 205, 211–212, 248, 251
Modulation, 19, 35, 124, 137
Molecules, 248–249, 266
Monochord, 13, 21, 106, 116–117
Monodromy, 236, 306n27
Monteverdi, Claudio (1567–1643), 48, 70, 72
Moreno, Jairo, 6
Motet, 2, 31–32, 39–40, 43–45, 61, 69, 74, 78–82, 86, 102, 106
Mousikē, 6–10, 14
Mozart, Wolfgang Amadeus (1756–1791), 170, 271
Multiverse, 282
Murner, Thomas (1475–ca. 1537), 69

Muses, 9, 28
Music
  "ancient," 2, 4, 34–35, 45, 49, 67, 92, 158–159
  and mathematics, 2, 18, 55, 133–140, 144–153
  "modern," 4, 63, 67, 79–80, 158–160
  practical, 55–56, 60, 67, 73, 79, 90, 109, 112, 134, 158, 256
  theory of, 3–6, 18, 20–21, 25, 56, 77, 92, 109, 117, 124, 137, 144, 229
Musica instrumentalis, 78
Musica mundana, 49, 78, 109, 282
Musicians
  amateur, 70, 76, 82, 257
  professional, 75–76, 256–258
Muslim music, 75, 77
Myograph, 217–218

Nambu, Yoichiro, 279–280
Napoleon Bonaparte (1769–1821), 179, 185–186
Natural philosophy, 1–5, 21, 25, 89–90, 94–97, 100, 103, 109, 131, 134, 156, 158, 164, 176–177, 185, 198, 233, 282–284
Nature, unity of, 4, 181–182, 188, 190, 193, 263, 282
Naturphilosophie, 4, 181, 185, 188–189, 194, 196
Neoplatonism, 73, 105–106
Neutron, 46
New Music (Athenian), 19, 38
New philosophy, 1–2, 5, 6, 21, 89, 94, 96, 103
Newton, Isaac (1642–1727), 3, 6–7, 20, 46, 73, 121–131
  and Descartes, 121
  Newton's rings, 127–131, 156, 164, 174, 176, 301n21
  theory of light, 7, 125–131
Nicomachus (ca. 60–ca. 120 C.E.), 10–11
Nielsen, Holger, 279
Node, 185, 275
Notre Dame, school of, 31
Novalis (Friedrich von Hardenberg; 1772–1801), 188, 190
Number. *See also* Arithmetic
  amicable, 144–145
  concept of, 2–4, 9–17, 29–30, 56–72, 82–83
  irrational, 2–3, 9, 15–17, 28–32, 55–87, 137, 141, 143, 282
  perfect, 144–145
  theory, 4, 133–137, 143–145, 233

Occult arts, 74–75, 93, 189
One, concept of, 14–16
Ophelia, 1–2
Ophthalmoscope, 217
Ophthalmotrope, 217, 220
Optics–music analogy, 124–131, 133–134, 151–157, 161, 164, 170–179

Oresme, Nicole (ca. 1320–1382), 2–3, 21–35, 45–46, 57, 87, 145
Organ, 74, 105, 112, 118–119, 134, 164, 174, 203, 205, 210
Organic forms, 190–191
Orpheus, 106, 109
Ørsted, Hans Christian (1777–1851), 4, 181, 185–192, 200, 207, 209
Oscillator, 201, 266–267, 279–280
Ossicles (hammer, anvil, and stirrup), 226, 234–236
Ouranos, 99
Overtone, 3–4, 93, 116–120, 137, 156–157, 164, 170, 200, 223, 225, 248–252, 275, 277

Painting, 56, 93, 135, 185, 195, 222–223, 229, 239, 260
Palisca, Claude (1921–2001), 6, 85
Panofksy, Erwin (1892–1968), 6
Paradox, 24–25, 46, 53, 66–68, 95, 140, 154, 179, 245, 269
Parhelia (mock suns), 93–94
Particle, 7, 127, 135, 151–156, 164, 173, 176–178, 191, 198, 207–209, 214, 247, 273, 275, 279, 280
Pendulum, 93–96, 114, 167, 203, 209, 248, 266
Pentagon, 77, 83
Perpendicular effect
  of currents, 190–191
  of sounds, 209, 211
Philolaus (ca. 470–ca. 385 B.C.E.), 9–12, 14, 48
Phosphorescence, 160
Physicist, 5, 90, 196, 255, 258
Physics
  Aristotelian, 21, 38, 46, 99
  Maxwellian, 232, 245, 248, 266, 269
  Newtonian, 121–131, 133–134, 177, 196, 245, 280, 282
Physis, 90, 99
Piano, 157, 268, 200–201, 206, 210, 222, 256–257, 260–261, 271
Pictet, Marc Auguste (1752–1825), 188–189
Pipe, vibrations of, 11–13, 66, 112, 119, 134, 163–168, 174, 203, 210
Planck, Max (1858–1947), 5, 255–271, 273, 279–282
  and anthropomorphism, 255–256, 269, 282, 309n22
  and Helmholtz, 255–258, 262, 264–265
  music experiments, 258–263
  Planck constant, 267–268
  Planck units, 268, 309n34
  and thermodynamics, 256, 263, 265–268
Planets, 2, 6, 19, 22, 27, 30, 33–36, 46–49, 73, 78–88, 99, 101–106, 116, 127, 131, 279
Plato (429–347 B.C.E.), 2, 9, 14–22, 29, 35, 47, 66, 76, 82, 92, 104, 120, 123, 272, 282
  Republic, 14, 19, 123
  Timaeus, 14, 120, 272

Pleasure, 29–30, 38, 59, 84, 86, 90, 92, 120, 135, 139–140, 158, 189
Poetry, 9, 111, 121, 143, 234–235, 238
Poincaré, Henri (1854–1912), 242
Polyhedra, 4, 146–149
Polyphony, 3, 31, 38, 48, 76, 78, 80–83, 158, 205, 279
Praetorius, Johannes (1537–1616), 47–48, 53
Proclus, 105
Protestantism, 74
Proton, 46
Pseudosphere, 240, 242, 307n52
Ptolemy, Claudius (ca. 90–ca. 168 C.E.)
 astronomical writings, 30, 47, 105
 harmonics, 13, 18–19, 76
Pulse theory of sound, 97, 135, 137, 151–156, 164
Pythagoras (ca. 570–ca. 495 B.C.E.), 9–12, 17, 32, 35, 73, 131, 280, 282
Pythagorean
 cosmology, 14, 22, 24
 distance relation, 232
 ratios, 9–17, 135, 203, 250
 theorem, 232, 237
 thought, 2, 5, 9–24, 47–48, 73, 135, 245, 251, 278–279, 284
 tuning, 60, 77

Quadrivium, 2, 19–22, 29, 46, 49, 54, 58, 76, 121, 282, 284
Quakers (Society of Friends), 4, 161–162, 300n2, 301n6
Quantization, 269, 275, 278

Radioactivity, 46
Rainbow, 93, 176
Rameau, Jean-Philipe (1683–1764), 135, 156–158
Ratio, 4, 10–18, 26–33, 47, 69, 73, 77, 82, 115–116, 127, 135, 141, 149, 155, 159, 174, 182, 203, 221, 229
 superparticular, 32–33, 135
Rational, 3, 10, 28, 32–33, 55–66, 70, 123, 141
Raumproblem. See Space, problem of
Ray, 100, 151–156, 188–190, 214, 222, 247, 276–277
Rayleigh, Lord (John William Strutt; 1842–1919), 245, 251–252, 255
Recorde, Robert (ca. 1512–1558), 55–56
Recurrence, cosmic, 22, 31
Reed, 13, 205, 266–267
Regiomontanus (Johannes Müller von Königsberg; 1436–1476), 107
Rehding, Alexander, 90, 296n15, 297n39
Relativity, 206, 229, 243, 255, 264, 273, 282
Renaissance, 21, 75, 90
Republic of Letters, 103, 296n3
Resonance, 6, 168, 182, 200, 205–206, 247, 279–281
 sympathetic, 152, 200, 203

Resonator, 5, 200, 203, 223, 225, 266–268
Revolution, scientific, 5, 90, 284
Rheticus (Georg Joachim de Porris; 1512–1574), 47
Rhetoric, 19, 24, 47–49, 68, 74, 77–78, 172, 234–235
Rhythm, 14–15, 32, 92, 166, 168, 238
Riemann, Bernhard (1826–1866), 5, 145, 231–243, 273
 and Helmholtz, 231–234
 on foundations of geometry, 231–232
 on mechanics of the ear, 233–235
Riemann, Hugo (1849–1919), 157
Ritter, Johann (1776–1810), 4, 181, 188–193
Roman Catholic Church, 38, 40, 87, 104–105
Roman thought, 13, 15, 18–19, 105
Romanticism, 4, 170, 188–190, 272
Rotation, 205–206, 237–238, 306
 of Earth, 25
 electrodynamic, 197–198
 of planets, 86
 of sirens, 223, 226
Royal Institution, 176–177, 195, 198, 205–207
Rudolf II (Holy Roman Emperor; 1552–1612), 74–75
Rumford, Count (Benjamin Thompson; 1753–1814), 176–177

Sadness, 4, 92, 139–140, 282
Saturn, 35, 80
Sauveur, Joseph (1653–1716), 112
Savart, Félix (1791–1841), 181, 185, 187, 194, 207, 210
Schelling, Friedrich Wilhelm Joseph von (1775–1854), 191
Schrödinger, Erwin (1887–1961), 5, 272–278, 280
Schumann, Robert (1810–1856), 190
Schütz, Heinrich (1585–1672), 260, 262, 264
Schwartz, Hillel, 6
Sedimentation, 255, 271, 284
Sexuality, 3, 82–84, 88, 294n59
Shakespeare, William (1564–1616), 161, 179, 195
Shēng, 205
Singing, 31, 70, 74, 77, 81, 109, 168, 261
 tubes, 196
Siren, 81, 205, 223, 226–229
Skeleton, melodic, 76–78
Skepticism, 25, 30, 198, 256
Smith, Robert (1689–1768), 164, 172–173
Socrates (469–399 B.C.E.), 9, 14–16, 18
Sodium, 249
Sommerfeld, Arnold (1868–1951), 245
Sonification, 279
Sound, speed of, 3, 112–113, 155, 214

Space
  empty, 93, 95, 99, 119 (*see also* Vacuum)
  n-dimensional, 232
  problem of, 4, 191, 193, 217, 219, 226, 228, 231–243
Space-time, 232, 241, 243, 273
Spectrum
  elemental, 247, 266
  mass, 279
  solar, 3, 5, 121, 125, 127, 188, 190, 220–224, 245–250, 266, 269, 274
Spheres, celestial, 22, 81
  music of the, 27–28, 30–37, 46–50, 78, 131, 245, 279, 282
Spitta, Philipp (1841–1894), 260
Sterne, Jonathan, 6
Stifel, Michael (1487–1567), 3, 55–60, 69–70, 77
Stoney, G. Johnstone (1826–1911), 5, 248–251, 266, 275
String, vibrating, 3, 45, 182, 187, 249, 251, 266, 275–276
  ancient accounts, 11–13, 17
  and Descartes, 93–98
  and Euler, 134–135, 152–157
  and Faraday, 208, 210
  and Kepler, 77
  and Mersenne, 107–120
  and Newton, 125
  Renaissance, 5, 47, 56–57, 66–68
  and Young, 163, 166, 168, 174
String theory, 5, 279–282
Sun, 24, 33–40, 46–49, 53–54, 73, 80–84, 96, 99, 102, 105–106, 127, 245, 247
Superposition, 173, 175, 226
Susskind, Leonard, 279
Sweetness, 123–124, 136, 268
Symmetry, 46–47, 157, 185, 188–189, 193, 261

Tachistoscope, 217, 219
Tartini tones, 166, 226
Telegraphy, 4, 207, 213–214
Temperament
  equal, 5, 59–60, 70–71, 137, 169, 257, 264, 267–269
  just intonation, 60, 77, 117, 137, 258, 291n29
  Pythagorean, 104–107, 116
  Young, 168–169
Tempo, 114, 170
Tennis, 99, 111
Thermodynamics, 255–256, 263, 265–266
Thompson, Emily, 6
Thornton, Robert (1898–1982), 243
Thunderstorms, 134, 193
Timbre, 164, 168, 235, 240–241, 248
Time, 86–87, 223, 226, 228, 231, 240–243, 248
Tinctoris, Johannes (ca. 1435–1511), 35
Topology, 4, 133, 145–149

Tragedy, joy of, 139–140
Translation, 179, 181
Transmutation, 45–46
Triads, 78, 117, 136–139, 170, 203, 258, 260–264
Trissino, Giovanni Giorgio (1478–1550), 61
Trumpet, 75, 117–118, 195
Tuning, 5, 47, 59–60, 63, 70, 74–75, 77, 84, 168, 256, 258–269, 272, 282. *See also* Temperament
Tuning fork, 203, 206, 223, 225
Tyndall, John (1820–1893), 13, 184, 214

Undertone, 4, 156–157, 203
Unity of nature, 4, 173, 182, 188, 193, 263
Universe, 3, 22, 33–36, 46, 75, 98–99, 102, 104–111, 282
Urania, 79, 81

Vacuum, 3, 94–98, 119
Valla, Giorgio (1447–1500), 35–36
Van Wymeersch, Brigitte, 6, 295n10
Veneziano amplitude, 279–280
Ventriloquism, 168, 198, 200, 214
Venus, 80–81, 84
Vibration
  longitudinal, 97, 152–153, 178, 191, 193, 198, 207, 209, 215
  of plates, 175, 178, 181–182, 185, 190–193, 203, 207, 308n18
  transverse, 4, 153, 178–179, 193, 196, 198, 206–209, 215
Vicentino, Nicola (1511–ca. 1575), 55, 60–70, 76, 123
Viète, François (1540–1603), 55–56
Violence, 44–45, 77
Violin, 75, 166, 182, 185, 18–188, 191–195, 206, 208, 249, 256–258, 271–272
Virgil (70–19 B.C.E.), 74, 84, 113
Voice, physiology of, 162, 168
Volmar, Axel, 6
Vortex, 89, 99–102, 152–153

Wagner, Richard (1813–1883), 229
Walker, D. P. (1914–1985), 6, 81, 83, 100
Wardhaugh, Benjamin, 6
Wave, 111, 136
  electrical, 213, 232
  equation, 232, 272–273, 276–277
  light, 3–4, 7, 125, 127, 131, 151–155, 161–179, 221
  mechanics, 5, 275–277
  vs. pulse, 151
  sound, 97, 137, 188, 198, 200
Wavelength, 127, 155, 173–175, 178, 222, 248, 250, 268, 275
Weil, André (1906–1998), 145
Weyl, Hermann (1885–1955), 278, 310n6
Wheatstone, Charles (1802–1875), 4, 195, 198–207, 210, 213–215, 279–280

Whewell, William (1794–1866), 5
Whitehead, Alfred North (1861–1947), 1
Widmann, Erasmus (1572–1634), 74
Willaert, Adrian (ca. 1490–1562), 61
Williams, L. Pearce, 198, 205, 207
Wolff, Christian (1679–1754), 146, 151
Wollaston, William (1766–1828), 245
Wonder, 29, 93, 280
Wunderkammer, 205

Xerography, 182

Young, Thomas (1773–1829), 4, 161–181, 198, 200,
  205, 210, 276
 and color vision, 172–173, 217–219, 223, 231, 273
 and ether, 172, 178–179, 214–215
 and Euler, 131, 154

Zarlino, Gioseffo (1517–1590), 48–53, 70–71,
  77–78, 85, 137
Z boson, 281
Zero, concept of, 14–16, 148
Zeta function, 145, 232

WITHDRAWN

WITHDRAWN